QSO HOSTS AND THEIR ENVIRONMENTS

QSO HOSTS AND THEIR ENVIRONMENTS

Edited by

Isabel Márquez
Josefa Masegosa
Ascensión del Olmo
Lucas Lara
Emilio García
and
Josefina Molina
Instituto de Astrofísica de Andalucía (CSIC)
Granada, Spain

Kluwer Academic / Plenum Publishers
New York, Boston, Dordrecht, London, Moscow

Library of Congress Cataloging-in-Publication Data

QSO hosts and their environments/edited by Isabel Márquez ... [et al.].
    p.   cm.
Includes bibliographical references and index.
ISBN 0-306-46662-7
  1. Radio sources (Astronomy)—Congresses. I. Márquez, Isabel. II. International Workshop on QSO Hosts and their Environments (20001: Granada, Spain)

QB475.A1 Q66 2001
523.1'12—dc21

2001038596

Proceedings of an International Workshop on QSO Hosts and their Environments, held January 10–12, 2001. in Granada, Spain

ISBN 0-306-46662-7

©2001 Kluwer Academic/Plenum Publishers, New York
233 Spring Street, New York, New York 10013

http://www.wkap.com

10   9   8   7   6   5   4   3   2   1

A C.I.P. record for this book is available from the Library of Congress

All rights reserved

No part of this book may be reproduced, stored in a retrieval system, or transmitted in any form or by any means, electronic, mechanical, photocopying, microfilming, recording, or otherwise, without written permission from the Publisher

Printed in the United States of America

Preface

In the last few years, a number of observational and theoretical developments have dramatically improved our understanding of AGNs. *HST* and large, ground-based telescopes with adaptive optic systems have brought new insights into the nature of QSO and BlLac objects and their environments. In particular, it has become clear that quasars reside in the center of galaxies and that it was possible to study the morphology and nature of the host galaxy.

To understand the origin and nature of these objects, various questions should be addressed in more detail: what are the stellar and gaseous content of the underlying galaxy; is it possible to characterize the population of companions and the nature of their interaction with the host galaxy; is there a connection between radio-loud QSO and radio-galaxies; between QSOs and ULIRGs galaxies; what is the evolution with redshift of both the host galaxy and its environments, and the main implications in theories of galaxy formation and evolution.

The use of IR facilities (SCUBA,ISO, GEMINI) is a promising tool to analyse the dust/obscured AGN connection in the case of FIR luminous QSOs. Moreover, 10m-class telescopes (Keck, VLT) are providing new and valuable data, since they allow to reach fainter objects. High resolution X-ray observations are expected to provide a unique information on the cluster environment of these objects. The combination of multi-wavelength data from different telescopes, together with the advent of new observational facilities like ALMA, FIRST, AXAF, GRANTECAN, SIRTF, NGST, will be essential for future progress.

This 3 days workshop, held at the Instituto de Astrofísica de Andalucía (Granada), from 10-12th January 2001, has been devoted to the revision of the *state of the art* on QSO hosts and their environments, the multi-wavelength properties of these objects till look back times, and their implications in galaxy formation and evolution. The workshop involved more than 70 participants from different countries. The invited speakers presented excellent review papers and the talks and posters completed the programme. Thanks to the participants, the contributions were of a high scientific level and the discussions were lively and exciting. We are specially grateful to Prof. Miley for conducting the final discussion. The present proceedings gather almost all the invited, contributed and poster papers presented during the seven sessions of the meeting.

We acknowledge Deborah Dultzin-Hacyan, who encouraged us to organize the meeting. It wouldn't have run well without a hard core of dedicated people. We are specially indebted to our workshop secretary, Mariano Domenicone, to our reception secretary, Susana Gómez, and the logistic services offered by Francisco Navarro. We are very gratefull to Sven Müller, who took all the pictures.

It is also a great pleasure to thank the numerous institutions that sponsored this meeting. Our special thanks to the City Council of Granada and its representative, Asunción Jódar, for the opening wellcome and the splendid evening at the Carmen de los Mártires. Financial help was given by the Instituto de Astrofísica de Andalucía, the Spanish National Research Council (CSIC), the Spanish Ministry of Science and Technology (MCYT), and the Junta de Andalucía.

The Organizing Committee

SCIENTIFIC

J.N. Bahcall (USA)
M. Bremer (UK)
D. Dultzin-Hacyan (México)
J. Dunlop (UK)
J. Heidt (Germany)
J. Hutchings (Canada)
I. Márquez (Spain)
P. Petitjean (France)

LOCAL (IAA)

A. del Olmo
E. García
L. Lara
I. Márquez
J. Masegosa
J. Molina

The Local Organizing Committee operated under the auspices of:

Instituto de Astrofísica de Andalucía
Spanish Ministry of Science and Technology (MCYT)
Consejería de Educación y Ciencia de la Junta de Andalucía
Spanish National Research Council (CSIC)

Pepa Masegosa, Mariano Domenicone and Ascensión del Olmo at the conference dinner

The Alhambra at night

List of participants

Antxon Alberdi	antxon@iaa.es
Eduardo Battaner	battaner@ugr.es
Joanne Baker	jcb@astron.Berkeley.EDU
Xavier Barcons	barcons@ifca.unican.es
Carlton Baugh	c.m.baugh@durham.ac.uk
Malcolm Bremer	m.bremer@bris.ac.uk
Alessandro Bressan	bressan@pd.astro.it
Andrea Cattaneo	cattaneo@ast.cam.ac.uk
José Antonio de Diego	jdo@astroscu.unam.mx
Antonio Delgado	delgado@iaa.es
Ascensión del Olmo	chony@iaa.es
Mariano Domenicone	dae@arrakis.es
James Dunlop	j.dunlop@roe.ac.uk
Wolfgang J. Duschl	wjd@ita.uni-heidelberg.de
Andreas Eckart	eckart@1ph.uni-koeln.de
Aaron Evans	aevans@mail.astro.sunysb.edu
Heino Falcke	hfalcke@mpifr-bonn.mpg.de
Renato Falomo	falomo@pd.pastro.it
David Floyd	djef@roe.ac.uk
Emilio Garcia	garcia@iaa.es
Rosa M. González Delgado	rosa@iaa.es
Matteo Guainazzi	mguainaz@xmm.vilspa.esa.es
Martin Hardcastle	m.hardcastle@bristol.ac.uk
Jochen Heidt	jheidt@lsw.uni-heidelberg.de
John Hutchings	John.Hutchings@hia.nrc.ca
Chris Impey	cimpey@as.arizona.edu
Knud Jahnke	kjahnke@uni-hamburg.de
Matt Jarvis	mjj@astro.ox.ac.uk
Elena Jiménez Bailón	elena@laeff.esa.es
Monique Joly	monique.joly@obspm.fr
Birgit Kelm	Kelm@astbo3.bo.astro.it
Johan H. Knapen	knapen@ing.iac.es
Jari Kotilainen	jarkot@deneb.astro.utu.fi

Yair Krongold	yair@astroscu.unam.mx
Bjoern Kuhlbrodt	BKuhlbrodt@uni-hamburg.de
Marek Kukula	mjk@roe.ac.uk
Lucas Lara	lucas@iaa.csic.es
Stephane Leon	sleon@asiaa.sinica.edu.tw
Wen-shuo Liao	wsliau@asiaa.sinica.edu.tw
Jeremy Lim	jlim@asiaa.sinica.edu.tw
Isabel Márquez	isabel@iaa.es
Paolo Marziani	marziani@pd.astro.it
Josefa Masegosa	pepa@iaa.es
Ross McLure	rjm@astro.ox.ac.uk
George Miley	miley@strw.leidenuniv.nl
Josefina Molina	fina@iaa.es
Sven Müller	smueller@astro.ruhr-uni-bochum.de
Kari Nilsson	kani@astro.ku.dk
Matthew O'Dowd	modowd@physics.unimelb.edu.au
Ken Ohsuga	ohsuga@rccp.tsukuba.ac.jp
Eva Orndahl	eva@astro.uu.se
William Percival	wjp@roe.ac.uk
Sebastian Rabien	srabien@mpe.mpg.de
Joseph Rhee	rhee@astro.ucla.edu
Susan Ridgway	ridgway@pha.jhu.edu
José Miguel Rodriguez Espinosa	jre@ll.iac.es
Michael Rowan-Robinson	m.rrobinson@ic.ac.uk
Stanislaw Rys'	strys@oa.uj.edu.pl
Dave Sanders	sanders@uhifa.ifa.hawaii.edu
Riccardo Scarpa	scarpa@stsci.edu
David Schade	David.Schade@hia.nrc.ca
Andrew Sheinis	sheinis@ucolick.org
Aimo Sillanpää	aimosill@oj287.astro.utu.fi
Jason Surace	jason@ipac.caltech.edu
Clive Tadhunter	C.Tadhunter@sheffield.ac.uk
Ichi Tanaka	itanaka@optik.mtk.nao.ac.jp
Dario Trèvese	trevese@roma1.infn.it
Masayuki Umemura	umemura@rccp.tsukuba.ac.jp
Fausto Vagnetti	vagnetti@roma2.infn.it
Sylvain Veilleux	veilleux@astro.umd.edu
José M. Vilchez	jvm@iaa.es
Chris Willott	cjw@astro.ox.ac.uk
Lutz Wisotzki	lutz@astro.physik.uni-potsdam.de
Margrethe Wold	wold@astro.su.se
Toru Yamada	yamada@optik.mtk.nao.ac.jp

Valentina Zitelli zitelli@bo.astro.it

At the conference dinner: from left in clockwise sense, James Dunlop, Mrs. Hutchings, John Hutchings, Jochen Heidt, David Sanders, Isabel Márquez, Michael Rowan–Robinson and his wife, Pepa Masegosa, Mariano Domenicone, Ascensión del Olmo, Eduardo Battaner and Malcolm Bremer.

Pepa Masegosa, Mariano Domenicone (backwards) and Jochen Heidt

Contents

Part I Radio quiet/loud dichotomy. Unification

The hosts galaxies of radio-loud and radio-quiet quasars 3
J. Dunlop

The Nucleus-Host Galaxy Connection in Radio-Loud AGN 13
M. O'Dowd, C.M. Urry, R. Scarpa, R. Falomo, J.E. Pesce, and A. Treves

The host galaxies of luminous radio-quiet quasars 21
W.J. Percival, L. Miller, R.J. McLure and J.S. Dunlop

The radio loudness dichotomy: environment or black-hole mass? 27
R. McLure and J. Dunlop

QSO environments at intermediate redshifts 33
M. Wold, M. Lacy, P.B. Lilje and S. Serjeant

Host galaxies and cluster environment of BL Lac objects at $z > 0.5$ 39
J. Heidt, J. Fried, U. Hopp, K. Jäger, K. Nilsson and E. Sutorius

Associated absorption and radio source growth 45
J.C. Baker

Host galaxies of RGB BL Lacertae objects 51
K. Nilsson, L. Takalo, T. Pursimo, A. Sillanpää, and J. Heidt

On The Parent Population of Radio Galaxies and the FR I–II Dichotomy 55
R. Scarpa and C.M. Urry

The real difference between radio-loud and radio-quiet AGNs 59
A. Sillanpää

Broadband optical colours of intermediate redshift QSO host galaxies 61
E. Örndahl and J. Rönnback

Three peculiar objects from a new sample of radio galaxies 65
L. Lara, W.D. Cotton, L. Feretti, G. Giovannini, J.M. Marcaide, I. Márquez and T. Venturi

Part II QSO redshift evolution and their cluster environments

QSO hosts and companions at higher redshift 71
J.B. Hutchings

The luminosity function of QSO host galaxies 83
L. Wisotzki, B. Kuhlbrodt and K. Jahnke

QSO host galaxy star formation history from multicolour data 89
K. Jahnke, B. Kuhlbrodt, E. Örndahl and L. Wisotzki

Near-infrared imaging of steep spectrum radio quasars 95
J.K. Kotilainen and R. Falomo

Adaptive-optics imaging of low and intermediate redshift quasars 101
I. Márquez, P. Petitjean, B. Theodore, M. Bremer, G. Monnet and J.L. Beuzit

Subaru Observations of the Host Galaxies and the Environments of the Radio Galaxy 3C324 at $z = 1.1$ 107
T. Yamada

Extremely red radio galaxies 113
C. J. Willott, S. Rawlings and K. M. Blundell

The environments of radio-loud quasars 119
J.M. Barr, M.N. Bremer, and J.C. Baker

Extended X-ray emission around radio-loud quasars 127
M. Hardcastle

WFPC2 Imaging of Quasar Environments 133
R. A. Finn, C. D. Impey and E. J. Hooper

Deceleration and asymmetry in QSO radio map 137
S. Ryś

Spatially resolved spectroscopy of emission-line gas in QSO Host galaxies 141
A.I. Sheinis

Host Galaxies and the Spectral Variability of Quasars 145
F. Vagnetti and D. Trèvese

Part III Tidal Interactions/Mergers. ULIRGS

The AGN-Starburst Connection in Ultraluminous Infrared Galaxies 151

Contents

M. Rowan-Robinson

The QSO – Ultraluminous Infrared Galaxies Connection 165
S. Veilleux, D.-C. Kim, and D.B. Sanders

Recent star formation in very luminous infrared galaxies 171
A. Bressan, B. Poggianti, and A. Franceschini

A Molecular Gas Survey of $z < 0.2$ Infrared Excess, Optical QSOs 177
A. S. Evans, D. T. Frayer, J. A. Surace, and D.B. Sanders

Molecular gas in nearby powerful radio galaxies 185
S. Leon, J. Lim, F. Combes and D. Van-Trung

HI Imaging of low-z QSO Host Galaxies 191
J. Lim, H. Chuo, S. Wen, W. Liao, and P.T.P. Ho

Dust in Quasars and Radiogalaxies as seen by ISO 199
S. A. H. Müller, R. S. Chini, M. Haas, K. Meisenheimer, U. Klaas, D. Lemke, and E. Kreysa

Imaging of a Complete Sample of IR-Excess PG QSOs 205
J.A. Surace and D.B. Sanders

Interaction patterns in a complete sample of compact groups 209
V. Zitelli, P. Focardi, B. Kelm and C. Boschetti

Part IV Galaxies hosting lower level AGN activity

AGN Host galaxies: HST at $z \sim 0.1$ and Gemini at $z \sim 2$ 215
D. Schade, S. Croom, B. Boyle, M. Letawsky, T. Shanks, L. Miller, N. Loaring and R. Smith

Host galaxies and nuclear structure of AGN with H_2O megamasers as seen with HST 223
H. Falcke, C. Henkel, A.S. Wilson, and J.A. Braatz

Statistics of Seyfert galaxies in clusters 229
B. Kelm and P. Focardi

More bars in Seyfert than in non-Seyfert galaxies 235
J.H. Knapen, I. Shlosman, R.F. Peletier, and S. Laine

Stellar Populations in the Host galaxy of AGNs 241
M. Joly, C. Boisson and D. Pelat

The Stellar Population of powerful Seyfert 2 galaxies: Implications for QSOs 247
R.M. González Delgado

Anisotropy in the mid-IR emission of Seyfert galaxies 255
A.M. Pérez Garcia and J.M. Rodriguez Espinosa

FIR and Radio emission in star forming galaxies 261
A. Bressan, G.L. Granato and L. Silva

No double nucleus at the center of Arp 220 ? 265
A. Eckart and D. Downes

Starburst Activity in Seyfert 2 Galaxies: UV–X-ray emission in NGC1068 269
E. Jiménez Bailón, J.M. Mas Hesse and M. Santos Lleó

Host Galaxies and Environment of Narrow Line Seyfert 1 Nuclei 273
Y. Krongold, D. Dultzin-Hacyan and P. Marziani

The Circum-Galactic Environment of LINERs and Bright IRAS Galaxies 277
Y. Krongold, D. Dultzin-Hacyan and P. Marziani

HI Imaging of Seyfert Galaxies 281
W. Liao, J. Lim, and P.T.P. Ho

Starburst-AGN Connection; Regulated by Radiatively-Supported Obscuring Walls 285
K. Ohsuga and M. Umemura

Rotation curve and mean stellar population of the I Zw 1 QSO host 289
J. Scharwächter, A. Eckart and S. Pfalzner

Part V QSO at high z as tracers of structures. Galaxy Formation

Semi-analytic galaxy formation: understanding the high redshift universe 295
C. M. Baugh, A. J. Benson, S. Cole, C.S. Frenk and C.G. Lacey

Radiation-Hydrodynamical Model for the QSO Formation 307
M. Umemura

Gravitationally Lensed Quasar Host Galaxies 313
C. Impey, H.-W. Rix, B. McLeod, C. Peng, C. Keeton, C. Kochanek, E. Falco, J. Lehár, and J.A. Muñoz

Host Galaxies of Low Luminosity Quasars at Intermediate Redshift 321
S. Rabien and M. Lehnert

A NICMOS imaging study of quasar host galaxy evolution 327
M. Kukula, J.S. Dunlop, R.J. McLure, L. Miller, W. J. Percival, S. A. Baum and C. P. O'Dea

The radio galaxy $K-z$ relation to $z \sim 4.5$ 333
M.J. Jarvis, S. Rawlings, S. Eales, K.M. Blundell and C.J. Willott

What fuels AGNs? 339
A. Cattaneo

Contents   xvii

VLT–ISAAC imaging of three radio loud quasars at $z \sim 1.5$ 343
R. Falomo, J.K. Kotilainen, and A. Treves

Two-dimensional modeling of AGN host galaxies   347
B. Kuhlbrodt, L. Wisotzki and K. Jahnke

Superclustering of Galaxies Traced by a Group of QSOs at $z = 1.1$ 351
I. Tanaka, T. Yamada, E.L. Turner and Y. Suto

Part VI **Future prospects with new instrumentation**

Uncovering AGN with X-ray observations   357
X. Barcons

New Ground Based facilities in QSO research   367
J.M. Rodriguez Espinosa

Author Index   375

QSO HOSTS AND THEIR ENVIRONMENTS

I

RADIO QUIET/LOUD DICHOTOMY. UNIFICATION

James Dunlop

Matthew O'Dowd

THE HOSTS GALAXIES OF RADIO-LOUD AND RADIO-QUIET QUASARS

James S. Dunlop
Institute for Astronomy, Royal Observatory, Edinburgh EH9 3HJ, U.K.

Abstract   I review our knowledge of the properties of the host galaxies of radio-loud and radio-quiet quasars, both in comparison to each other and in the context of the general galaxy population. It is now clear that the hosts of radio-loud *and* radio-quiet quasars with $M_V < -23.5$ are virtually all massive elliptical galaxies. The masses of these spheroids are as expected given the relationship between black-hole and spheroid mass found for nearby quiescent galaxies, as is the growing prevalence of disc components in the hosts of progressively fainter AGN. There is also now compelling evidence that quasar hosts are practically indistinguishable from normal ellipticals, both in their basic structural parameters and in the old age of their dominant stellar populations; at low z the nuclear activity is *not* associated with the formation of a significant fraction of the host galaxy. While the long-held view that quasar radio power might be a simple function of host morphology is now dead and buried, I argue that host-galaxy studies may yet play a crucial role in resolving the long-standing problem of the origin of radio loudness. Specifically there is growing evidence that radio-loud objects are powered by more massive black holes accreting at lower efficiency than their radio-quiet couterparts of comparable optical output. A black-hole mass $> 10^9 M_\odot$ appears to be a necessary (although perhaps not sufficient) condition for the production of radio jets of sufficient power to produce an FRII radio source within a massive galaxy halo.

Introduction

Studies of the host galaxies of low-redshift quasars are of crucial importance for defining the subset of the present-day galaxy population which is capable of producing quasar-level nuclear activity. They are also of value for constraining physical models of quasar evolution, for exploring the extent to which radio-loudness might be connected with host-galaxy properties, and as a means to estimate the masses of the central black holes which power the active nuclei.

Our view of low-redshift quasar hosts has been clarified enormously over the last five years, primarily due to the angular resolution and dynamic range offered by the Hubble Space Telescope. In this overview I have therefore chosen to concentrate on the results of recent, primarily HST-based studies of low-redshift quasars, and will only briefly mention the latest results at higher redshift which are discussed in detail elsewhere in these proceedings. I have also chosen to centre the discussion around our own, recently-completed, HST imaging study of the hosts of quasars at $z \simeq 0.2$. Preliminary results from this programme can be found in McLure et al. (1999) and final results from the completed samples are presented by Dunlop et al. (2001). Here I focus on a few of the main results from this study and discuss the extent to which other authors do or do not agree with our findings.

1. Host galaxy luminosity, morphology and size

After some initial confusion (e.g. Bahcall et al. 1994), recent HST-based studies have now reached agreement that the hosts of all luminous quasars ($M_V < -23.5$) are bright galaxies with $L > L^*$ (Mclure et al. 1999, McLeod & McLeod 2001, Dunlop et al. 2001). However, it can be argued, (with some justification) that this much had already been established from earlier ground-based studies (e.g. Taylor et al. 1996).

In fact the major advance offered by the HST for the study of quasar hosts is that it has enabled host luminosity profiles to be measured over sufficient angular and dynamic range to allow a de Vaucouleurs $r^{1/4}$-law spheroidal component to be clearly distinguished from an exponential disc, at least for redshifts $z < 0.5$. In our own study this is the reason that we have been able to establish unambiguously that, at low z, the hosts of both radio-loud quasars (RLQs) *and* radio-quiet quasars (RQQs) are undoubtedly massive ellipticals with (except for one RQQ in our sample) negligible disc components (McLure et al. 1999, Dunlop et al. 2001). This result is illustrated in Fig. 1.

Fig. 1 confirms that the hosts of radio-loud quasars and radio galaxies all follow essentially perfect de Vaucouleurs profiles, in good agreement with the results of other studies. The perhaps more surprising aspect of Fig. 1 is the extent to which our radio-quiet quasar sample is also dominated by spheroidal hosts. At first sight this might seem at odds with the results of some other recent studies, such as those of Bahcall et al. (1997) and Hamilton et al. (2001) who report that approximately one third to one half of radio-quiet quasars lie in disc-dominated hosts. However, on closer examination it becomes clear that there is no real contradiction provided one compares quasars of similar power. Specifically, if atten-

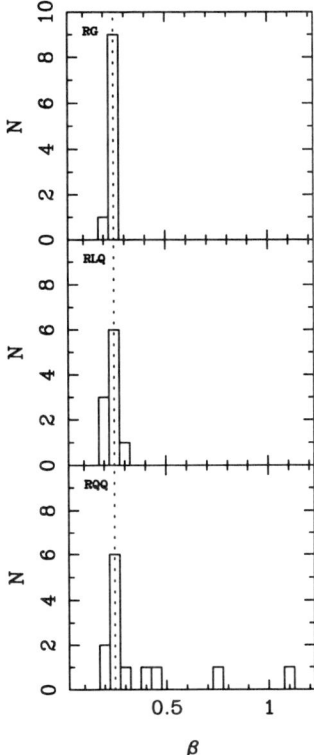

Figure 1. Histograms of the best-fit values of β, where host-galaxy surface brightness is proportional to $exp(-(r)^\beta)$, shown for the radio-galaxy, radio-loud quasar and radio-quiet quasar sub-samples imaged with the *HST* by Dunlop et al. (2001). The dotted line at $\beta = 0.25$ indicates a perfect de Vaucouleurs law, and all of the radio-loud hosts are consistent with this within the errors. Two of the three RQQs with hosts for which $\beta > 0.4$ transpire to be the two least luminous nuclei in the sample, and should really be reclassified as Seyferts.

tion is confined to quasars with nuclear magnitudes $M_V < -23.5$ we find that 10 out of the 11 RQQs in our sample lie in ellipticals, Bahcall et al. find that 6 of their 7 similarly-luminous quasars lie in ellipticals, while an examination of the data in Hamilton et al. shows that in fact at least 17 out of the 20 comparably-luminous RQQs in their archival sample also appear to lie in spheroidal hosts.

It is thus now clear that above a given luminosity threshold we enter a regime in which AGN can only be hosted by massive spheroids, regardless of radio power. It is also clear that, within the radio-quiet population, significant disc components become more common at lower nuclear luminosities. This dependence of host-galaxy morphology on

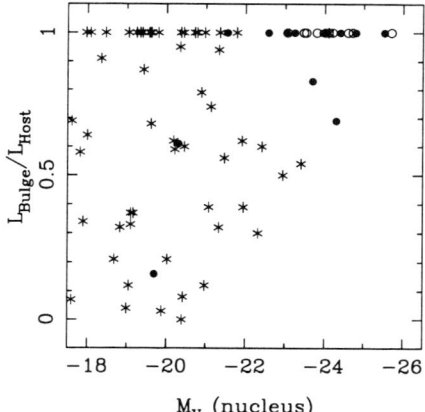

Figure 2. The relative contribution of the spheroidal component to the total luminosity of the host galaxy plotted against absolute V-band luminosity of the nuclear component. The plot shows the results for our own HST sample (RLQs as open circles, RQQs as filled circles) along with the results from Schade et al. (2000) for a larger sample of X-ray selected AGN spanning a wider but lower range of optical luminosities (asterisks). This plot illustrates very clearly how disc-dominated host galaxies become increasingly rare with increasing nuclear power, as is expected if more luminous AGN are powered by more massive black holes which, in turn, are housed in more massive spheroids.

nuclear luminosity is nicely demonstrated by combining our own results with those of Schade et al. (2000) who have studied the host galaxies of lower-luminosity X-ray selected AGN. This I have done in Fig. 2 where the ratio of bulge to total host luminosity is plotted as a function of nuclear optical power. Fig. 2 is at least qualitatively as expected if black-hole mass is proportional to spheroid mass (Magorrian et al. 1998, Merritt & Ferrarese 2001), and black-hole masses $> 5 \times 10^8 M_\odot$ are required to produce quasars with $M_R < -23.5$.

In concluding this discussion of host morphology I should note that there is at least some (albeit yet tentative) evidence that the hosts of some of the most luminous quasars may in fact have a significant disc contribution (Percival et al. 2001). At first sight this would appear to be at odds with the appealingly simple picture presented in Fig. 2, and it will certainly be interesting to see if this result survives the scrutiny of HST imaging currently underway. However, if confirmed, such a result need not contradict the universality of elliptical hosts, but rather might mean that some of the most luminous quasars arise from the merger of the elliptical host with a massive gas-rich disc galaxy, in which case the underlying massive elliptical might (at least temporarily) appear to have acquired a significant disc component.

In our *HST* study we have also been able to break the well-known degeneracy between host galaxy surface-brightness and size. This point is illustrated by the fact that we have, for the first time, been able to demonstrate that the hosts of RLQs and RQQs follow a Kormendy relation (Fig.3). Moreover the slope (2.90 ± 0.2) and normalization of this relation are identical to that displayed by normal quiescent massive ellipticals. The average half-light radii of the host galaxies in our sub-samples are 11 kpc for the RGs, 12 kpc for the RLQs, and 8 kpc for the RQQs ($H_0 = 50$ km s^{-1} Mpc^{-1}, $\Omega_m = 1.0$). For comparison the average half-light radius of brightest-cluster galaxies observed by Schneider et al. (1983) is 13 kpc.

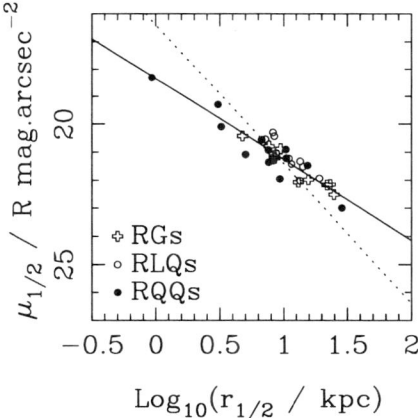

Figure 3. The Kormendy relation followed by the hosts of all 33 powerful AGN studied by Dunlop et al. (2001) with the *HST* . The solid line is the least-squares fit to the data which has a slope of 2.90 ± 0.2, in excellent agreement with the slope of 2.95 found by Kormendy (1977) for inactive ellipticals. For the few RQQs which have a disc component the best-fitting bulge component has been plotted.

2. Host galaxy ages

It is well known from simulations that the merger of two disc galaxies can produce a remnant which displays a luminosity profile not dissimilar to a de Vaucouleurs $r^{1/4}$-law. This raises the possibility that the apparently spheroidal nature of the quasar hosts discussed above might be the result of a recent major merger which could also be responsible for stimulating the onset of nuclear activity. This would also be the natural prediction of suggested evolutionary schemes in which ULIRGs are presumed to be the precursors of RQQs. Could a recent merger of two massive gas-rich discs be simultaneously responsible for the triggering of nuclear activity and the production of an apparently spheroidal host?

The answer appears to be no. One piece of evidence against such a picture comes from the fact that, as mentioned above, the Kormendy relation displayed by quasar hosts appears to be indistinguishable from that of quiescent, well-evolved massive ellipticals. However, a more direct test comes from attempts to determine the ages of the dominant stellar populations in the quasar hosts. Within our own sample we have attempted to estimate the ages of the host galaxies both from optical-infrared colours (now possible for the first time by combining our *HST* images with our pre-existing UKIRT data; Taylor et al. 1996) and from deep optical off-nuclear spectroscopy (Nolan et al. 2000). The results of this investigation are summarized in Fig. 4, which shows that the hosts of both radio-loud and radio-quiet quasars are dominated by old well-evolved stellar populations (with typically less than 1% of stellar mass involved in recent star-formation activity). There are currently no comparably-extensive studies of host-galaxy stellar populations with this result can be compared. However, Canalizo & Stockton (2000) have published results from a more detailed spectroscopic study of three objects, one of which, Mkn 1014, is also in our RQQ sample. This is in fact the only quasar host for which we have found clear spectroscopic evidence of A-star features and a significant (albeit still only $\simeq 2\%$ by mass) young stellar population. It is presumably no coincidence that this is also the only quasar in our sample which was detected by IRAS, and the only host which displays spectacular tidal-tail features comparable to those commonly found in images of ULIRGs (see Sanders, this proceedings). However, even for this apparently star-forming quasar host, Canalizo & Stockton agree that $\simeq 95\%$ of the host is dominated by an old well-evolved stellar population (although they argue that $5-8\%$ of the galaxy has been involved in recent star formation).

In summary, at least for low-redshift quasars, the timescale of the *primary* star-formation epoch in the host appears to be completely disconnected from that of the more recent nuclear activity which has resulted in the object featuring in quasar catalogues. The production of a low-redshift quasar only seems to require the massive, well-evolved spheroid housing the massive black hole to undergo a relatively minor interaction. In contrast the production of a ULIRG seems to require a major merger between two massive galaxies at least one of which is gas rich. Present evidence suggests that the overlap between these two phenomena is rather limited at low redshift, and that the ULIRG $\rightarrow$ quasar evolutionary route can only apply to a fairly small subset

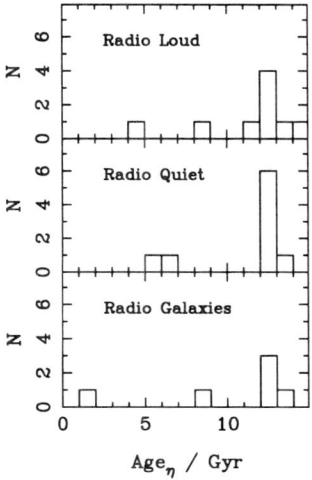

Figure 4. The age distribution of the dominant stellar populations in the sub-samples of host galaxies studied by Nolan et al. (2001). The ages were derived by fitting a 3-component model (comprising scattered quasar light, a young (0.1Gyr) stellar population, and an underlying stellar population of age ranging from 0.1 to 14 Gyr) simultaneously to off-nuclear optical spectra and the $R - K$ colours of the host galaxies. The dominant populations in the hosts of both radio-loud and radio-quiet AGN are predominantly old (12-14 Gyr) as is found for quiescent elliptical galaxies.

of objects (e.g. Mkn 1014). Of course at high redshift the prospect for star-formation and nuclear activity having completely disconnected timescales is much more limited, and it seems likely that the first epoch of quasar activity in a massive galaxy is closely connected with massive (possibly dust-enshrouded) star-formation activity in the host (e.g. Fabian 1999, Archibald et al. 2001).

3. Black hole masses and radio loudness

Having established that the hosts of quasars are massive spheroids one can estimate the mass of their central black holes using the relationship between spheroid luminosity and black-hole mass recently derived from dynamical studies of nearby galaxies (e.g. Magorrian et al. 1998, Merritt & Ferrarese 2001). While undoubtedly uncertain to within a factor of a few, the attractiveness of this approach is that it allows an estimate of the central black-hole mass which is independent of any of the observed properties of the active nucleus. This estimate can then be compared with, for example, an estimate based on the assumption that the nucleus is accreting at the Eddington limit.

Using one of the most recent determinations of the black-hole:spheroid mass relationship, $M_{bh} = 0.0013 M_{spheroid}$ (Merritt & Ferrarese 2001), we find average black-hole mass estimates of $\langle M_{bh} \rangle = 1.5 \times 10^9 M_\odot$ for the RLQs in our sample, and $\langle M_{bh} \rangle = 0.9 \times 10^9 M_\odot$ for the RQQs. This subtle but apparently persistent difference (see below) arises directly from the fact that, although perfectly matched in optical *nuclear* luminosity, the hosts of our RQQs are, on average, $\simeq 1.5 - 2$ times less luminous than the hosts of their radio-loud counterparts.

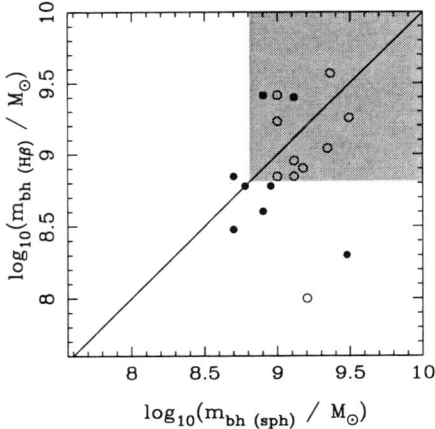

Figure 5. A comparison between the black-hole masses of quasars as predicted from host-galaxy spheroidal luminosity by Dunlop et al. (2001), and the corresponding values determined from $H\beta$ line-width by McLure & Dunlop (2001). The shaded area is shown to demonstrate that there is a region in which both approaches agree that $M_{bh} \gtrsim 10^9 M_\odot$, and that this region contains all except one of the RLQs (open circles), but excludes all except 2 of the RQQs (filled circles).

A comparison of the resulting predicted Eddington luminosities with the actual observed output of the quasar nuclei leads to the conclusion that most of the RLQs in our sample are emitting at $\simeq 5 - 10\%$ of their potential Eddington limit, while the radio-quiet objects span a wider range in efficiency, from $\simeq 10\%$ to 100% of the Eddington limit.

The above black-hole mass estimates can also be compared with values derived, completely independently, from an analysis of the velocity width of the H_β lines in the quasar nuclear spectra under the assumption that the broad-line region is gravitationally bound. This has been a growth industry in recent years (e.g. Wandel 1999, Laor 2000), bolstered by estimates of the size of the broad-line region from reverberation mapping of Seyfert galaxies. Recently Ross McLure and I have applied this technique to estimate the masses of the black holes which power the quasars we have imaged with the *HST*. The results are remarkably

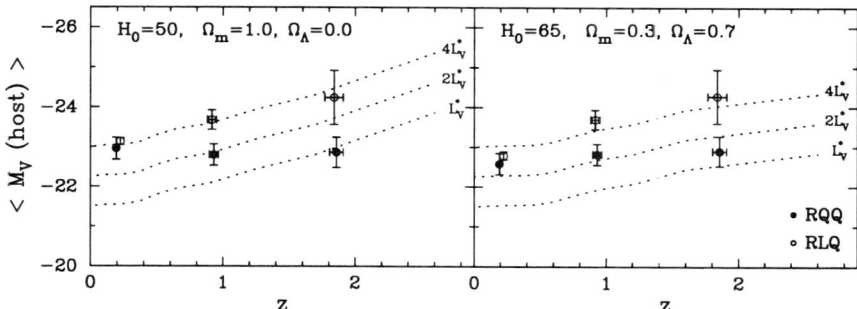

Figure 6. Mean absolute V-band magnitude versus mean redshift for the host galaxies of the RLQs (open circles) and RQQs (filled circles) in the NICMOS study of Kukula et al. (2001). Also shown is the subset of 5 RLQs and 7 RQQs from the Dunlop et al. (2001) WFPC2 study of quasars at $z \sim 0.2$ which have total (host + nuclear) luminosities in the same range as the high-redshift samples ($-24 \geq M_V \geq -25$). Error bars show the standard error on the mean. The dotted lines show the luminosity evolution of present day L^*, $2L^*$ and $4L^*$ elliptical galaxies, assuming a formation epoch of $z = 5$ with a single rapid burst of star formation followed by passive evolution thereafter. LH panel: assuming $H_0 = 50$ km s^{-1} Mpc^{-1}, $\Omega_m = 1.0$ and $\Omega_\Lambda = 0.0$. RH panel: $H_0 = 65$ km s^{-1} Mpc^{-1}, $\Omega_m = 0.3$ and $\Omega_\Lambda = 0.7$.

similar to the values described above, with the H_β line-width yielding $\langle M_{bh} \rangle = 1 \times 10^9 M_\odot$ for the RLQs, and $\langle M_{bh} \rangle = 5 \times 10^8 M_\odot$ for the RQQs.

Such agreement (to within a factor of two - Fig. 5) suggests that these mass estimates should be taken seriously, and of special interest is the fact that the apparent mass offset between the black holes which power radio-loud and radio-quiet objects persists (Fig. 5). Indeed, given the uncertainties involved, the division in mass appears fairly clean, at least in the sense that the radio-loud objects all lie above a certain mass threshold. Black-hole mass estimation from host spheroid luminosity leads to the conclusion that 9 out of the 10 RLQs have $M_{bh} > 10^9 M_\odot$ while only 4 out of the 11 RQQs lie above this threshold. From the H_β analysis 11 out of 13 RLQs have $M_{bh} > 10^{8.8} M_\odot$, while only 4 out of 17 RQQs lie in this regime (see McLure, these proceedings). A similar conclusion has recently been reached by Laor (2000).

There are a number of possible explanations for this apparent black-hole mass difference between radio-loud and radio-quiet objects. Interestingly Blandford (2000) argues that highly-collimated jets might only be produced by sub-Eddington accretion. Thus it may simply be the case that by selecting RLQs and RQQs of comparable optical output, we are guaranteed to find sub-Eddington accreters in the radio-loud sample, whereas the radio-quiet sample can contain at least some less massive holes emitting at close to maximum efficiency.

4. The connection to high redshift

The effective study of quasar hosts at high redshift is still in its infancy. However, already it is becoming clear that the mass offset between RQQ and RLQ hosts described above appears to grow with increasing redshift (see Fig. 6, plus Kukula et al. (2001), and contributions from Kukula, Ridgway, Impey and Rix in these proceedings), lending additional credence to its reality. Specifically, for the same nuclear luminosity, RQQ hosts at $z \simeq 2$ appear to be a factor of 2–3 less massive than either their low-z counterparts or their $z \simeq 2$ radio-loud counterparts. It is too early to say whether this is due to changes in the host population, or simply due to (on average) more efficient black-hole fueling revealing more clearly the mass threshold required for radio-loud activity. Over the next few years it will be extremely interesting to see if high-resolution infrared imaging with 8-m class telescopes can clarify our picture of high-z quasar hosts in the same way as has been achieved with the *HST* at low redshift.

References

Archibald E. et al. 2001, MNRAS, in press (astro-ph/0002083)
Bahcall J.N., Kirhakos S., Schneider D.P., 1994, ApJ, 435, L11
Bahcall J.N., Kirhakos S., Saxe D.H., Schneider D.P., 1997, ApJ, 479, 642
Blandford R.D., 2000. Phil. Trans. Roy. Soc. A., in press (astro-ph/0001499)
Canalzo G., Stockton A., 2000. AJ, 120, 1750
Dunlop J.S. et al. 2001, MNRAS, submitted
Fabian A.C., 1999, MNRAS, 308, 39
Hamilton T.S., Casertano S., Turnshek D.A., 2001, (astro-ph/0011255)
Kormendy J., 1977, ApJ, 217, 406
Kukula M.J. et al., 2001, MNRAS, in press (astro-ph/0010007)
Laor A., 2000, ApJ, 543, L111
Magorrian J. et al., 1998, AJ, 115, 2285
McLeod K.K., McLeod B.A., 2001, ApJ, in press (astro-ph/0010127)
McLure R.J., Dunlop J.S., 2001, MNRAS, in press (astro-ph/0009406)
McLure R.J. et al., 1999, MNRAS, 308, 377
Merritt D., Ferrarese L., 2001, MNRAS, 320, L30
Nolan L.A. et al., 2001, MNRAS, in press (astro-ph/0002027)
Percival W.J. et al., 2001, MNRAS, in press, astro-ph/0002199
Schade D., Boyle B.J., Letawsky M., 2000, MNRAS, 315, 498
Schneider D.P., Gunn J.E., Hoessel J.G., 1983, ApJ, 268, 476
Taylor G.L., Dunlop J.S., Hughes D.H., Robson E.I., 1996, MNRAS, 283, 968
Wandel A., 1999, ApJ, 519, L39

THE NUCLEUS-HOST GALAXY CONNECTION IN RADIO-LOUD AGN

Matthew O'Dowd,* C. Megan Urry
Space Telescope Science Institute, 3700 San Martin Dr., Baltimore, MD 21218, USA

Riccardo Scarpa
European Southern Observatory, Alonso de Cordova 3107, Vitacura, Casilla 19001, Santiago, Chile

Renato Falomo
Osservatorio Astronomico di Padova, Vicolo dell'Osservatorio 5, 35122 Padova, Italy

Joseph E. Pesce
Department of Astronomy, Pennsylvania State University, 525 Davey Lab, University Park, PA 16802

Aldo Treves
Universita' dell'Insubria, via Lucini 3, 22100 Como, Italy

Abstract We studied a moderate-redshift ($0.15 < z < 0.5$) sample of *HST* [1]-imaged radio-loud (RL) AGN spanning a large range of nuclear power. The total sample consists of 57 objects, divided into sub-samples of 40 BL Lac objects and 17 RL quasars. Essentially all the AGN nuclei lie in luminous elliptical galaxies (Es), which follow the Kormendy relation for normal Es. The host galaxies occupy a narrow range of ~ 1 magnitude, compared to >5 orders of magnitude in nuclear power, and the two sub-samples have indistinguishable luminosity–redshift distributions over the redshift range $0.15 < z < 0.5$. No correlation between nuclear brightness (corrected for beaming) and galaxy luminosity is seen. Assuming the host galaxy–black hole mass correlation for local galaxies, the derived Eddington ratios range from 2×10^{-4} for low-power AGN (BL Lac objects) to 2×10^{-1} for high power AGN (RL quasars). This implies a considerable difference in rate of fueling. At least for RL

---

*Also at School of Physics, University of Melbourne, Parkville, Victoria, Australia 3052
[1] Based on observations with the NASA/ESA Hubble Space Telescope, obtained at the Space Telescope Science Institute, which is operated by the Association of Universities for Research in Astronomy, Inc. under NASA contract No. NAS5-26555.

AGN, the host galaxy properties appear to have no close connection to the power of the jet.

Introduction

Whether there is a link between the intrinsic power of Active Galactic Nuclei (AGN) and their host galaxies is not known. It seems plausible that more massive host galaxies might form in high density regions that also support the formation of more massive nuclear black holes (Small & Blandford 1992, Haehnelt & Rees 1993, Kauffmann & Haehnelt 2000), and/or that more massive host galaxies could support an increased rate of fueling. Locally there is observational evidence that the mass of the central supermassive black hole is correlated with bulge mass (Magorrian et al. 1998, van der Marel 1999) and with bulge velocity dispersion (Ferrarese & Merrit 2000, Gebhardt et al. 2000).

There are some observational clues that suggest AGN nuclear luminosity might be related to host galaxy mass (McLeod et al. 1999, Schade et al. 2000, Hooper et al. 1997, Hutchings et al. 1984); however, other studies find no relation (Urry et al. 2000, McLure et al. 1999, Bahcall et al. 1997, Smith et al. 1986). Certainly among RL AGN no trend has been detected, though only a small range of (high) nuclear power has been probed.

Here we describe the relationship between host galaxy properties and nuclear power in a sample of RL AGN ($F_{5GHz}/F_B > 10$; Kellerman et al. 1989). We restrict ourselves to RL objects, which have relativistic jets formed near the central supermassive black hole, in order to probe a similar kind of physics near the black hole. The range of observed nuclear power reflects a continuum of accretion powers, which in turn arises from some variation in the process of fueling and/or jet formation near the black hole.

Using BL Lac objects and RL quasars it is possible to define redshift-matched samples that span a broad range of intrinsic nuclear luminosity. Our goal is to probe the link between nuclear power (processes near the black hole) and environment (host galaxy properties).

1. The Low-Power Sample

The difficulty in performing this analysis over a wide range of nuclear power lies with a redshift selection bias. Most low-power AGN, namely FR I radio galaxies or Seyferts, are not found in complete samples beyond about $z \sim 0.2$. In contrast, samples with large numbers of quasars extend to $z \sim 0.5$ and beyond. Indeed, in any flux-limited sample, there is an induced correlation between redshift and luminosity.

Even in the range $0 < z < 0.5$ galaxies can undergo significant luminosity evolution – up to 0.5 mag. in R band for a passively evolving E (Bressan et al. 1994). The coupling of this evolution with the induced redshift-nuclear luminosity correlation can in principle lead to a spurious correlation between host galaxy and nuclear luminosities.

Using BL Lac objects as representatives of the low nuclear power regime we can overcome this problem. BL Lacs are intrinsically low-power blazars whose close alignment with the line of sight results in strong relativistic beaming of the jet emission. This magnification means that BL Lacs can be found in large numbers beyond $z \sim 1$.

Our Cycle 6 HST snapshot survey of 132 BL Lac objects from seven complete samples (4 radio-, 1 X-ray-, 1 optically-selected) yielded WFPC2 R-band observations of 110 BL Lacs; a complete sub-sample in the redshift range $0.027 \leq z \leq 1.34$. The full details of the image reduction and host galaxy fitting can be found in Scarpa et al. 2000 and the host galaxy results are in Urry et al. 2000.

Host galaxies were resolved in 65% of the sample, with 95% resolved for $z < 0.5$, and none resolved for $z > 0.7$. For all resolved host galaxies with sufficient signal-to-noise ratios, a de Vaucouleurs profile (i.e., a bulge-dominated host) was clearly preferred over an exponential profile (i.e., a disk-dominated host). This strongly supports the idea that RL AGN reside in E galaxies rather than spirals.

The average K-corrected absolute magnitude of the host galaxies is $M_R = -23.7 \pm 0.6$ mag. ($H_0 = 50$, $q_0 = 0$). This makes them somewhat brighter than normal Es, for which $L^* = -22.4$ mag. (Efstathiou et al. 1988), and of similar magnitude to brightest cluster galaxies (-23.9 mag; Thuan & Paschell 1989).

2. The High-Power Sample

For the high-power comparison sample we use RL Quasars (RLQs) with comparable published imaging data. Specifically, we limit the sample to quasars that satisfy the following selection criteria:

• **Published HST imaging data are available.** The superior, stable resolution of HST is critical for accurate, consistent determination of host morphologies.

• **Low redshift range, $z \leq 0.5$.** For this redshift range our detection rate of BL Lac host galaxies was $> 95\%$, i.e., there is minimal bias against faint hosts with bright nuclei.

• **Host galaxies detected for the full sample.** That is, $\sim 100\%$ of the RLQs with $z \leq 0.5$ in the published sample must have resolved host galaxies. As above, this prevents bias against faint low host-nuclear luminosity ratios.

We identified three quasar studies meeting these criteria: McLure et al. 1999, 6 objects; Boyce et al. 1998, 5 objects; Bahcall et al. 1997, 6 objects. The total RLQ sample contains 17 quasars. The lowest RLQ redshift is $z = 0.15$. We therefore restrict the statistical comparison of the BL Lac and RLQ samples to $0.15 \leq z \leq 0.5$.

To verify the difference in intrinsic power of the two samples, we look at extended radio power, which is a rough bolometer of time-integrated nuclear power. The median extended power of BL Lac sub-sample is $P_{5GHz} = 24.3$ W Hz^{-1} sr^{-1}, compared to $P_{5GHz} = 27.2$ W Hz^{-1} sr^{-1} for the RLQs: a difference of ~ 3 orders of magnitude in median nuclear power. Over 5 orders of magnitude separate the least radio-powerful BL Lacs ($P_{5GHz} \sim 23$ W Hz^{-1} sr^{-1}) from the most radio-powerful RLQs ($P_{5GHz} > 28$ W Hz^{-1} sr^{-1}).

3. Results

The distribution of host galaxy luminosity versus redshift appears identical for the low- and high-power RL AGN samples, as can be seen in Fig. 1. In the redshift range studied, $0.15 \leq z \leq 0.5$, the average absolute R-band magnitude of the BL Lac host galaxies is $M_R = -23.79 \pm 0.6$ mag, and for RLQs is $M_R = -23.9 \pm 0.6$ mag. A KS test gives a 88% probability that these two populations are drawn from the same distribution.

Normal (non-active) Es exhibit a tight relationship between effective radius (R_e) and surface brightness at that radius (μ_e), the so-called Kormendy relation (Kormendy 1977), whose parameters depend on stellar dynamics and galaxy morphology. The Kormendy relations for BL Lac and RLQ host galaxies are indistinguishable (Fig. 2), suggesting that these galaxies are dynamically and morphologically very similar. Other samples of RL AGN, both FR IIs and FR Is (from Smith & Heckman 1990), follow the same general locus as well. Furthermore, this universal RL Kormendy relation is consistent with that for normal Es (Hamabe & Kormendy 1987).

To first order, it is clear that large differences in nuclear power do not correspond to large differences in host galaxy properties. To evaluate this more quantitatively, we compared nuclear luminosity to host galaxy luminosity. In the BL Lac objects, strong beaming (which allowed us to select this matched redshift sample) affects the perceived brightness. To correct for this we use published estimates of BL Lac Doppler factors from two studies: Ghisellini et al. 1993, in which lower limits on the

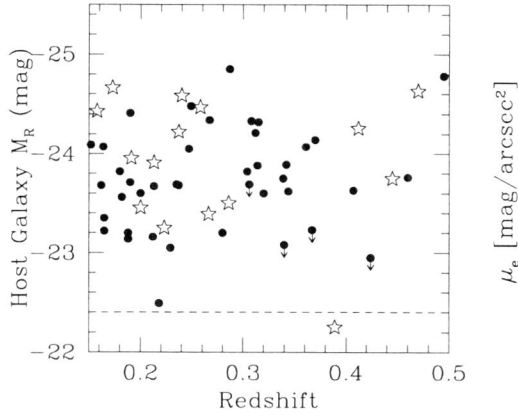

Figure 1. M_R(host) vs. z for BL Lacs (circles) and RLQs (stars). A K-S test in the range $0.15 \leq z \leq 0.5$ gives an 88% probability that these sub-samples are drawn from the same distribution. The dashed line shows L^* (-22.4 mag) for normal Es (Efstathiou et al. 1988).

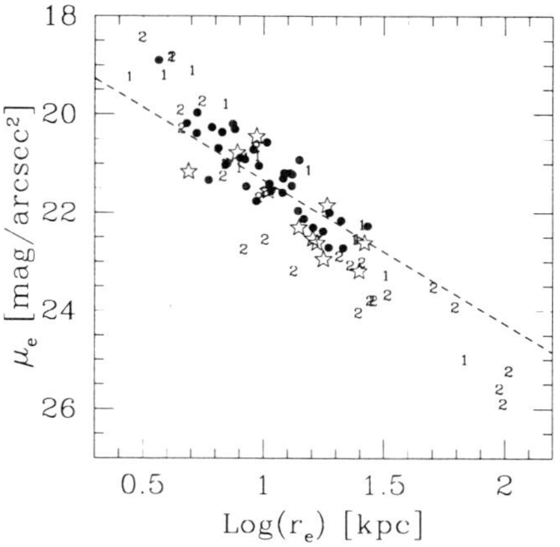

Figure 2. Kormendy relation for BL Lac objects (circles), RLQs (stars) and FRI and FRII radio galaxies (1s and 2s). The best-fit Kormendy relation for normal Es (dashed line; Hamabe & Kormendy 1987) suggests that RL AGN are dynamically and morphologically similar to normal Es.

Doppler factor are calculated from measurements of bulk motion in the radio jet, and Dondi & Ghisellini (1995), in which lower limits in the γ-ray emission region are calculated from measurements of the ratio of γ- to X-ray photons assuming a synchrotron self-Compton model. These two studies give lower limits of Doppler factors for 9 BL Lacs; we use the average of these values for the remaining sources, averaging separately for high-frequency peaked and low-frequency peaked BL Lacs (HBLs and LBLs; Padovani & Giommi 1995).

The lower limits in the beaming estimates translate to upper limits for the BL Lac nuclear luminosities. The median corrected R-band nuclear magnitude for the entire BL Lac subsample is -19.7 mag, while the median for the RLQ sub-sample is -25.2 mag. At least 5 orders of magnitude in power separate the least powerful BL Lacs ($M_R \sim -15$ mag) and the most luminous RLQs ($M_R \sim -27$ mag). Yet over this range there is essentially no difference in host galaxy luminosity – the hosts all fall in a tight range ($RMS \sim 1$ mag) near the luminosity of brightest cluster galaxies. There is no significant correlation between the two quantities.

Since these host galaxies seem to be normal Es, it is plausible that central black hole mass and bulge luminosity are correlated, as observed in nearby non-active galaxies (Magorrian et al. 1998, van der Marel 1999). If so, we can estimate upper limits to the Eddington ratios for

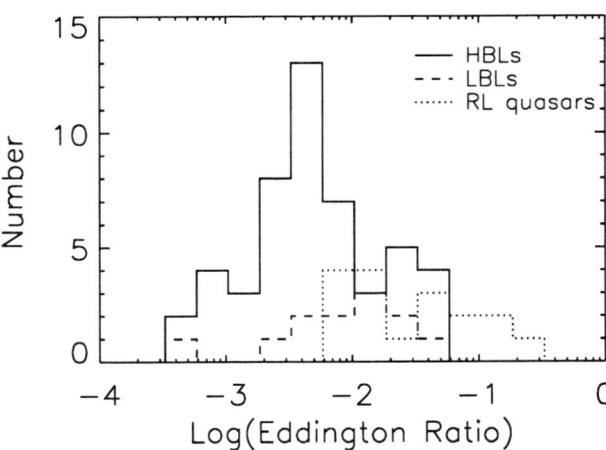

Figure 3. Histogram of Eddington ratios for BL Lacs and RLQs, assuming a constant M_{bh} to M_{bulge} ratio (Magorrian et al. 1998, Ho 1999). The values for BL Lacs represent upper limits. These are divided into high frequency- and low frequency-peaked BL Lacs (HBLs and LBLs). A progression in Eddington ratio is apparent from HBLs to LBLs to RLQs.

these AGN. Fig. 3 shows the histogram of Eddington ratios, assuming $M_{black\ hole} = 0.002 * M_{bulge}$ (Ho 1999), and a mass-to-light ratio of 4 (in R-band; following McLeod et al. 1999) for the host galaxies. The BL Lac Eddington ratios are upper limits, so the range of Eddington ratios for the sample as a whole spans at least ~ 4 orders of magnitude. We also note a steady increase in Eddington ratio from the least radio-powerful high-frequency peaked BL Lacs, to low-frequency peaked BL Lacs, to the most radio-powerful RLQs.

4. Conclusions

We find that essentially all BL Lac objects reside in luminous Es, as appears to be the case for RL AGN in general. The host galaxies of the BL Lac objects and RLQs in our study span a tight range (~ 1 mag) in optical luminosity, while both extended radio power and beaming-corrected nuclear luminosity span a much larger range of ~ 5 orders of magnitude. The two sub-samples have indistinguishable luminosity–redshift distributions over the redshift range $0.15 \leq z \leq 0.5$. We also find that these (and all) RL AGN follow Kormendy relations consistent with non-active Es.

Under the assumption that the host galaxies follow the same bulge luminosity–central black hole mass correlation observed in nearby, non-active galaxies, we find a range of > 4 orders of magnitude in Eddington ratio from the least luminous BL Lacs to the most luminous RLQs. We suggest that the difference in nuclear power between BL Lac objects and RLQs is due to rate of fueling, rather than black hole mass.

The nuclear power of RL AGN appears to have little dependence on global host galaxy properties. The property of "radio loudness" may select for host galaxies above a certain luminosity, but otherwise the host galaxy properties have no close connection to the power of the jet.

Acknowledgments

We thank Laura Maraschi for helpful discussion of this issue.
Support for this work was provided by NASA through grant number GO-06363.01-95A from the Space Telescope Science Institute, which is operated by AURA, Inc., under NASA contract NAS5-26555.
This work was also supported in part by NASA grant NAG5-9327.

References

Bahcall, J. N., Kirhakos, S., Saxe, D. H., Schneider. D. P., 1997, ApJ, 479, 642
Bressan, A., Chiosi, C., Fagotto, F., 1994, ApJS, 94, 63
Boyce, P. J., Disney, M. J., Blades, J. C. et al., 1998, MNRAS, 298, 121
Dondi, L., Ghisellini, G., 1995, MNRAS, 273, 583
Efstathiou, G., Ellis, R. S., Peterson, B. A., 1988, MNRAS, 232, 431
Ferrarese, L., Merritt, D., 2000, ApJ, 539, L9
Gebhardt, K., Bender, R., Bower, G. et al., 2000, ApJ, 539, L13
Ghisellini, G., Padovani, P., Celotti, A., Maraschi, L., 1993, ApJ, 407, 65
Haehnelt, M. G., Rees, M. J., 1993, MNRAS, 263, 168
Hamabe, M., Kormendy, J., 1987, in IAU Symp. 127, Structure and Dynamics of Elliptical Galaxies, ed. P. T. de Zeeuw (Dordrecht: Kluwer), 379
Ho, L., 1999, in Observational Evidence for Black Holes in the Universe, ed. S. K. Chakrabarti (Dordrecht: Kluwer), 157
Hooper, E. J., Impey, C. D., Foltz, C. B., 1997, ApJ, 491, 146
Hutchings, J. B., Crampton, D., Campbell, B., 1984, ApJ, 280, 41
Kauffmann, G., Haehnelt, M. G., 2000, MNRAS, 311, 576
Kellermann, K. I., Sramek, R., Schmidt et al., 1989, AJ, 98, 1195
Kormendy, J., 1977, ApJ, 218, 333
Magorrian, J., Tremaine, S., Richstone, D., 1998, AJ, 115, 228
McLeod, K. K., Rieke, G. H., Storrie-Lombardi, L. J., 1999, ApJ, 511, L67
McLure, R. J., Kukula, M. J., Dunlop, J. S., 1999, MNRAS, 308, 377
Padovani, P., Giommi, P., 1995, ApJ, 444, 567
Scarpa, R., Urry, C. M., Falomo, R., Pesce, J., Treves, A., 2000, ApJ, 532, 740
Schade, D. J., Boyle, B. J., Letawsky, M., 2000, MNRAS, 315 & 498
Small, T. A., Blandford, R. D., 1992, MNRAS, 259, 725
Smith, E. P., Heckman T. M., 1990, ApJS, 69, 365
Smith, E. P., Heckman, T. M., Bothun, G. D. et al., 1986, ApJ, 306, 64
Thuan, T. X., Paschell, J. J., 1989, ApJ, 346, 34
Urry, C. M., Scarpa, R., O'Dowd, M. et al., 2000, ApJ, 532, 816
van der Marel, R. P., 1999, AJ, 117, 744

William Percival

Ross McLure

THE HOST GALAXIES OF LUMINOUS RADIO-QUIET QUASARS

W.J. Percival,[1] L. Miller,[2] R.J. McLure[2] and J.S. Dunlop[1]
[1] *Institute for Astronomy, University of Edinburgh, Royal Observatory, Blackford Hill, Edinburgh EH9 3HJ, U.K.*
[2] *Dept. of Physics, University of Oxford, Nuclear & Astrophysics Laboratory, Keble Road, Oxford OX1 3RH, U.K.*

Abstract The results from a deep K-band imaging study, which reveals the host galaxies around a sample of luminous radio-quiet quasars, are shown to be important for models of the quasar population. The K-band images, obtained at UKIRT, were of sufficient quality to allow accurate 2D modelling of the underlying host galaxy. The derived average K-band absolute K-corrected host galaxy magnitude for these luminous radio-quiet quasars is $\langle M_K \rangle = -25.15 \pm 0.04$, comparable to luminosities derived from samples of quasars of lower total luminosity. Nuclear-to-host ratios therefore break the lower limit previously suggested from studies of lower nuclear luminosity quasars and Seyfert galaxies. We show how simple theoretical models of quasar activity predict such a limit, and suggest how this picture must be changed in light of our results.

1. Revealing the Host Galaxies

This Section summarises the observation and analysis of 13 luminous ($M_V \leq -25.0$) radio-quiet quasars, which resulted in determination of their host galaxy properties. The sample and its analysis are discussed in detail in Percival et al. (2001). The observations were all taken using the 256×256 pixel InSb array camera IRCAM 3 on the 3.9 m UK Infrared Telescope (UKIRT). The pixel scale is $0.281\,\mathrm{arcsec\,pixel^{-1}}$ which gives a field of view of $\sim 72\,\mathrm{arcsec}$. The sample was observed during three observing runs in 09/1996, 09/1997 and 05/1998. For the later two runs the tip-tilt AO system provided exceptional image quality and consistency, with FWHM $\sim 0.45\,\mathrm{arcsec}$. The variation in FWHM of the images taken on 07/09/1997 is presented in Fig. 1, and demonstrates the consistency of the images through the night.

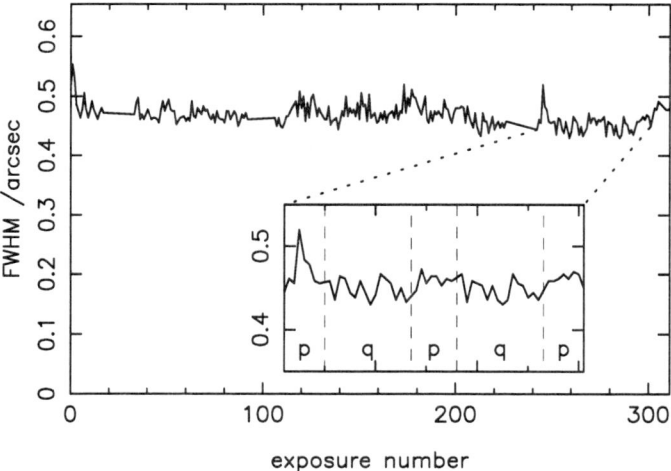

Figure 1. Plot showing the seeing variation quantified by the FWHM of the images taken on 07/09/1997 with tip-tilt operational. The inset shows a 'blow-up' of the seeing variations for the tail end of the night when quasar 0244−012 was observed. The dashed lines represent changes between quasar and PSF observations, denoted by 'q' or 'p' respectively. The repeated pattern for each of the qso mosaics is due to undersampling, as the quasar is centred on a different part of the chip for each observation, but a consistent position for corresponding images in different mosaics.

Obtaining an accurate PSF is vital for the correct removal of the nuclear component. With these ground-based observations the PSF varies with seeing conditions and telescope pointing, so an empirical PSF was determined for each quasar by observing a nearby bright star. For three of the quasars, there was a nearby star which could be placed on the frame with the quasar. For the remaining quasars the telescope was offset to a nearby bright star to use as the psf, before and after each quasar integration (which lasted a maximum of 1600 secs). To ensure consistent adaptive optics correction, properties of the tip-tilt guiding were matched between quasar and PSF measurements. To do this PSF stars were selected to enable tip-tilt guiding from a star of a similar magnitude, distance from the object, and position angle to that used for the quasar image. By examining fine resolution contour plots of the PSF images, it was found that the PSF was stable over the course of each night, but varied between nights at the telescope and for different telescope pointing. Because of this, the final stacking of PSF images was performed with the same weighting between days as for the quasar images.

Preliminary removal of the nuclear point source revealed two quasars with extended structure disturbed from an elliptical profile, and they were excluded from subsequent analysis. The luminosity and morphology of the host galaxies of the remaining 11 luminous quasars were estimated by fitting 2-dimensional model images to the data. The surface brightness profile fitted to the hosts was

$$\mu = \mu_o \exp\left[-\left(\frac{r}{r_o}\right)^{1/\beta}\right] \tag{1}$$

where μ_o, r_o, and β are fitted parameters. Additionally we fitted the axial ratio, angle on the sky and brightness of the nuclear component. A χ^2 minimisation technique was used to estimate the goodness of fit of each 6-parameter model, and to find the 'best-fit' host.

2. Luminosities of the hosts

For three of the quasars, analysis of how χ^2 varies within the parameter space revealed that the best-fit integrated host luminosity, L_{int}, was not well constrained. A host galaxy was determined as being present in that a lower limit was determined in all cases. However, the maximum light which could have come from the host was not clear because the shape of the host was not sufficiently resolved. The morphology of the best-fit galaxy at large L_{int} could alter to place the majority of the host light in the central region. This effect could have been avoided by placing limits on r_o or, for instance, using the near-infrared Fundamental Plane (Pahre, Djorgovski & de Carvalho 1998), although these upper limits would have been highly dependent on the criteria set. The host luminosity is ultimately limited by the total light in the image, and it is expected that the host luminosities for these quasars do have upper bounds at high values of L_{int}, but these high values would not be of any use in determining the actual host light. For the remaining eight luminous quasars, the minima were sufficiently constrained to provide 68.3% confidence intervals.

Host and nuclear luminosities for our quasars are compared with the results of other infrared studies in Fig. 2. The McLeod & Rieke (1994a,b) data were converted from the H-band to the K-band using observed colours for the total light, and assuming an evolved stellar population for the host (see Percival et al. 2001 for details). The study of Taylor et al. (1996) was performed in the K-band, and the apparent K-band magnitudes of host and nuclear components were taken directly from this work. The data from the different infrared samples were converted to absolute magnitudes by applying the K-correction of Glazebrook et al.

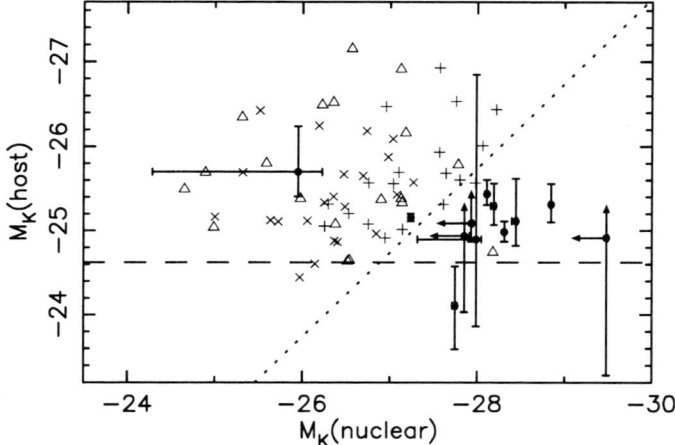

Figure 2. Nuclear vs. integrated host absolute K-band magnitudes for our sample of quasars (solid circles) with 68.3% error bars. The errors in the measured nuclear component are derived from these and consequently the errors are strongly correlated. Plotted for comparison are the calculated host magnitudes for the radio-quiet quasars imaged by Taylor et al. (1996) (open triangles), McLeod & Rieke (1994a) (crosses) and McLeod & Rieke (1994b) (plus symbols). The luminosity of an L^* galaxy, $M_K^* = -24.6$ (Gardner et al. 1997) is also plotted (dashed line), as is the locus of points with a rest-frame K-band nuclear-to-host ratio of 8 (dotted line).

(1995) for the host galaxy and assuming the nuclear component follows a standard power law spectrum $f(\nu) = \nu^{-0.5}$.

The average integrated host galaxy magnitude for the quasars of Percival et al. (2001) was found to be $\langle M_K \rangle = -25.15 \pm 0.04$. For comparison, when converted for cosmology exactly as our data, the sample of Taylor et al. (1996) gives $\langle M_K \rangle = -25.68$, McLeod & Rieke (1994a) $\langle M_K \rangle = -25.42$ and McLeod & Rieke (1994b) $\langle M_K \rangle = -25.68$.

3. The importance of these data for AGN models

The triangular shape of the McLeod & Rieke points in Figs. 2 and 3 found for low redshift ($0 < z < 0.3$) Seyferts and quasars of lower luminosity than those in the sample of Percival et al. (2001), has been shown to be in accord with a lower limit to the host luminosity which increases with nuclear luminosity (McLeod & Rieke 1995). This cut-off in the host luminosity is roughly equivalent to an upper limit to quasar nuclear-to-host ratios, which, rather worryingly, would also be the case if there was an observational limit to the possibility of host detection. In the K-band, this corresponds to a nuclear-to-host ratio of ~ 8.

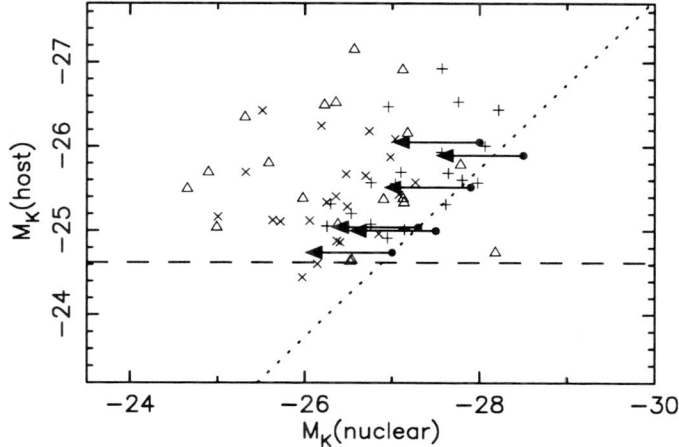

Figure 3. Figure showing why the 'triangular shape' of the McLeod & Rieke data (symbols as in Fig. 2) is favoured by models with a decreasing light curve, and a peak luminosity proportional to the black hole mass (which is expected in turn to be proportional to the spheroid luminosity). The locus of points traced out by such quasars are shown by the arrows.

This 'triangular' distribution of host and nuclear luminosities can be simply explained by a theoretical picture of quasar activity that includes:

1. A decreasing quasar light curve for each AGN (e.g. adopted by Kauffmann & Haehnelt 1999, $L = L_{\max} \exp(-t/t_{\rm acc})$)

2. A peak nuclear luminosity proportional to the mass of the black hole.

3. A linear relation between black hole mass and spheroid luminosity (e.g. Magorrian et al. 1998)

The locus of points expected to be traced out by AGN following these 'rules' is shown by the arrows in Fig. 3. In particular, Kauffmann & Haehnelt (2000) used these ideas to obtain a triangular shape for the distribution of quasars in the nuclear-host luminosity plane.

However, the integrated host luminosities derived from the K-band images of Percival et al. (2001) exhibit a low dispersion around a mean similar to that calculated in studies of less luminous quasars. The simple model of quasar emission with a peak nuclear-to-host ratio of ~ 8 described above does not hold for these quasars.

A possible fix would be to invoke the scatter of the Magorrian et al. (1998) relations to explain high luminosity quasars (and high mass black holes) within low luminosity structures, and invoke a steeply-declining host mass function to explain the lack of really massive hosts. Further work on this model would then be required, particularly with regard to explaining the slope of the high luminosity tail of the quasar luminosity function.

Alternatively, factors other than black-hole mass, such as nuclear obscuration, accretion processes, etc. could be the cause of differing nuclear luminosities within reasonably similar galaxies (with similar black hole masses). Here, to fit in with the simple model described above, the upper limit to allowed nuclear-to-host ratios would have to be ~ 20. Note that this would place the peak emission close to the predicted Eddington limit for the black holes expected to be powering these quasars.

In order to help constrain these models, a more detailed, and statistically robust analysis of the distribution of quasar host and nuclear luminosities is required. This would then constrain quasar models not only using peak luminosities, but also by the shape of the host mass function and how that ties in with the quasar luminosity function. Further work to test and further constrain the luminosities of the hosts of luminous quasars using HST and ground-based telescopes is underway.

References

Gardner J.P., Sharples R.M., Frenk C.S., Carrasco B.E., 1997, ApJ, 480, L99
Glazebrook K., Peacock J.A., Miller L., Collins C.A., 1995, MNRAS, 275, 169
Kauffmann G., Haehnelt M., 2000, MNRAS, 311, 576
Magorrian J. et al., 1998, AJ, 115, 2285
McLeod K.K., Rieke G.H., 1994a, ApJ, 420, 58
McLeod K.K., Rieke G.H., 1994b, ApJ, 431, 137
McLeod K.K., Rieke G.H., 1995, ApJ, 441, 96
Pahre M.A., Djorgovski S.G., de Carvalho R.R., 1998, AJ, 116, 1591
Percival W.J., Miller L., McLure R.J., Dunlop J.S., 2001, MNRAS, in press, astro-ph/0002199
Taylor G.L., Dunlop J.S., Hughes D.H., Robson E.I., 1996, MNRAS, 283, 968

THE RADIO LOUDNESS DICHOTOMY: ENVIRONMENT OR BLACK-HOLE MASS?

Ross McLure[1] and James Dunlop[2]
[1] *University of Oxford, UK*
[2] *Institute for Astronomy, University of Edinburgh, UK*

Abstract The results of a comprehensive study of the cluster environments and black-hole masses of an optically matched sample of radio-loud and radio-quiet quasars are presented. No evidence is found for a difference in large-scale environments, with both quasar classes found to be located in clusters of Abell class ~ 0. Conversely, virial black-hole mass estimates based on H_β line-widths show a clear difference in the quasar black-hole mass distributions. Our results suggest that a black-hole mass of $\sim 10^9$ $M_\odot$ is required to produce a powerful radio-loud quasar, and that it is black-hole mass and accretion rate which hold the key to the radio-loudness dichotomy.

1. Cluster environments of powerful AGN at $z \simeq 0.2$

The sample for this study consists of 44 powerful AGN with redshifts in the range $0.1 < z < 0.3$. The full sample is comprised of three matched sub-samples of 10 radio galaxies (RGs), 13 radio-loud quasars (RLQs) and 21 radio-quiet quasars (RQQs), the majority of which are drawn from the *HST* host-galaxy study of McLure et al. (1999) and Dunlop et al. (2001), with further objects taken from the host-galaxy study of Bahcall et al. (1997). The sub-samples are matched such that the $L_{5GHz} - z$ distribution of the two radio-loud sub-samples, and the $M_V - z$ distribution of the two quasar sub-samples are statistically indistinguishable.

Counts of excess galaxies surrounding each AGN were used to calculate the spatial clustering amplitude (B_{gq}) in order to provide a quantitative estimate of the richness of their cluster environments (Longair & Seldner 1979). A full description of the analysis and results of this study can be found in McLure & Dunlop (2001a).

1.1. Do RLQs have richer environments than RQQs?

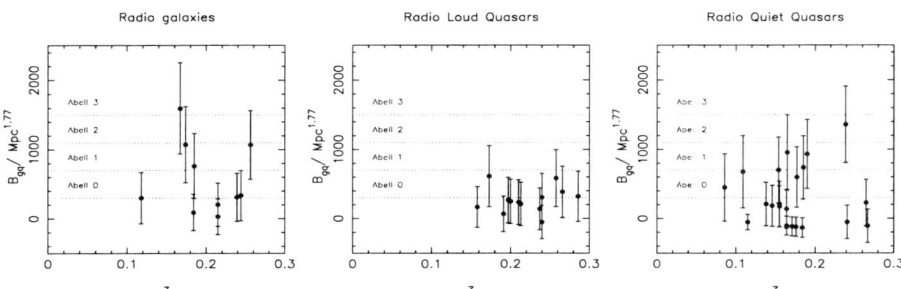

Figure 1. The $B_{gq} - z$ distribution for the three sub-samples. Also shown are approximate Abell cluster classifications according to the linear scheme proposed by Yee & López-Cruz (1999), renormalized such that $B_{gq} = 300\text{Mpc}^{1.77}$ corresponds to Abell class 0.

The spatial clustering amplitude results for the three sub-samples are shown in Fig. 1. It can be seen that there is no indication that the RLQs occupy systematically richer environments than the RQQs, with their respective mean clustering amplitudes of 267 ± 51 and $326\pm94\text{Mpc}^{1.77}$ indicating that both quasar classes occupy cluster environments comparable to Abell class 0. We note here that our finding that RQQs occupy similar environments to RLQs is not in general agreement with the literature, with the majority of previous studies finding RQQs to inhabit poorer environments than RLQs. For example, both Smith, Boyle & Maddox (1995) and Ellingson, Yee & Green (1991) found that at $z < 0.3$ RQQs have environments perfectly consistent with those of field galaxies. Although the exact cause of this discrepancy is difficult to determine, it is probably due to the fact that the RQQs and RLQs in our study are well matched in terms of nuclear luminosity.

Further support for this argument is provided by the good agreement between our results and those of the recent study by Wold et al. (2000, 2001, these proceedings) which involved a similar analysis of an optically matched sample of RQQs and RLQs at redshifts in the range $0.5 < z < 0.8$. Wold et al. also conclude that there is no difference in the average cluster richness of RQQs and RLQs of similar optical luminosity, suggesting that our results persist at higher redshift. In Fig. 2 we show a comparison of the clustering results for our RLQ sample, and those for a sub-sample of the Wold et al. RLQs which display the same range in radio luminosity. It is clear from Fig 2 that there is no evidence for evolution in RLQ cluster environments, at least out to $z \sim 0.8$.

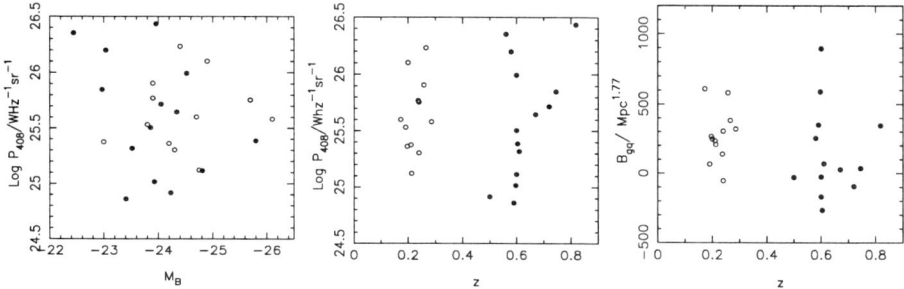

Figure 2. The left-hand and middle panels show the matching between our RLQ sample (open circles) and the sub-sample of the objects studied by Wold et al. (2000) discussed in the text (filled circles). The right-hand panel shows a comparison of their respective spatial clustering amplitudes.

The lack of any significant difference in the cluster environments of optically-matched samples of RQQs and RLQs suggests that cluster environment cannot be the primary cause of the quasar radio-loudness dichotomy. In the following section we move on from studying the large-scale environments of quasars, to explore what role (if any) is played by black-hole mass in determining the radio luminosity of AGN.

2. The black-hole masses of quasars and Seyfert galaxies

In this study (McLure & Dunlop 2001b) the quasar sample from host-galaxy study of McLure et al. (1999) and Dunlop et al. (2001) was combined with a sample of 15 Seyfert 1 galaxies from the reverberation mapping study of Wandel, Peterson & Malkan (1999). This combined sample provided the opportunity to investigate the relation between black-hole mass (M_{bh}) and host-galaxy bulge mass (M_{bulge}) over a wide range in AGN nuclear luminosity, and was assembled with the aim of addressing two important questions. Firstly, is the ratio of M_{bh}/M_{bulge} in Seyfert galaxies really a factor of twenty lower than seen in nearby galaxies and quasar hosts, as claimed by Wandel (1999), or can the apparent discrepancy be explained by a systematic over-estimate of the Seyfert bulge luminosities? Secondly, do differences in black-hole masses and gas accretion rates play a crucial role in determining the radio properties of AGN?

Host-galaxy bulge luminosities for the vast majority (39/45) of the sample were determined via full two dimensional disc/bulge decomposition of high resolution *HST* data, with the luminosities of the remaining objects being taken from literature disc/bulge decompositions. Consequently, our revised Seyfert bulge luminosities should be more accurate than the L_{bulge}/L_{tot} morphology-based estimates adopted by Wandel (1999). The black-hole masses are virial estimates (Wandel, Peterson & Malkan 1999) where it is assumed that the velocities of the broad-line regions clouds (as estimated from the FWHM of the H_β emission line) are due to the gravitational potential of the central black-hole. We make the further assumption that the broad-line region has a flattened disc-like geometry, and that the H_β line-widths are therefore orientation dependent, as suggested by the results of Wills & Browne (1986). The success of this model in reproducing the observed line-width distribution is illustrated in Fig. 3.

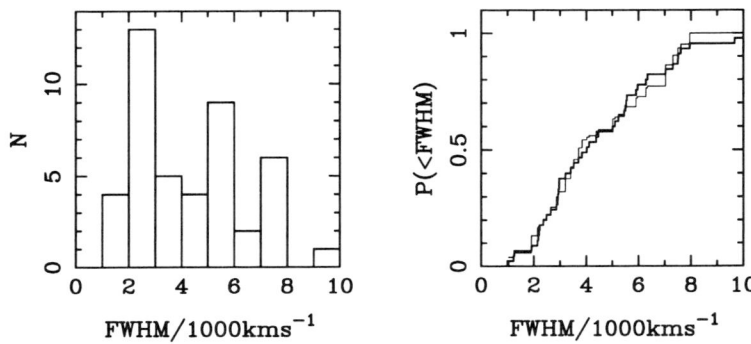

Figure 3. The left-hand panel shows the distribution of H_β FWHM measurements for the 45 objects in the combined quasar+seyfert sample. The right-hand panel shows the cumulative FWHM distributions displayed by the data (thick line) and that of the disc BLR model discussed in the text (thin line). The two cumulative distributions are indistinguishable, with a KS test probability of $p = 0.99$

3. The $M_{bh} - M_{bulge}$ relation

The black-hole mass vs. bulge luminosity relation for the combined quasar and Seyfert galaxy sample is plotted in Fig. 4. The two quantities can be seen to be well correlated, with the rank-order coefficient of $r_s = -0.66$ having a significance of 4.4σ. It can be seen immediately that, with our revised bulge luminosity estimates, there is no evidence that the Seyfert galaxies follow a different relation to the more luminous quasars. The best-fitting relation ($\chi^2 = 64.8$ for 45 d.o.f.) is found to be:

$$\log(M_{bh}/M_\odot) = -0.61(\pm 0.08)M_R - 5.41(\pm 1.75) \qquad (1)$$

and is shown as the solid line in the left-hand panel of Fig 4. It is noteworthy that the slope expected from a linear $M_{bh} - M_{bulge}$ relation is -0.52, which is not inconsistent with the best-fitting value of

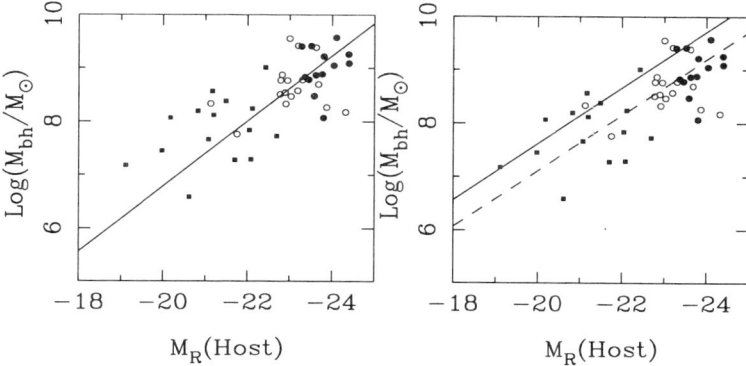

Figure 4. Both panels show black-hole mass vs. host galaxy R-band magnitude. The quasars are shown as open (radio quiet) and filled (radio loud) circles, while the Seyfert galaxies are shown as filled squares. In the left-hand panel the solid line is the best fit to the data. The solid line in the right-hand panel is the predicted relation from Magorrian et al. (1998). The dashed line in the right-hand panel is best fit to the data forcing a constant M_{bh}/M_{bulge} ratio, and corresponds to $M_{bh}/M_{bulge} = 0.0025$.

-0.61 ± 0.08. Consequently, a least-squares fit of the data was undertaken with an enforced slope of -0.52, and is shown as the dashed line in the right-hand panel of Fig. 4. This can be seen to be a reasonable representation of the data, and corresponds to a relationship of the form $M_{bh} = 0.0025\,M_{bulge}$. For reference, the solid line in right-hand panel of Fig. 4 shows the predicted relation from Magorrian et al. (1998), which can be seen to systematically over predict the black-hole masses by a factor of ~ 2.5.

The black-hole mass estimates for the 17 RQQs and 13 RLQs quasars provide an opportunity to determine the influence (if any) of black-hole mass on quasar radio luminosity. The continuum luminosity distributions of the two quasar sub-samples are indistinguishable, implying that any difference in their black-hole mass distributions are presumably linked to the difference in radio properties.

Our results suggests that a difference does exist between the RQQ and RLQ black-hole mass distributions, with the median black-hole mass of the RLQs being a factor of three larger than their radio-quiet counterparts. A natural division between the quasar sub-samples appears to occur at $M_{bh} \sim 10^{8.8}\,M_\odot$. Only 2/13 of the radio-loud quasars have $M_{bh} < 10^{8.8}\,M_\odot$, while only 4/17 of the radio-quiet have $M_{bh} > 10^{8.8}\,M_\odot$. This difference in black-hole mass distributions is shown to be significant at the 2.9σ ($p = 0.004$) level by a KS test. The implication from the quasar sample is therefore that (albeit with substantial overlap) for a given nuclear luminosity, the probability of a source being radio-loud increases with black-hole mass, or alternately decreases with L/L_{Edd}. The same conclusion was recently reached by Laor (2000) from his study of the virial black-hole masses of PG quasars.

4. Conclusions

- We find no detectable difference in the average cluster environments of optically matched RQQs and RLQs at $z \sim 0.2$, implying that cluster richness is not the primary cause of the radio-loudness dichotomy.

- Our results suggest that on average, luminous quasars occupy cluster environments as rich as Abell class 0, although a large scatter is present.

- We find that the bulges of both Seyfert galaxies and quasar host galaxies follow the same relation between black-hole and bulge mass, with a best-fitting linear relation of $M_{bh} = 0.0025 \, M_{bulge}$.

- Our black-hole mass estimates suggest that black-hole mass is a key parameter in determining an AGN's radio luminosity, and that a black-hole of mass $\simeq 10^9 \, M_\odot$ is required to produce a powerful radio-loud quasar.

References

Bahcall J.N., Kirhakos S., Saxe D.H., Schneider D.P., 1997, ApJ, 479, 642
Dunlop J.S. et al., 2001, MNRAS, submitted
Ellingson E., Yee H.K.C., Green R.F., 1991, ApJ, 371, 49
Longair M.S., Seldner M., 1979, MNRAS, 189, 433
Laor A., 2000, ApJ, 543, L111
Magorrian J. et al., 1998, AJ, 115, 2285
McLure R.J., Dunlop J.S., 2001, MNRAS, in press (astro-ph/0007219)
McLure R.J., Dunlop J.S., 2001, MNRAS, submitted (astro-ph/0009406)
McLure R.J., Kukula M.J., Dunlop J.S., Baum S.A., O'Dea C.P., Hughes D.H., 1999, MNRAS, 308, 377
Smith R.J., Boyle B.J., Maddox S.J., 1995, MNRAS, 277, 270
Wandel A., 1999, ApJ, 519, L39
Wandel A., Peterson B.M., Malkan M.A., 1999, ApJ, 526,579
Wills B.J., Browne I.W.A., 1986, ApJ, 302, 56
Wold M., Lacy M., Lilje P.B., Serjeant S., 2000, MNRAS, 316,267
Wold M., Lacy M., Lilje P.B., Serjeant S., 2001, MNRAS, in press (astro-ph/0011394)
Yee H.K.C., López-Cruz O., 1999, AJ, 117, 1985

QSO ENVIRONMENTS AT INTERMEDIATE REDSHIFTS

Margrethe Wold[1], Mark Lacy[2], Per B. Lilje[3] and S. Serjeant[4,5]
[1] *Stockholm Observatory, University of Stockholm, Sweden*
[2] *IGPP, Lawrence Livermore National Laboratory and UC Davis, California*
[3] *Institute of Theoretical Astrophysics, University of Oslo*
[4] *Astrophysics Group, Imperial College London*
[5] *Unit for Space Sciences and Astrophysics, University of Kent at Canterbury*

Abstract   We have made a survey of quasar environments at $0.5 \leq z \leq 0.8$, using a sample of both radio-loud and radio-quiet quasars matched in B-band luminosity. Our observations include images of background control fields to provide a good determination of the field galaxy counts. About 10 per cent of the quasars appear to live in rich clusters, whereas approximately 45 per cent live in environments similar to that of field galaxies.

The richness of galaxies within a 0.5 Mpc radius around the radio-quiet quasars is found to be indistinguishable from the richness around the radio-loud quasars, corresponding on average to groups or poorer clusters of galaxies. Comparing the galaxy richness in the radio-loud quasar fields with quasar fields in the literature, we find no evidence of an evolution in the environment with epoch. Instead, a weak, but significant correlation between quasar radio luminosity and environmental richness is present. It is thus possible that the environments of quasars, at least the powerful ones, do not evolve much between the present epoch and $z \approx 0.8$.

1.   Sample selection and observations

We report on a survey of the galaxy environments of intermediate redshift radio-loud and radio-quiet quasars (RLQs and RQQs, hereafter), described in more detail by Wold et al. (2000a,b). The quasar sample was pulled from complete catalogues with different flux limits, in order to cancel the correlation between redshift and luminosity known to exist in flux-limited surveys. It contains 20 RQQs and 21 steep-spectrum RLQs within a narrow redshift range ($0.5 \leq z \leq 0.8$), but with a wide range in AGN luminosity, both optical and radio. The two quasar samples are matched in both redshift and optical luminosity, and the distribution of sources in the redshift–optical/radio luminosity plane is shown in Fig. 1. Our assumed cosmology has $H_0 = 50$ km s^{-1} Mpc^{-1}, $\Omega_0 = 1$ and $\Lambda = 0$.

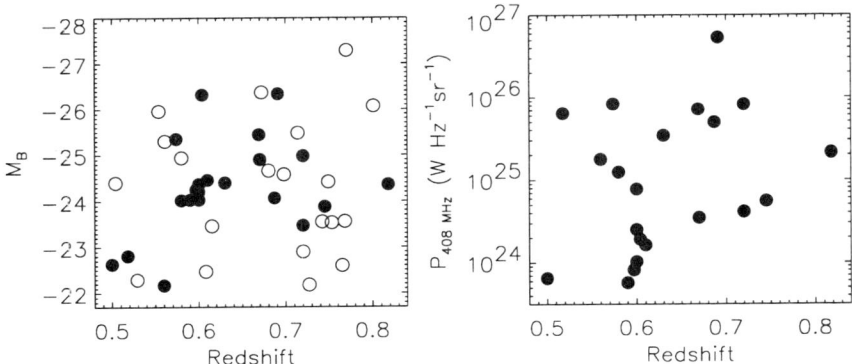

Figure 1. Left: quasar absolute magnitudes as a function of redshift. Open circles correspond to RQQs, and filled circles correspond to RLQs. Right: radio luminosity as a function of redshift for the RLQs.

The RLQs were selected from two complete surveys limited both in the optical and in the radio; low radio luminosity quasars from the 7CQ survey (Riley et al. 1999) and high radio luminosity sources from the MRC-APM survey (Serjeant 1996, Maddox et al. in prep., Serjeant et al. in prep.) The RQQs were selected from three different optically selected surveys, the faint UVX survey by Boyle et al. (1990), the intermediate-luminosity LBQS (Hewett et al. 1995) and the bright BQS (Schmidt & Green 1983).

Most of the quasar fields were imaged with the 2.56-m Nordic Optical Telescope on La Palma, Spain, but there are also some data from the *HST* and the 107-in telescope at the McDonald Observatory. In addition to the quasar fields, we also imaged several background control fields by offsetting the telescope 5–10 arcmin away from the quasar targets. Our observing strategy was to use filters that target emission longwards of the rest-frame 4000 Å break at the quasar redshifts, and thereby give preference to galaxies with evolved stellar populations physically associated with the quasars. In some cases we also used two filters in order to straddle the redshifted 4000 Å break. Depending on whether the quasar had a redshift $z < 0.67$ or $z \geq 0.67$, we used V and R, or R and I, respectively.

Using the galaxy counts from the background control fields, we evaluated the number of excess galaxies within a radius of 0.5 Mpc centered on the quasar, and thereafter converted to the 'clustering amplitude', B_{gq}, the amplitude of the spatial galaxy–quasar cross correlation function. The analysis involving the clustering amplitude is discussed elsewhere (e.g. Longair & Seldner 1979, Yee & Green 1987, Yee & López-Cruz 1999).

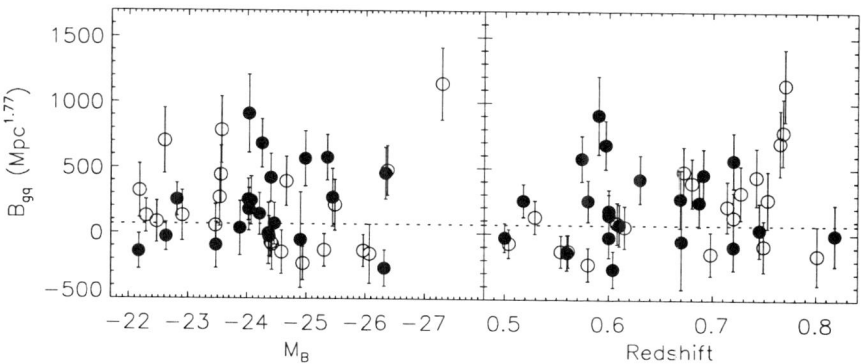

Figure 2. Clustering amplitudes as a function of quasar absolute magnitude (left) and redshift (right). The open circles correspond to the RQQs, and the filled circles are the RLQs. The dotted line indicates the clustering amplitude for local field galaxies (Davis & Peebles 1983).

2. Radio-loud vs radio-quiet qsos

The clustering amplitude for each quasar field is plotted as a function of quasar absolute B magnitude and redshift in Fig. 2. As seen from the figure, there is a wide spread in the amplitudes. About 10 per cent of the quasars seem to live in very rich environments ($B_{gq} \gtrsim 700$ Mpc$^{1.77}$), in some cases perhaps corresponding to Abell class 1–2 clusters. Another 10 per cent live in fairly rich environments, with B_{gq} between 500 and 700, and 45 per cent in environments similar to the field, $B_{gq} \lesssim 100$ Mpc$^{1.77}$.

Interestingly, we find that the average environment of the RQQs is indistinguishable from that of the RLQs. The mean clustering amplitudes for the RLQ and the RQQ samples are 210±82 and 213±66 Mpc$^{1.77}$, respectively. We thus find that the mean clustering amplitudes for the RLQ and the RQQ samples are statistically indistinguishable, implying that the RLQs and the RQQs live in similar environments at these redshifts. This result disagrees with Ellingson, Yee & Green (1991) who found that at $0.3 < z < 0.6$, the RLQs exist more often in rich environments than the RQQs, perhaps due to subtle selection effects in their somewhat heterogeneous sample. But our result is consistent with recent host galaxy studies finding that powerful RQQs exist in luminous, massive elliptical galaxies similar to the RLQs (e.g. Bahcall et al. 1997, McLure et al. 1999). Other investigators are also finding similar environments for powerful RLQs and RQQs at $z \approx 0.2$ (Fisher et al. 1996, McLure & Dunlop 2000, these proceedings).

Note that both the quasar fields and the background comparison fields were obtained in exactly the same manner, and that we have applied the same analysis to both samples. The comparison between the RLQ and the RQQ environment is therefore direct and internally consistent. Straight number counts in the quasar and background fields show a clear excess of galaxies at faint magnitudes (see Wold et al. 2000a,b), and we also find tentative evidence for clusters at the quasar redshifts in the form of a red sequence present in the richest RQQ fields, see Fig. 3.

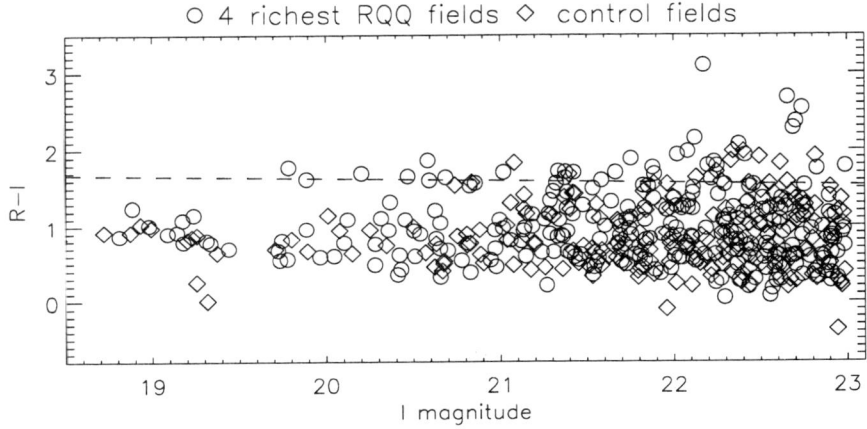

Figure 3. Colour-magnitude diagram of galaxies in the four richest RQQ fields (circles) and in the background control fields (diamonds). There is a hint of a red sequence (dashed line) in the quasar fields which is not present in the background control fields. The mean redshift of the four RQQs is 0.74, and the colour of the red sequence corresponds to the expected colour of galaxies at $z \approx 0.7$–0.8.

3. A link between radio luminosity and environmental richness?

This section treats the environment of the RLQs. As discussed in the beginning, our aim with selecting sources from different surveys with different flux-density limits was to overcome the luminosity-redshift degeneracy.

In our data, we found a hint of a correlation between the clustering amplitude and the radio luminosity of the quasars. To further investigate this, we added 51 steep-spectrum quasars from the work by Yee & Green

(1987), Ellingson et al. (1991) and Yee & Ellingson (1993), mostly at redshifts $0.2 < z < 0.6$. These data are plotted in Fig. 4, where it can be seen that the clustering amplitude correlates more with radio luminosity than with redshift.

To analyze this, we used Spearman's partial rank correlation coefficients giving the correlation coefficient between two variables holding the third constant. The correlation coefficient between B_{gq} and z, holding L_{408} constant, is 0.10 with a 0.8σ significance, whereas the correlation coefficient between B_{gq} and L_{408}, holding z constant, is 0.4 with a 3.4σ significance. We thus find no evidence for the cosmic evolution in B_{gq} as has been claimed for RLQs and radio galaxies (Yee & Green 1987, Ellingson et al. 1991, Hill & Lilly 1991).

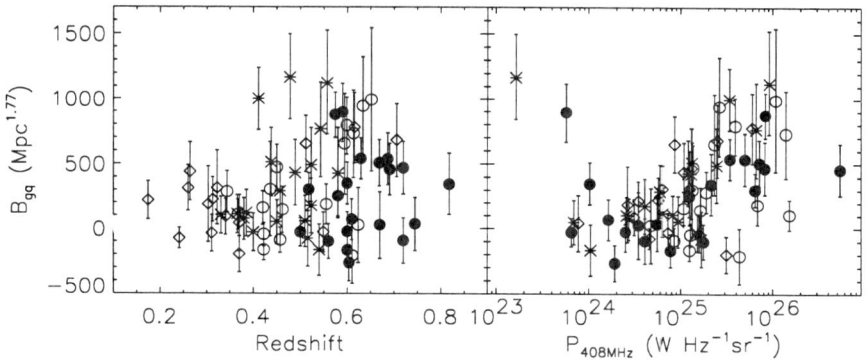

Figure 4. Clustering amplitudes in RLQ fields as a function of redshift (left plot) and radio luminosity at 408 MHz (right plot). Filled circles: this work, asterisks: Ellingson et al. (1991), open circles: Yee & Green (1987), diamonds: Yee & Ellingson (1993).

The correlation between B_{gq} and L_{408} is weak with much scatter, nevertheless, it is significant. Does this imply that the large-scale radio emission in the RLQs is affected by the environment? Models of radio sources certainly suggest this, where the minimum energy density in the radio lobes scales with the ram pressure at the working surface, implying that the synchrotron luminosity scales with the external gas density (Miller, Rawlings & Saunders 1993). Wold et al. (2000a) compare the data with a model which assumed *all* the variation in L_{408} is due to the differences in the environments, but find that the predicted correlation is much steeper than observed. It is thus possible that both the environmental density and the bulk kinetic power in the radio jets determine the radio luminosity. Alternatively, the relation between B_{gq} and L_{408} may just reflect an increasing mass of the host. This is possible if the radio luminosity is determined mostly by black hole mass, and if galaxies with massive black holes prefer rich environments.

4. Summary

1 Both the radio-loud and the radio-quiet quasars studied in this survey live in a diversity of environments, from field-like environments to what appears to be rich galaxy clusters. Only about 10 per cent of the quasars live in relatively rich clusters of Abell richness class 1–2, and approximately 45 per cent live in field-like environments.

2 The average environmental richness in the RLQ and the RQQ fields is statistically indistinguishable, corresponding to groups or poorer clusters. We therefore find that on scales of 0.5 Mpc there is no difference in the environments of the RLQs and the RQQs.

3 We find no evidence of an evolution with epoch in the environments of RLQs. Instead, the claimed evolution with redshift might have been caused by selection effects in flux-limited samples. The true underlying correlation may be that of environmental richness with radio luminosity.

References

Bahcall J.N., Kirhakos S., Saxe D.H., Schneider D.P., 1997, ApJ, 479, 642
Boyle B.J., Fong R., Shanks T., Peterson B.A., 1990, MNRAS, 243, 1
Davis M., Peebles P.J.E., 1983, ApJ, 267, 465
Ellingson E., Yee H.K.C., Green R.F., 1991, ApJ, 371, 49
Fisher K.B., Bahcall J.N., Kirhakos S., Schneider D.P., 1996, ApJ, 468, 469
Hewett P.C., Foltz C.B., Chaffee F.H., 1995, AJ, 109, 1498
Hill G.J., Lilly S.J., 1991, ApJ, 367, 1
Longair M.S., Seldner M., 1979, MNRAS, 189, 433
McLure R.J., Dunlop J.S., 2000, MNRAS, submitted (astro-ph/0007219)
McLure R.J., Kukula M.J., Dunlop J.S. et al. 1999, MNRAS, 308, 377
Miller P., Rawlings S., Saunders R., 1993, MNRAS, 263, 425
Riley J.M., Rawlings S., McMahon R.G. et al. 1999, MNRAS, 307, 293
Serjeant S.B.B., 1996, DPhil Thesis, Univ. Oxford
Schmidt M., Green R.F., 1983, ApJ, 269, 352
Wold M., Lacy M., Lilje P.B., Serjeant S., 2000a, MNRAS, 316, 267
Wold M., Lacy M., Lilje P.B., Serjeant S., 2000b, MNRAS, accepted (astro-ph/0011394)
Yee H.K.C., Ellingson E., ApJ, 411, 43
Yee H.K.C., Green R.F., 1987, ApJ, 319, 28
Yee H.K.C., López-Cruz O., 1999, AJ, 117, 1985

HOST GALAXIES AND CLUSTER ENVIRONMENT OF BL LAC OBJECTS AT $Z > 0.5$

Jochen Heidt[1], Josef Fried[2], Ulrich Hopp[3], Klaus Jäger[4], Kari Nilsson[5] and Eckhard Sutorius[1]

[1] *Landessternwarte Heidelberg, Königstuhl, 69117 Heidelberg, Germany*
[2] *Max-Planck-Institut für Astronomie, Königstuhl 17, 69117 Heidelberg, Germany*
[3] *Universitäts-Sternwarte, Scheinerstr. 1, 81679 München, Germany*
[4] *Universitäts-Sternwarte, Geismarlandstr. 11, 37083 Göttingen, Germany*
[5] *Tuorla Observatory, 21500 Piikkiö, Finland*

Abstract First results from an extensive study of the properties of the host galaxies and cluster environment of BL Lac objects at $z > 0.5$ are presented. The main aim of this study is to confirm the apparent evolution previously claimed. A host galaxy was detected in all BL Lac objects at $z < 0.7$. At higher redshifts our search was not successful. An exception might be PKS 0426-380 ($z > 1.03$). Here we detected a galaxy, which might be either along the line of sight or the host galaxy of the BL Lac itself. If the latter is correct, the upper limit to the redshift is wrong. A clear overdensity of galaxies have been found for two BL Lac objects, a slight overdensity for two others. In the remaining sources no overdensity of galaxies could be detected. Spectroscopy of the environment of PKS 0537-441 ($z = 0.894$) confirmed the physical association of at least 4 galaxies with the BL Lac. A previously discussed lensing scenario for this source can be ruled out with high confidence.

Introduction

In the last years, the host galaxies and the cluster environment of BL Lac objects have been extensively studied. The host galaxies of BL Lac objects are luminous ($M_R = -23.5$) and large ($r_e = 10$ kpc) with some evidence for evolution (e.g. Wurtz et al. 1996, Falomo & Kotilainen 1999). However, only a few BL Lac hosts above $z = 0.5$ are most likely resolved, whereas BL Lac hosts above $z = 0.7$ remain undetected so far. This is mainly due to the short integration times used and/or bad seeing conditions (partly old data, c.f. Stickel et al. 1993). Therefore, evolutionary trends are currently somewhat uncertain.

Table 1.

Object	z	T [min]	FWHM ["]	Filt	Tel.
OV 236	0.352	36	0.70	I	VLT
MS 2347.4+1924	0.515	20	0.95	I	NTT
PKS 1538+149	0.605	4	1.45	R	VLT
PKS 1519-273	~ 0.7	32	0.80	I	VLT
PKS 2240-260	0.774	54	0.75	I	NTT
PKS 0057-338	0.875	32	0.95	I	VLT
PKS 0537-441	0.894	123	0.85	I	NTT
PKS 0426-380	> 1.03	88	0.75	I	NTT
PKS 2029+121	1.215	64	1.20	I	VLT
PKS 2131-021	1.285	16	1.00	I	VLT
MH 2133-449	?	16	1.00	R	VLT
PKS 1349-439	?	16	1.05	R	VLT

The situation is similar for the cluster environment. While BL Lac objects are found in relatively poor cluster environments (Abell 0-1 at best) at $z < 0.3$, there seems to be an evolution (denser environments) towards higher redshifts (Fried et al. 1993, Wurtz et al. 1997). Contrary, the parent population of the BL Lac objects, the FR I radio galaxies do not show evidence for evolution of their cluster environment (Hill & Lilly 1991), but FR II radio galaxies and radio-loud QSO do (Yee & Green 1987, Hill & Lilly 1991, but see also Wold et al. this proceedings for contrary results). If this trend is correct, the Unified Scheme should be modified, and FR II as the parents be included at least in part (Wurtz et al. 1997). As for BL Lac host galaxies, sufficiently deep, high quality data are entirely lacking at $z > 0.6$.

Therefore we started a project to study the host galaxies and cluster environment of BL Lac objects at redshifts $z > 0.5$ by means of very deep imaging with the best resolution possible in order to confirm/disprove the apparent evolution of BL Lac hosts and their cluster environment. This study is accompanied by multi-object spectroscopy of a few selected fields. In this contribution the initial results of this study for a sample of 12 BL Lac objects are presented. In the following $H_0 = 50$ and $q_0 = 0$ is assumed.

1. The data

An overview about the data set is given in Table 1. The images have been acquired mostly through an I filter using SUSI2 at the NTT and FORS1 at the VLT1 (Antu) with exposure times ranging from 4 up to 120 min. We enjoyed subarcsecond seeing for most of the sources. Unfortunately, the sources with highest or unknown redshift have the worst seeing.

2. Host galaxies

The host galaxies of the BL Lac objects have been studied by means of a fully 2-dimensional decomposition (see Heidt et al. 1999a and Nilsson et al. 1999 for details). Three different types of models were fitted to the images. A model consisting of a bulge (representing the host) and a scaled PSF (representing the AGN), a model of a disk and a scaled PSF and a pure PSF model. The bulge/disk models were convolved with the PSF. The PSF was generated from a combination of several well exposed, but unsaturated stars on the same frame. This was possible for all sources except for PKS 2240-260 and PKS 0537-441, where fainter stars had to be used. The analysis was also hampered for PKS 0057-338, PKS 1519-273 and PKS 2029+121, which have relatively bright, nearby stars or galaxies overlapping with the isophotes of the BL Lac. In these cases an iterative procedure was used, which allowed to model the nearby galaxy/star as well (see Heidt et al. 1999b for details about this procedure).

The results are not very encouraging. In all 3 BL Lac objects at $z < 0.7$ a host galaxy could be resolved. Their morphological parameters derived are similar to those obtained by others (Falomo & Kotilainen, 1999, Scarpa et al. 2000). However, with one possible exception (PKS 0426-380), we do not succeed in resolving a host in the BL Lac objects at higher redshift.

The case of PKS 0426-380 is puzzling. Its lower limit to the redshift based on Mg II in absorption is $z < 1.03$ (Stickel et al. 1993). After subtraction of a scaled PSF residuals are clearly present, which can not be due to an imperfect PSF alone (see Fig. 1). Invoking a galaxy results in a much better fit to the image. If the galaxy detected is along the line of sight to the BL Lac at $z = 1.03$ and thus be responsible for Mg II in absorption in the BL Lac spectrum, it would be a very bright galaxy ($M_I = -26.6$(bulge), $M_I = -25.2$(disk)). Alternatively, the (limit to the) redshift of the BL Lac is wrong. In that case, the BL Lac could be at a much lower redshift, and thus the galaxy detected the BL Lac host galaxy. Spectroscopy of this field is required.

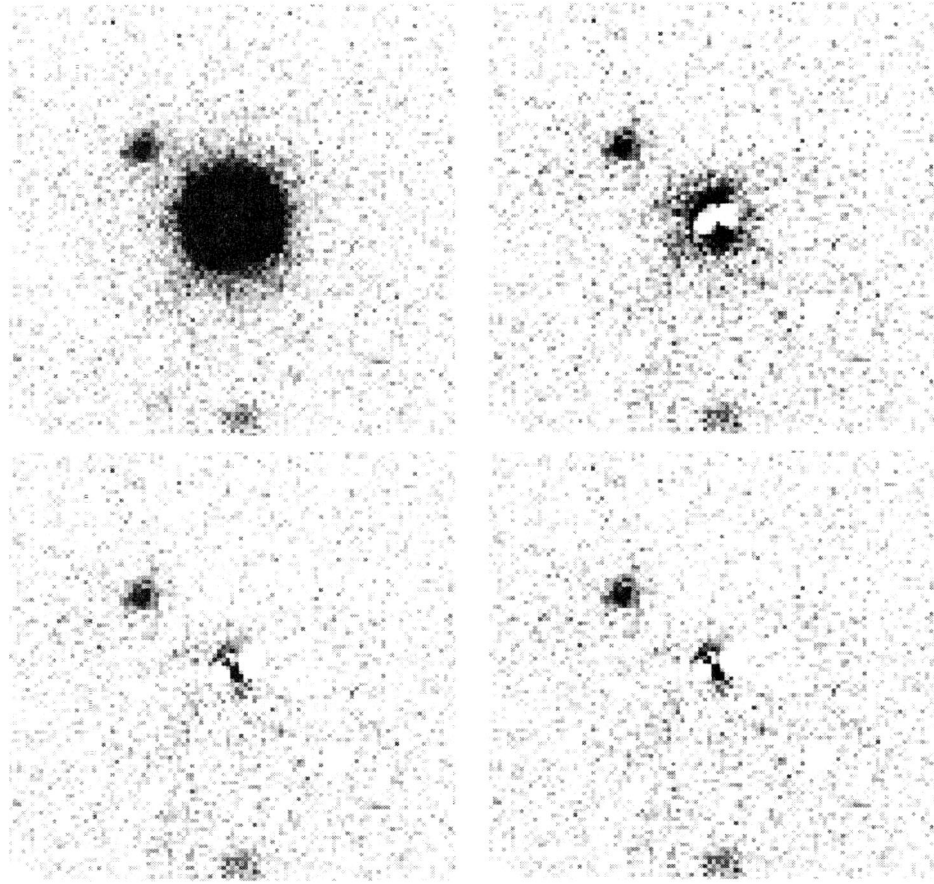

Figure 1. Top left) Image of PKS 0426-380 (z > 1.03). Top right) Same image after subtraction of best-fit PSF. Bottom left) Same image after subtraction of best-fit PSF + disk model. Bottom right) Same image after subtraction of best-fit PSF + bulge model. Field of view is 15" × 15".

3. Cluster environment

The cluster environment of the BL Lac objects was investigated by comparing the number density of galaxies within 250 kpc projected distance to the BL Lac to the number density beyond a radius of 500 kpc from the BL Lac. Due to the large field of view of SUSI2 and FORS1 a sufficiently large area was available to properly determine the number counts in the field. For PKS 0426-380 $z = 1.03$ was assumed, whereas $z = 0.7$ was chosen for the BL Lac objects with unknown redshift. Object detection and classification was done using SExtractor (Bertin &

Arnouts 1996). The completeness limits were estimated from the peaks of the differential number-count distribution. Except for the BL Lac objects with the highest redshifts (PKS 2029+121, PKS 2131-021), the completeness limits are 2–3 mag. fainter than a M_I^* galaxy (K-correction and extinction included) thus measuring the galaxy luminosity function sufficiently.

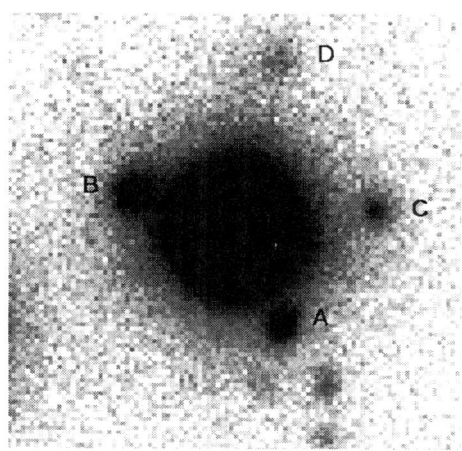

Figure 2. Image of PKS 0537-441 (z = 0.894). Field of view is 15" × 15". The four companion galaxies A-D forming an "Einstein cross" around he BL Lac are clearly visible.

In only two BL Lac objects (PKS 0537-441, PKS 2240-260) a clear overdensity of galaxies was detected ($\leq$ Abell 1), whereas two others (MS 2347.4+1924, PKS 0057-338) have a marginal excess of galaxies within 250 kpc projected radius as compared to the field. For the remaining 8 sources no overdensity was found. Thus our preliminary analysis do not confirm the apparent evolution found by Fried et al. (1993) and Wurtz et al. (1997).

4. PKS 0537-441

PKS 0537-441 is one of the most controversary discussed BL Lac objects. Stickel et al. (1988) claimed the detection of a galaxy along the line of sight to the BL Lac and concluded that its properties might be influenced by gravitational lensing effects. This was not confirmed, however, by Falomo et al. (1992) and Scarpa et al. (2000) by high resolution images from ground and the HST. On the other hand, Lewis & Ibata (2000) analysed the same HST image by Scarpa et al. (2000) and found 3 companion galaxies within a few arcseconds of the BL Lac. They argued that these 3 companion galaxies might be distant sources, which are gravitationally lensed by a small group or cluster harbouring PKS 0537-441.

In our deep image of PKS 0537-441 we also do not see the galaxy along the line of sight claimed by Stickel et al. (1988). The 3 companion galaxies studied by Lewis & Ibata (2000) are clearly present (Fig. 2). Additionally, a fourth galaxy at a similar distance as the other companion galaxies can bee seen as well. The optical appearance of these 4 galaxies is striking and reminiscent of an "Einstein cross". Their apparent magnitudes range from I = 21 up to 21.3. It is possible to make a crude lens model under the assumption that these 4 galaxies are at redshifts between $z = 2 - 3$ (Seitz priv. com.). In that case the lensing system could be a small group of galaxies or cluster at the redshift of PKS 0537-441.

To test this idea, we acquired spectra of the 4 companion galaxies using FORS1 at the VLT1 (Antu). The preliminary analysis suggest that all of the 4 galaxies are at redshifts similar to the BL Lac, their projected distances are $\leq$ 30 kpc from the BL Lac. They form together at least a group of galaxies, thus confirming the results of our analysis of the cluster environment of PKS 0537-441. Since they are so close to PKS 0537-441, some of them might be located in the halo of its host galaxy. Based on our imaging and spectroscopic results either lensing scenario (lensed or being lensed) can be ruled out with high confidence.

Acknowledgments

This work was supported by the Deutsche Forschungsgemeinschaft through SFB 328 and SFB 439.

References

Bertin E., Arnouts S., 1996, A&AS 117, 393
Falomo R., Melnick J., Tanzi E.G., 1992, A&A 255, L17
Falomo R., Kotilainen J., 1999, A&A 352, 85
Fried J.W., Stickel M., Kühr H., 1993, A&A 268, 53
Heidt J., Nilsson K., Sillanpää A., Takalo L.O., Pursimo T., 1999a, A&A 683, 692
Hill G.J., Lilly S.J., 1991, ApJ 367, 1
Lewis G.F., Ibata R.A., 2000, ApJ 528, 650
Nilsson K., Pursimo T., Takalo L.O. et al., 1999, PASP 111, 1223
Scarpa R., Urry C.M., Falomo R., Pesce J.E., Treves A., 2000, ApJS 532, 740
Stickel M., Fried J.W., Kühr H., 1988, A&A 206, L30
Stickel M., Fried J.W., Kühr H., 1993, A&AS 98, 393
Wurtz R., Stocke J.T., Yee H.K.C., 1996, ApJS 103, 109
Wurtz R., Stocke J.T., Ellingson E., Yee H.K.C., 1997, ApJ 480, 547
Yee H.K.C., Green R.F., 1987, ApJ 319, 28

ASSOCIATED ABSORPTION AND RADIO SOURCE GROWTH

Joanne C. Baker
University of California, Berkeley
jcb@astro.berkeley.edu

Abstract   Evidence is presented that C IV associated absorption is seen preferentially towards steep-spectrum radio sources of small linear size. A model for the clearing out of absorbing gas along the jet axis as the radio source grows is proposed. The general properties of the absorbers suggest that AGN environments are gas-rich.

Introduction

Quasar absorption systems provide a unique and powerful tool to probe tenuous gas clouds throughout the universe. The physical regimes probed in absorption may be quite different from those that dominate the emission from galaxies and active galactic nuclei (AGN), and hence the information gained is complementary to that obtained by other methods.

Relevant to this meeting, absorption lines close to the quasar redshift — associated absorbers — are likely to be valuable probes of quasar environments, as our sightline may intercept gas from a range of locations. For example, absorption may arise in gas outflowing from the quasar central engine, the ISM of the quasar host galaxy and/or ISM of nearby galaxies in clusters or groups near the quasar.

Previous studies have demonstrated that associated absorption is relatively common in quasar spectra and, moreover, is more frequent than expected by chance alignment of galaxies along the sightline to the quasar (Foltz 1988). The majority of quasars showing narrow absorption lines near the quasar redshift are radio-loud, steep-spectrum radio sources (Anderson et al. 1987, Foltz 1988), raising the question of whether the presence of absorbing material is in some way related to the formation of powerful radio jets in a fraction of AGN. Alternatively, optically-selected samples might be biased against objects with absorption. The presence of excess gas in the vicinity of quasars itself may play a role in the triggering and fuelling of AGN activity.

1. A study of C IV absorption in radio-loud quasars

To elucidate the relationship between the absorbing gas and quasar properties, spectroscopy of a complete sample of radio-loud quasars has been carried out. The quasars were drawn from the 408-MHz-selected Molonglo Quasar Sample (MQS; Kapahi et al. 1998) and comprise all MQS quasars with redshifts $1.5 < z < 3.0$ where C IV and/or Ly α is visible in the optical.

Spectroscopy at 1–2Å resolution has been obtained for 22 out of a total of 27 $z > 1.5$ MQS quasars with the Anglo-Australian Telescope, ESO 3.6m telescope and VLT UT1 telescope. We note that these observations are challenging due to the faintness of the high-redshift quasar targets, $19 < b_J < 22$. The data were reduced with FIGARO in the standard way.

To look for absorption trends with radio and optical properties, we focussed initially on the narrow C IV$\lambda\lambda$1548,1550 absorption-line doublet. For all quasars observed, the strongest C IV absorption system was identified within 5000 km s^{-1} of the C IV emission-line peak and its equivalent width measured (when visible).

The original motivation for this study came from noting that absorption was especially common in 'compact, steep-spectrum' (CSS) quasars in the MQS (Baker & Hunstead 1995, 1996). CSS radio sources are characteristically unresolved at the typical resolution of the VLA ($< 2''$), but have steep radio emission indicative of lobes rather than compact cores ($\alpha > 0.5$ where $S_\nu \propto \nu^\alpha$). Hence CSSs have until now been neglected in studies of the 'unified schemes' for radio sources which use core-to-lobe ratio as an orientation indicator (Barthel 1989). To assess the role of CSS sources within the AGN population, we have obtained $\sim 0\rlap{.}''1$ resolution MERLIN images for many CSS quasars in our sample at 1.5 and 5 GHz (de Silva et al., in preparation). Most of the CSSs are resolved in the MERLIN images, allowing measurement of linear sizes (or limits for those that remain compact).

2. Results

Fig. 1 shows the equivalent widths of the C IV absorption lines as a function of radio spectral index. Associated absorption is detected *exclusively* in steep-spectrum quasars in our sample (although we looked at only 3 flat-spectrum quasars). This pattern agrees with the results

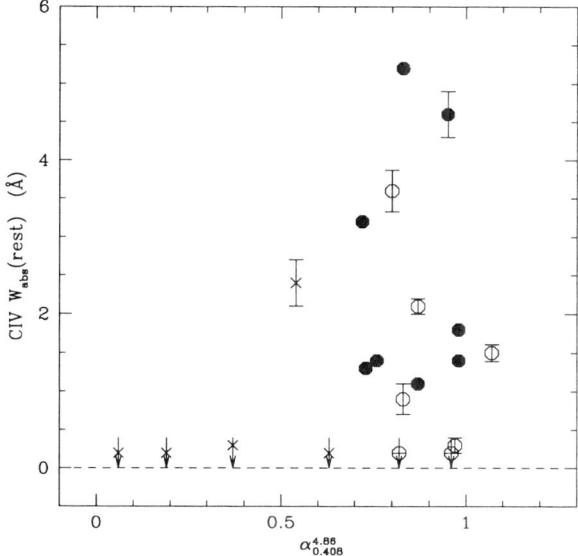

Figure 1. Equivalent width of C IV absorption as a function of radio spectral index, $\alpha_{0.408}^{4.86}$ (between 408 MHz and 4.8GHz). Symbols: filled circles are CSSs ($l < 25$ kpc), open circles are lobe-dominated quasars ($l > 25$ kpc and $R < 1$), crosses are core-dominated quasars ($R > 1$). Arrows indicate non-detections, with the limiting accuracy plotted.

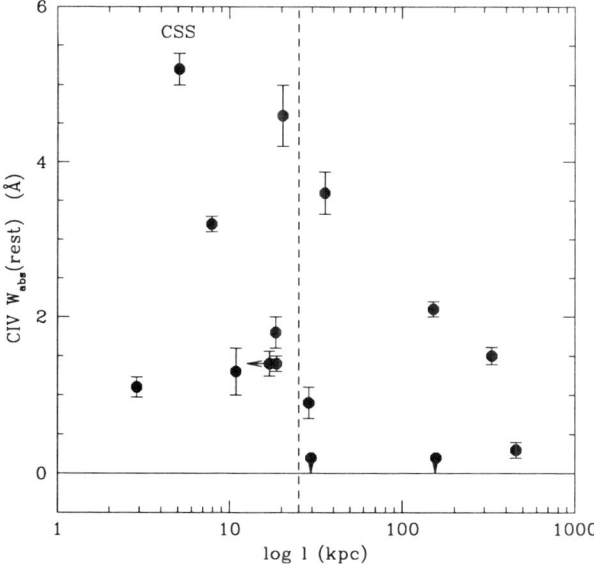

Figure 2. Equivalent width of C IV absorption as a function of radio source size, l. The dotted line at $l = 25$ kpc marks the size division between CSSs and non-CSSs.

for a larger inhomogeneous study by Foltz (1988), and the 3CR sample of Anderson et al. (1987).

Moreover, the strongest absorption is seen predominantly in the smallest radio sources. Fig. 2 — C IV equivalent width in absorption versus linear size of the radio source — clearly shows a tendency for the absorption strength to decrease in larger sources. Notably all the CSS sources show $z_a \approx z_e$ absorption stronger than $W_{abs} = 1$Å.

3. Interpretation

The simplest interpretation of the lack of absorption in flat-spectrum quasars is that the absorbing material avoids the radio jet axis. Flat-spectrum sources are thought to be viewed at small angles to the radio jet, such that their radio emission is dominated by synchrotron components in the compact radio core. Steep-spectrum quasars, on the other hand, have weaker cores due to being viewed at larger angles to the jet. Thus the absorbing clouds must lie preferentially away from the jet. Such a picture is in line with the 'unified schemes' for AGN (Barthel 1989, Antonucci 1993), and consistent with a trend for C IV absorption strength to decrease with increasing radio-core dominance (Barthel, Tytler & Vestergaard 1997).

The anticorrelation of C IV absorption strength and linear size, on the other hand, cannot be interpreted solely as an orientation effect. However, it may be indicative of an effect of radio source growth. If we make the reasonable assumption that radio sources expand steadily during their lifetime, then we might expect that small sources will tend to be younger on average than larger sources in a flux limited sample (although the relationships between age, size and luminosity are complex). VLBI observations have confirmed in a number of cases that compact double sources can be as young as hundreds to 10^4 years (Taylor et al. 2000). Alternatively, the jets in small sources might be confined by unusually dense ISM, but there is currently little evidence to support this picture (at least for the CSS class as a whole). CSSs are generally consistent with being scaled-down versions of larger sources, with their radio emission confined to within the host galaxy.

The steady decrease in absorbing column density as the source grows may be due to the lobes physically displacing gas as they expand, or due to ionisation by the quasar nucleus, or a combination of the two.

4. Discussion

The picture described above relates the drop in absorbing column density to radio source growth. The question of the origin of the gas remains. The spectroscopy shows that the gas contains a mixture of H I and other species with a range of ionisation, from O I and C II to C IV and N V. Dust is almost certainly also present. The presence of an H I reservoir might be related to prior merger activity, thought to play a role in triggering AGN, or may be primordial and enriched by galactic winds and AGN outflows.

Similar absorbers have been reported in high-redshift radio galaxies (HzRGs), where the absorbing clouds have been shown to extend over tens of kpc, at least over the extent of Ly α emission halos seen around many HzRGs (Röttgering et al. 1995, van Ojik et al. 1997). The Ly α emission halos around HzRGs are sometimes clumpy, and have even been compared with the sites of primordial galaxy formation (Pentericci et al. 2000, Miley, these proceedings).

However, massive H I halos have also been seen in low redshift radio sources (Conway 1996, Peck et al. 2000), notably in compact ones. Thus it is interesting to ask whether H I/Ly α halos are ubiquitous and perhaps fundamental to radio source and/or AGN formation.

Another prediction of the displacement model is that there should exist examples of compact radio sources which are completely enshrouded by the dusty absorber. Interestingly, Tadhunter (these proceedings) has reported finding such an object.

5. Summary

Evidence is presented that associated absorption is seen preferentially in steep-spectrum quasars, and is especially strong in small radio sources (such as CSSs). We present a model for the clearing out of the absorbing gas close to the jet axis by lobe expansion as the source grows. Future study of the absorbers should shed light on their relation to large-scale Ly α and H I halos, and perhaps yield evidence for a role in triggering AGN activity. Certainly, this adds to the evidence that quasar environments are gas-rich.

Acknowledgments

JCB acknowledges support for this work which was provided by NASA through Hubble Fellowship grant #HF-01103.01-98A from the Space Telescope Science Institute, which is operated by the Association of Universities for Research in Astronomy, Inc., under NASA contract NAS5-26555.

References

Antonucci R.R.J., 1993, ARA&A, 31, 473
Anderson S.F., Weymann R.J., Foltz C.B., Chaffee F.H. Jr., 1987, AJ, 94, 278
Baker J.C. 1997, MNRAS, 286, 23
Baker J.C., Hunstead R.W., 1995, ApJL, 452, L95
Baker J.C., Hunstead R.W., 1996, The second workshop on GPS and CSS radio sources, eds I. Snellen, R Schilizzi, H. Röttgering and Bremer M., Leiden, p166
Barthel P.D., 1989, ApJ, 336, 606
Barthel P.D., Tytler D.R., Vestergaard M., 1997, in 'Mass ejection from AGN', eds N. Arav, I. Shlosman and R.J. Weymann, ASP Conf. Series, Vol. 128, p48
Conway J. 1996, The second workshop on GPS and CSS radio sources, eds I. Snellen, R Schilizzi, H. Röttgering and Bremer M., Leiden, p198
Foltz C.B., Chaffee F.H., Weymann R.J., Anderson S.F., 1988, in Quasar Absorption Lines: Probing the Universe, eds. Blades et al., C.U.P., Cambridge, p.53
Kapahi V.K., Athreya R.M., Subrahmanya C.R., Baker J.C., Hunstead R.W., McCarthy P.J., van Breugel W., 1998, ApJS, 118, 327
Peck A.B., Taylor G.B., Fassnacht C.D., Readhead A.C.S., Vermeulen R.C., 2000, ApJ, 534, 104
Pentericci L. et al., 2000, A&A, 361, L25
Röttgering H., Hunstead R.W., Miley G., van Ojik R., Wieringa M., 1995, MNRAS, 277, 389
Taylor G.B., Marr J.M., Pearson T.J., Readhead A.C.S., 2000, ApJ, 541, 112
van Ojik R., Röttgering H.J.A., Miley G.K., Hunstead R.W., 1997, A&A, 317, 358

HOST GALAXIES OF RGB BL LACERTAE OBJECTS

K. Nilsson, L. Takalo, T. Pursimo, A. Sillanpää
Tuorla Observatory, FIN-21500 Piikkiö, Finland
kani@astro.utu.fi

J. Heidt
Landessternwarte Heidelberg, Königstuhl, D-69117 Heidelberg, Germany
jheidt@lsw.uni-heidelberg.de

Abstract We present the status of our ongoing program at the Nordic Optical Telescope (NOT) to determine the host galaxy properties of the ROSAT – Green Bank (RGB) BL Lacertae objects. This sample is by far the largest BL Lac sample (127 objects) obtained from a single survey. We give a description of the sample and discuss first results based on ~ 60 objects.

Keywords: Galaxies: active – Galaxies: BL Lacertae objects: General

1. The RGB sample

The ROSAT–Green Bank (RGB) sample of BL Lacertae objects (Laurent-Muehleisen et al. 1999) was formed by cross-correlating the ROSAT All-Sky Survey (RASS) with the Green Bank (GB87) 5 GHz radio survey, searching for optical counterparts in the POSS I photographic plates and classifying the radio–optical coincidences into normal galaxies, emission line AGN and BL Lacertae objects based on their optical spectra.

This procedure produced 127 BL Lacs, of which 38 were new discoveries. Of the 127 objects 100 are definite BL Lacs, i.e. they adhere to the widely used standard $W_\lambda < 5\text{Å}$, $Br_{4000} < 25\%$ (Stocke et al. 1991) and 27 are BL Lac "candidates", i.e. their line strengths W_λ and/or 4000 Å breaks Br_{4000} exceed the above limits, but are within the wider definition proposed by Marchã et al. (1996).

The RGB BL Lac sample is triply limited (radio, optical and X-ray), but not complete. However, a complete subsample of 33 objects covering 3970 deg^2 can be formed. The RGB BL Lacs exhibit a wider range of spectral broadband characteristics than previous BL Lac samples selected at a single frequency band. In the $\alpha_{ox} - \alpha_{ro}$ diagram the RGB BL Lacs fill the gap between the areas occupied by radio- and X-ray selected BL Lacs showing that BL Lacs exhibit a continuous range of broadband spectral characteristics.

2. Observations and analysis

We have obtained R-band images of ~ 100 sample objects using the Nordic Optical Telescope (NOT). Exposure times have been 300 – 3000 seconds depending on object brightness and the relative prominence of the host galaxy. Atmospheric conditions have been usually very good with FWHM $\sim 0.5 - 1.2$ arcsec.

To determine the host galaxy parameters we fit the observed surface brightness distribution with a two-dimensional two-component model consisting of an unresolved nuclear component and the host galaxy component. We make the fits using $\beta = 0.25$ (representing an elliptical galaxy), $\beta = 1.00$ (a disk galaxy) or let β to be a free parameter. The model is convolved with the observed PSF to mimic the atmospheric blurring. Using standard nonlinear fitting methods we find the combination of model parameters that minimizes the χ^2 between the model and the data. The fitting procedure also provides the final two-dimensional model and a residual image that can be used to study phenomena close to the nucleus.

3. First results

So far we have fitted 66 objects of the RGB sample using the above procedure. Of these 50 are resolved well enough for the determination of host galaxy parameters. The discussion below is limited to those 30 objects that have spectroscopic redshifts. Below is a summary of the first results :

- With one exception, the host galaxies appear to be relatively luminous (-25.2 < M_R < -22.0) and large (5 kpc < r_{eff} < 40 kpc) elliptical galaxies. One of the BL Lac candidates appears to be residing in a galaxy with a strong disk component and spiral arms.

- In the $r_{eff} - \mu_{eff}$ projection of the fundamental plane (Fig. 1) the RGB BL Lacs at low redshifts (z < 0.15) follow a relation similar to nonactive ellipticals, FR I and FR II host galaxies.

- We do not find any significant correlations between the host galaxy parameters (M_R, r_{eff}) and nuclear parameters (M_{nucl}, α_{ox}, α_{ro}).

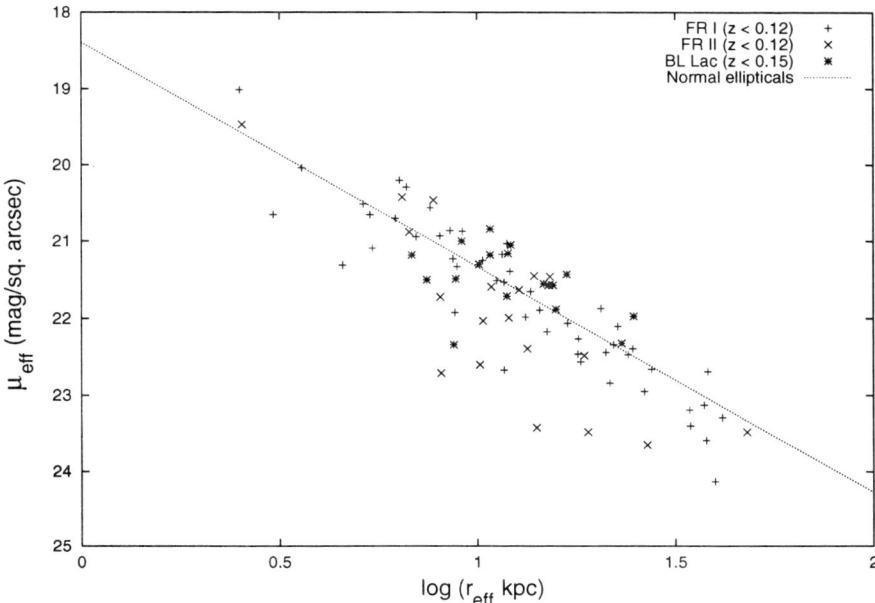

Figure 1. The Kormendy relation for various groups of elliptical galaxies. The data for FR I, FR II and normal galaxies are from Govoni et al. (2000).

Acknowledgments

This work has been supported by the Academy of Finland.

References

Govoni, F., Falomo, R., Fasano, G., Scarpa, R., 2000, A&A, 353, 507
Laurent-Muehleisen, S. A., Kollgaard, R. I., Feigelson, E. D., Brinkmann, W., Siebert, J., 1999, ApJ, 525, 127
Marchã, M. J. M., Browne, I. W. A., Impey, C. D., Smith, P. D., 1996, MNRAS, 281, 425
Stocke, J. T., Morris, S. L., Gioia, I. M., Maccacaro, T., Schild, R., Wolter, A., Fleming, T., Henry, J. P., 1991, ApJS, 76, 813

Renato Falomo (left), Riccardo Scarpa (center) and Alessandro Bressan (right)

From right end to left, counter-clockwise, Aimo Sillanpää, Kari Nilsson, Jari Kotilainen, Clive Tadhunter, Johan Knapen, Masayuki Umemura, his daughter and Mrs. Umemura.

ON THE PARENT POPULATION OF RADIO GALAXIES AND THE FR I–II DICHOTOMY

Riccardo Scarpa and C. Megan Urry

European Southern Observatory; Space Telescope Science Institute
rscarpa@eso.org; cmu@stsci.edu

The most promising explanation for the nuclear activity of galaxies is the presence of gas accretion around a massive black hole, and it seems clear now that all galaxies have a massive black hole in their center (Richstone et al. 1998, van der Marel 1999). This suggests that all elliptical galaxies have the basic ingredient for becoming active.

Here we test the possibility that all elliptical galaxies can host radio sources of any power and radio class. In particular, we test whether it is possible to link the optical luminosity function (LF) of non-radio and radio galaxies. To do that, we note that ellipticals of different luminosity might well have different probabilities of forming strong radio sources. Indeed in complete samples of radio sources (Ledlow & Owen 1996, Govoni et al. 2000) a roughly constant number of radio galaxies (RG) is observed between $-25 < M_R < -21$ mag, indicating the probability of observing radio emission increases strongly with the optical luminosity, L. To constrain this probability function, we start from the following general assumptions based on empirical result for RG:

(1) The optical LF of non-radio ellipticals is a Schechter function: $\Phi(L) = \frac{\Phi^*}{L^*}(\frac{L}{L^*})^\alpha e^{-(\frac{L}{L^*})}$. We set $L^* = 2.3 \times 10^{11} L_\odot$ (or $M^* = -22.8$ in the Cousins R band; $H_0 = 50$ km/s/Mpc; $q_0 = 0$) and $\alpha = +0.2$, as found for elliptical galaxies in the Stromlo-APM experiment (Loveday et al. 1992).

(2) All elliptical galaxies of all optical luminosities have the potential of being radio sources, with a probability $S(L) = S^*(\frac{L}{L^*})^h$. Where S^* sets the overall normalization of the function, and S(L) is dimensionless.

(3) Regardless of L, once activated, all ellipticals produce radio sources with the same power-law distribution $N(P) \propto P^{-2}$ (in units of P^{-1}; Toffolatti et al. 1987, Urry & Padovani 1995)

(4) In the radio-optical luminosity plane FR I and FR II are separated by a transition line roughly proportional to L^2, with normalization depending on the frequency under consideration.

From hypothesis 1 and 2, the normalized cumulative distribution of RG in luminosity L is given by the incomplete gamma function $\gamma(1 + \alpha + h, L/L^*)$. Both constants Φ^* and S^* cancel out, leaving h as the only free parameter. The best fit to the observations is obtained for $h = 2 \pm 0.4$ (Fig. 1).

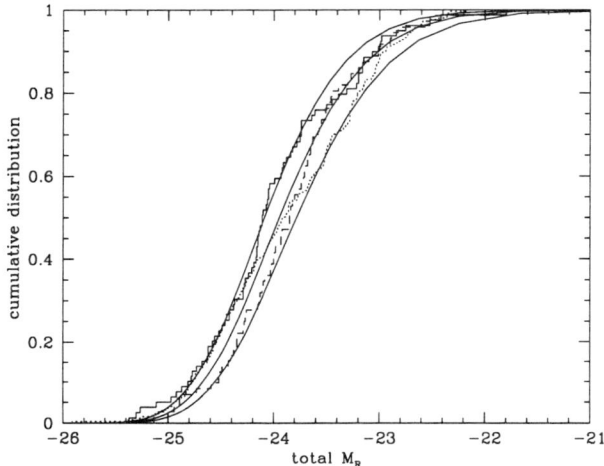

Figure 1. Cumulative distribution of optical magnitudes for RG from three different samples and $H_0 = 50$ km/s/Mpc. **solid line:** Govoni et al. (2000); **dotted line:** Ledlow & Owen (1996); **dashed line:** Smith & Heckman (1989). Superposed to the observed data, are the expected cumulative distribution for RG given by the incomplete gamma function $\gamma(1 + \alpha + h, L/L^*)$, for $h =$ 2.4, 2.0, and 1.6, from left to right, respectively.

Having fixed $h = 2$, we then use assumptions 3 and 4 to populate the radio-optical luminosity plane and see whether the introduction of this probability function can explain some known property of RG. In Figs. 2 and 3, it is shown that it is indeed possible to reproduce the observed distribution of RG in this plane starting from the LF of non-radio ellipticals. Moreover, our result is consistent with a picture in which FR I and FR II radio sources are hosted by galaxies extracted from the same parent population. No intrinsic differences are necessary to explain the well known difference of ~ 0.5 mag. in optical luminosity between the two classes of radio galaxies (Fig. 4). This is due to the transition region being a increasing function of the optical luminosity (Bicknell 1995).

On The Parent Population of Radio Galaxies and the FR I-II Dichotomy 57

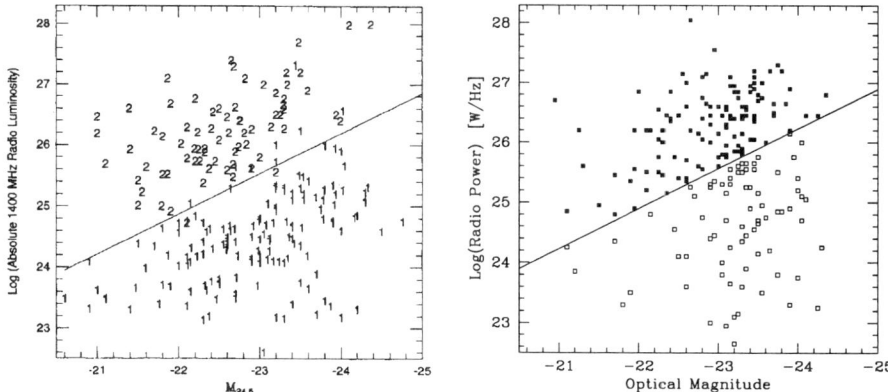

Figure 2. Radio power versus optical magnitude for RG from multiple radio surveys. **Left:** Observed distribution of FR I (symbol 1) and FR II (symbol 2) from the heterogeneous sample of Ledlow & Owen (1996). The solid line separating FR I from FR II was originally drawn by Ledlow & Owen. **Right:** Representative Monte Carlo simulation for a complete flux-limited sample matching the parameters of the Ledlow & Owen survey. Solid squares represent FR II, open squares FR I, defined entirely by their position with respect to the solid line (same as in left panel). Both source distribution and FR I - II relative population are well reproduced. For consistency with Ledlow & Owen (1996), this figure was computed with $H_0 = 75$ km/s/Mpc.

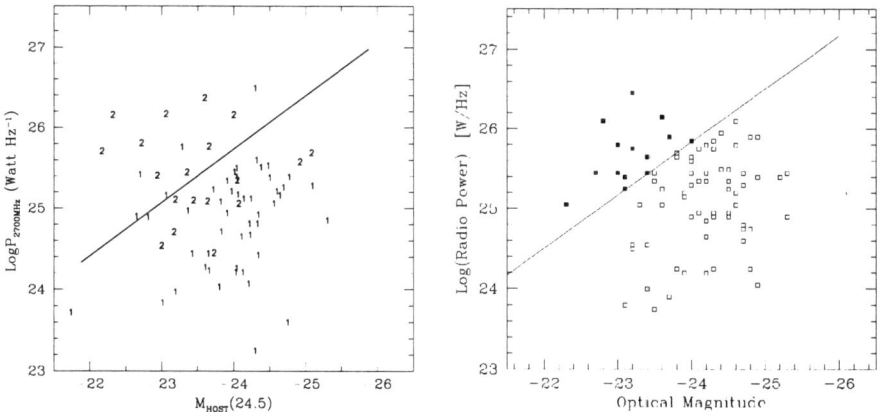

Figure 3. Radio power versus optical magnitude for RG from a well-defined volume-limited survey. **Left:** Observed data from Govoni et al. (2000). **Right:** Monte Carlo simulation matched to the Govoni et al. selection criteria. The agreement is excellent in both the distribution of sources in the radio-optical luminosity plane and the relative populations of FR I (open squares) and FR II (solid squares). For consistency, this figure is computed with $H_0 = 50$ km/s/Mpc.

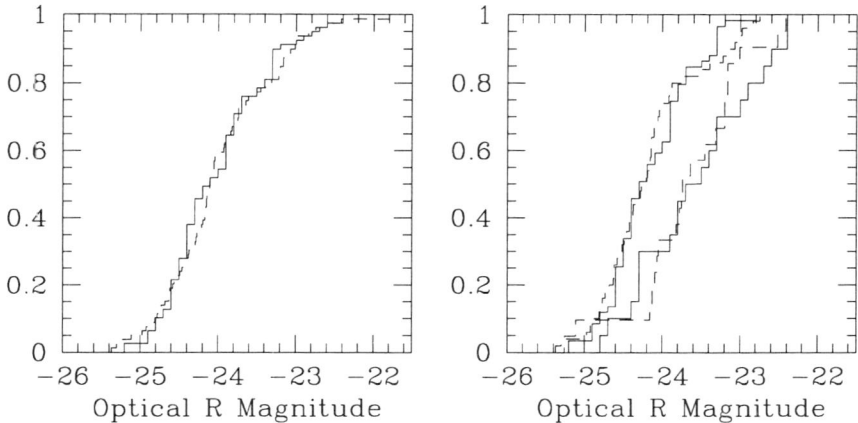

Figure 4. Cumulative distributions of R-band absolute magnitudes for RG data and simulations in Fig. 3. Solid line: simulated data; dashed line: observed data. **Left:** Cumulative distribution for the full data set of Govoni et al. (2000). **Right:** Separate cumulative distributions for FR I (left) and FR II (right). Simulated and observed data agree very well, indicating that the observed difference in average optical luminosity between FR I and FR II is essentially a selection effect.

The physical interpretation for this continuity of elliptical galaxy properties across all radio powers is that all ellipticals have a central black hole and therefore have the potential to generate radio sources. Once the radio source is created, its power should depend mainly on accretion rate, which should depend on the availability of gas and stage of development of the accretion activity. It is not too surprising, therefore, that the radio power is largely independent from L.

References

Bicknell G.V. 1995, ApJS, 101, 29
Govoni F., Falomo R., Fasano G., Scarpa R., 2000, A&A, 353, 507
Ledlow M.J., Owen F.N., 1996, AJ, 112, 9
Loveday J., Peterson B.A., Efstathious G.,
Maddox S.J. 1992, ApJ, 390, 338
Richstone D., Ajhar E.A., Bender R. et al., 1998, Nature ,395, 14
Smith E.P., Heckman T.M., 1989, ApJ, 341, 658
Toffolatti L., Franceschini A., Danese L & de
Urry C.M., Padovani P., 1995, PASP, 107, 803
van der Marel R., 1999, AJ, 117, 744

THE REAL DIFFERENCE BETWEEN RADIO-LOUD AND RADIO-QUIET AGNS

A. Sillanpää
Tuorla Observatory, FIN-21500 Piikkiö, Finland
aimosill@astro.utu.fi

Abstract It seems to me that AGN-astronomers can't see the "forest because of the trees" when they are fighting about the insignificant details between radio-loud and radio-quiet AGNs. It is quite clear that the real difference between these two classes is the JET. So the discussion should be: why some AGNs have a powerful radio jet and some other don't have. I will briefly discuss about the possible explanations for this fundamental difference.

Keywords: Galaxies: active – Galaxies: radio-loud and radio-quiet AGNs

Introduction

The standard "Unified Scheme" for AGNs can quite succesfully explain the differences between different types of RADIO-LOUD AGNs but this class is a very small minority in the whole AGN class. Most of the AGNs are radio-quiet even when all of the properties except radio emission between these two classes are just similar. So why this difference? It is widely believed that both objects have a supermassive black hole in the center of the host galaxy. Because of the unknown reason some black holes eject a powerful radio jet when some objects can produce only very weak uncollimated outflows. There are also remarkable similarities between galactic "microquasars" and extragalactic AGNs. In this paper I try to summarize some earlier explanations and also give the "FINAL TRUTH" for the bimodality.

1. Some fundamental properties of the AGNs (and microquasars)

- ALL radio-loud AGN-hosts are ellipticals.

- ALL quasars with spiral hosts are radio-quiet.

- There is no example of a radio-quiet AGN with a large-scale jet.

- The weak radio-ejecta of the radio-quiet objects are energetically insignificant.

- The space density of the radio-loud objects at the given optical luminosity is ~10 times lower than the radio-quiet objects.

- Despite the above mentioned differences the continua and lines from X-rays to IR are quite similar to both classes.

- Radio-loud AGNs are about 10000 times brighter in the radio than radio-quiet AGNs with the same [OIII] luminosity.

- In all cases of the galactic microquasars we have a BINARY SYSTEM.

2. Explanations and the "FINAL TRUTH"

- Massive galaxies have allways a supermassive black hole in their centers.

- One single black hole can't produce any significant radio jet because it can't collimate it.

- Thermal emission of the AGNs is produced in the accretion disks or very close to them so these properties are very similar.

- Elliptical galaxies are produced by merging of two spiral galaxies.

- When two massive galaxies merge, the two center black holes from a binary system.

- Because the life-time of the binary system is not VERY long (gravitational radiation) not all ellipticals harbour AGNs.

- ONLY SECONDARY BLACK HOLE CAN PRODUCE A MAGNETIC FIELD TO COLLIMATE THE EJECTING RADIO JET.

References

Laor, A. 2000, ApJ, L111-L114.
Koide, S. et al. 2000, ApJ, 536, 668-674.
McLure, R.J. et al. 1999, MNRAS, 308, 377-404.
Mirabel, F. 2000, astro-ph/0005203.
Wilson, A., Colbert, E. 1995, ApJ, 438, 62-71.

BROADBAND OPTICAL COLOURS OF INTERMEDIATE REDSHIFT QSO HOST GALAXIES

Eva Örndahl and Jari Rönnback
Department of Astronomy and Space Physics, Uppsala University, Box 515, S-751 20 Uppsala, Sweden
eva@astro.uu.se

1. The sample

We have obtained deep images for a sample of 100 radio-loud and radio-quiet quasars in the redshift range $0.4 < z < 0.8$. To facilitate a direct statistical comparison between radio-loud and radio-quiet objects, we selected pairs of objects with approximately the same redshift and V-magnitude. Data was obtained in 1994 at the ESO NTT and at the Nordic Optical Telescope, primarily in the R band, but some of them also in V and/or I.

Crucial for host galaxy analysis is the subtraction of the point spread function (PSF). The object frames contain at least one unsaturated star which we use in conjunction with model data (from Saglia et al. 1993) to construct a composite PSF. Empirical data forms the core of the PSF but in the wings we use model data, thus enabling us to reach fainter levels. Details on the point spread function subtraction can be found in Rönnback et al. (1996).

2. Host galaxy magnitudes

We identify a host galaxy in 90% of the radio-loud quasars (RLQ) and in 83% of the radio-quiet quasars (RQQ). The sample was split into three redshift bins and the mean absolute R magnitude for the host galaxies in RLQs and in RQQs computed, yielding the result presented in the histograms in Fig. 1. Here, $H_0 = 75$ km s^{-1} Mpc^{-1}, $\Omega_0 = 0.2$ and $\Lambda = 0$.

We conclude that there is a slight difference in the absolute magnitude of the hosts of radio-loud and radio-quiet QSOs already in the redshift range of 0.4 to 0.54, but that the RLQ host galaxies have become ~ 1

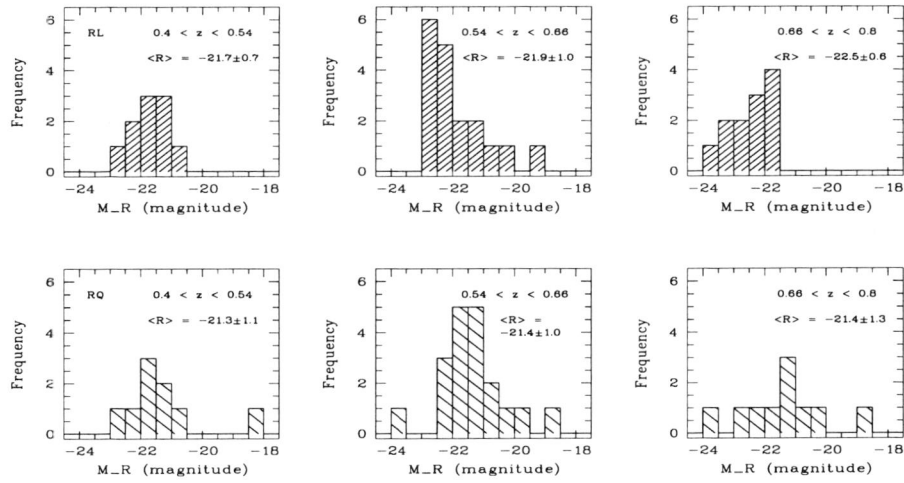

Figure 1. Distribution of host galaxy absolute R magnitudes for the three redshift bins. Top row: radio-loud objects, bottom row: radio-quiet objects.

mag brighter in the range 0.66 to 0.8 whereas the RQQ host galaxies' luminosities haven't changed significantly. The same trend of a widening gap between the magnitudes of RLQ host galaxies and RQQ host galaxies with redshift is seen in the *HST* investigations of Kukula et al. (2000, these proceedings), Dunlop et al. (2001) and McLure et al. (1999).

No K-corrections have been applied to the magnitudes above, but given that the corrections in this redshift interval are large for early-type galaxies but negligible for later types, this will not affect the above conclusions. With an equal morphological mix for the RLQ host galaxies and RQQ host galaxies the same broadening of the magnitude distribution will apply. If the host galaxies of RLQ more often tend to be of early-type than those of RQQ, any correction will only widen the difference in magnitudes.

3. Host galaxy colours

Colours have been determined for 14 host galaxies and are presented in Fig. 2. The colours of the template stellar populations shown in the panels were derived by integrating the redshifted filter profile convolved with spectral energy distributions adopted from Kinney et al. (1996).

There is a slight tendency for radio-loud host objects to have colours typical of (no-evolution) early-type galaxies, at least for lower redshifts, as seen in the $R - I$ versus z-diagram. At higher redshifts both these hosts and those of RQQ rather have the colours of late-type galaxies. However, due to the combination of filters and redshift range some contamination from emission lines cannot be excluded, mainly in I from [OIII] around a redshift of ~0.65. We also cannot rule out a contribution of scattered light from the QSO within the host galaxy in some of these cases.

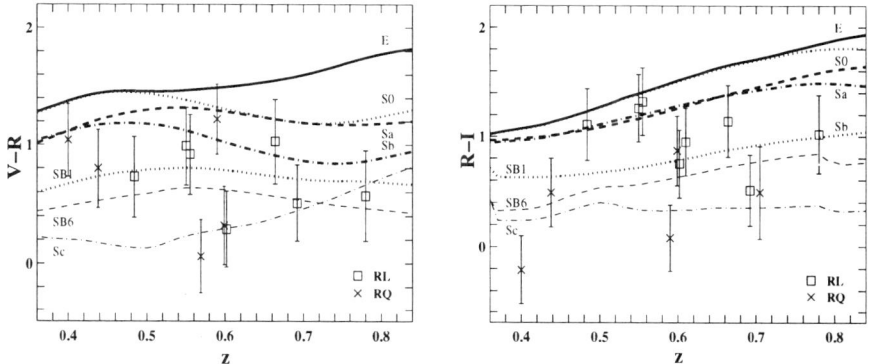

Figure 2. Colours versus redshift for the quasar host galaxies. Boxes are radio-loud hosts, crosses are radio-quiet hosts. The error bars include background as well as photometric errors. Lines represent different template stellar populations.

4. Summary

The mean absolute R magnitudes of the radio-loud host galaxies in our sample show a trend of brightening with redshift, as opposed to those of the radio-quiet host galaxies. Derived colours are in general typical of later-type galaxies.

References

Dunlop J.S. et al., 2001, MNRAS, submitted
Kinney A.L., Calzetti D., Bohlin R.C., McQuade K., Storchi-Bergmann T., Schmitt H.R., 1996, ApJ 467, 38
Kukula M.J. et al., 2000, MNRAS, submitted (astro-ph/0010007)
McLure R.J., Kukula M.J., Dunlop J.S., Baum S.A., O'Dea C.P., Hughes D.H., 1999, MNRAS, 308, 377
Saglia R.P., Bertschinger E., Baggley G., Burstein D., Colless M., Davies R.L., McMahan R.K., Wegner G., 1993, MNRAS, 264, 961
Rönnback J., van Groningen E., Wanders I., Örndahl E., 1996, MNRAS, 283, 282

Eva Örndahl

Lucas Lara

THREE PECULIAR OBJECTS FROM A NEW SAMPLE OF RADIO GALAXIES

L. Lara[1], W.D. Cotton[2], L. Feretti[3], G. Giovannini[3,4], J.M. Marcaide[5],
I. Márquez[1] and T. Venturi[3]

[1] *Instituto de Astrofísica de Andalucía (CSIC), Apdo. 3004, 18080 Granada, Spain*
[2] *NRAO, 520 Edgemont Road, Charlottesville, VA 22903-2475, USA*
[3] *Istituto di Radioastronomia (CNR), via P. Gobetti 101, 40129 Bologna, Italy*
[4] *Dipartimento di Fisica, Univ. di Bologna, via B. Pichat 6/2, 40129 Bologna, Italy*
[5] *Departamento de Astronomía, Universitat de València, 46100 Burjassot, Spain*

Abstract We have constructed a new sample of 84 large angular size radio galaxies selected from the NRAO VLA Sky Survey (NVSS). Radio sources with declination above $+60°$, total flux density greater than 100 mJy at 1.4 GHz and angular size larger then 4' have been selected and observed with the VLA at 1.4 and 4.9 GHz. A number of peculiar radio galaxies have been discovered, of which we present here results on J1835+620, J2114+820 and J2157+664.

Keywords: Radio galaxies, Samples, J1835+620, J2114+820, J2157+664

Introduction

The NVSS (Condon et al. 1998) covers the sky north of $-40°$ declination, at a frequency of 1.4 GHz and an angular resolution of 45''. Its high sensitivity and resolution render the NVSS a unique tool for the selection of samples of extended radio galaxies. In 1995 we undertook a project for the definition and study of a sample of large angular size radio galaxies. We selected objects from the NVSS with angular size larger than 4', flux density at 1.4 GHz larger than 100 mJy and declination above $+60°$. A total of 122 sources were pre-selected and observed with the VLA at 1.4 and 4.9 GHz between 1995 and 1998 in order to confirm large angular size radio galaxies, reject objects resulting from the superposition of two or three adjacent sources, and isolate the core emission to obtain accurate positions for subsequent optical identification and redshift determination. Out of the 122 pre-selected sources, 82 (+2 known giant radio galaxies) were finally identified beyond doubt

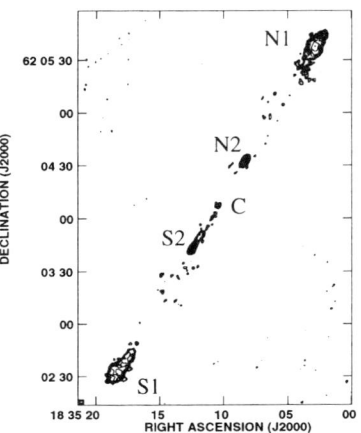

Figure 1. Radio map of J1835+620 made with the VLA at 4.9 GHz. Contours are separated by factors of $\sqrt{2}$, being the 1st one at 0.16 mJy beam^{-1}. The convolution beam is 1.79 arcsec ×1.41 arcsec, at P.A. −42.92°

with single radio galaxies and included in the sample. Optical spectroscopic observations of those objects with unknown redshift were carried out in 1997 and 1998 at the 2.2m telescope in Calar Alto (Spain). A detailed presentation of the sample and of the aims followed with the sample construction can be found in Lara et al. (2001a, 2001b).

1. Three peculiar objects

The selection criteria adopted for the sample construction guarantee that a large diversity of sources, with very different conditions, are selected. The sources present a large range in projected linear sizes[1] ($260 \leq L[\text{kpc}] \leq 3700$), in total emitting power at 1.4 GHz ($23.4 \leq \log P_t(1.4)[\text{W/Hz}] \leq 27.1$) and in morphological types (FR I and FR II type radio sources, wide angle tails, asymmetric sources, etc.). Among the radio galaxies belonging to the sample, we emphasize here three of them which deserve a special attention due to different peculiar properties:

1.1. J1835+620

J1835+620 presents an unusual radio structure consisting of a compact central core (C) and two radio lobes (N1 and S1) straddling two bright and symmetric components (N2 and S2; Fig 1). Given the ex-

---

[1] We assume that $H_0 = 50 \,\text{km}\,\text{s}^{-1}\,\text{Mpc}^{-1}$ and $q_0 = 0.5$

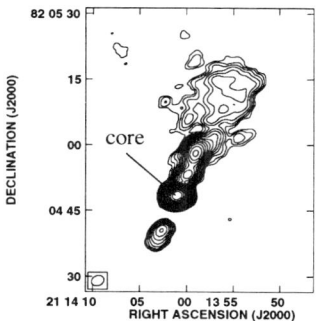

Figure 2. Radio map of J2114+820 made with the VLA at 4.9 GHz. Contours are separated by factors of $\sqrt{2}$, being the 1st one at 0.12 mJy beam^{-1}. The convolution beam is 2.85 arcsec ×2.25 arcsec, at P.A. $-67.50°$

traordinary symmetry of J1835+620, we interpret its radio structure as the result of an episode of restarting activity. Two distinct phases of core activity would be responsible for the observed morphology. The optical counterpart of J1835+620 lies in a group of at least three galaxies and shows strong narrow emission lines (z=0.518). This radio source belongs to the group of giant radio galaxies (see Lara et al. 1999a, Schoenmakers et al. 2000).

1.2.   J2114+820

J2114+820 is a low power radio source with a FR-I type structure which basically consists of a prominent core, a jet and counter-jet directed in NW direction (Fig. 2) and two extended S-shaped lobes (not shown in Fig. 2). The optical counterpart of J2114+820 is a bright elliptical galaxy with a strong unresolved central component. The spectrum shows prominent broad emission lines in H$_\alpha$, H$_\beta$ and H$_\gamma$ (z=0.084). However, according to current unification schemes of radio loud AGNs (eg. Urry & Padovani 1995), FR-I sources are the parent population of BL-Lac type objects, which are characterized by the absence of broad emission lines (at most only weak broad lines have occasionally been observed in BL-Lac type objects). In consequence, FR-I sources should not show such emission lines, contrary to what we observe in this radio source. More observations are needed in order to bring J2114+820 into a consistent scenario (see Lara et al. 1999b).

1.3.   J2157+664

J2157+664 is a strongly asymmetric radio source. High resolution VLA observations at 1.4 and 4.9 GHz show a bright core with a jet

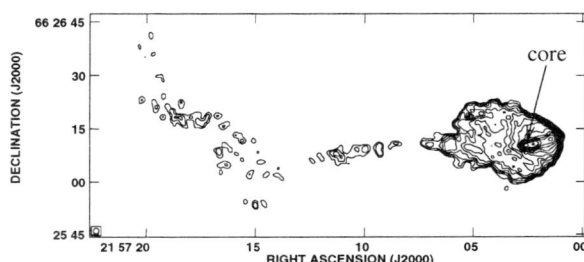

Figure 3. Radio map of J2157+664 made with the VLA at 4.9 GHz. Contours are separated by factors of $\sqrt{2}$, being the 1st one at 0.23 mJy beam^{-1}. The convolution beam is 1.6 arcsec ×1.6 arcsec

directed in west direction, which seems to interact abruptly with the external medium, producing a prominent bow shock (Fig. 3). Observations of the polarized radio emission (not presented) show evidence of such strong interaction of the jet, with an ordered magnetic field compressed along the bow shock edge. A long and extended tail is also observed in this source towards the east.

Acknowledgments

This research is supported in part by the Spanish DGICYT (PB97-1164). The NRAO is a facility if the National Science Foundation operated under cooperative agreement by Associated Universities, Inc.

References

Condon J.J., Cotton W.D., Greisen E.W., Yin Q.F., Perley R.A., Taylor G.B., Broderick J.J., 1998, AJ, 115, 1693

Lara L., Cotton W.D., Feretti L., Giovannini G., Marcaide J.M., Márquez I., Venturi T., 2001a, A&A, 370, 409

Lara L., Márquez I., Cotton W.D., Feretti L., Giovannini G., Marcaide J.M., Venturi T., 2001b, A&A, submitted

Lara L., Márquez I., Cotton W.D., Feretti L., Giovannini G., Marcaide J.M., Venturi T., 1999a, A&A, 348, 699

Lara L., Márquez I., Cotton W.D., Feretti L., Giovannini G., Marcaide J.M., Venturi T., 1999b, NewAR, 43, 643

Schoenmakers A., de Bruyn A.G., Röttgering H.J.A., van der Laan H., 2000, MNRAS, 315, 395

Urry C.M., Padovani P., 1995, PASP, 107, 803

II

QSO REDSHIFT EVOLUTION AND THEIR CLUSTER ENVIRONMENTS

John Hutchings

QSO HOSTS AND COMPANIONS AT HIGHER REDSHIFTS

John. B. Hutchings
Herzberg Institute of Astrophysics, NRC of Canada
Victoria, B.C., Canada

Abstract   This review presents the current state of work on QSO hosts and companions at redshifts above 1. This includes the properties of QSO host galaxies, such as size, scale length, and luminosity, and morphology, as they appear to change with redshift and radio activity. This leads to a view of how the properties of galaxies that host QSOs change with cosmic time. I also review studies of the galaxy companions to QSOs at higher redshifts, and studies of the emission line gas in and around higher redshift QSOs. These topics should see great progress in the next decade.

1.   Host galaxy detection

In the past decade we have made significant progress in detecting and measuring the host galaxies of QSOs at redshifts in the range 1 to 2.5, and even above 3 in a few instances. This is partly due to the fact that the hosts at higher redshifts are in very active star-forming stages of their lives, and are thus bright in the observed rest-frame UV. The first significant paper on high redshift host galaxies came from KPNO data with resolution only about 1.3 arcsec, showing dramatically that radio-loud hosts were both bright and large (Heckman et al. 1991). Since then the availability of NIR imaging and higher resolution imaging (via *HST* or adaptive optics) have been responsible for the bulk of the higher z investigations to date, and which I attempt to review here (see e.g. Lowenthal et al. 1995, Lehnert et al. 1992, Hutchings 1995b, Aretxaga et al. 1998, Hutchings 1998, Hutchings et al. 1999, Lehnert et al. 1999, Rix et al. 2000, Kukula et al. 2000, Ridgway et al. 2001). It is clear that the use of 8-10m class telescopes will help us reach to fainter details and higher redshifts where redshift dimming dominates, and will be a key to future investigations in this field (Falomo et al. 2000).

There are thus a number of investigations that have reported on high redshift host galaxies. Their interest as samples of high redshift galaxies in general has somewhat been overshadowed by the recent major progress in recognising and studying high redshift non-QSO galaxies, via photometric redshifts and 10m telescope spectroscopy. However, the QSO environment is still of considerable interest in the context of the general galaxy population at higher redshifts.

As well as the redshifts above 1, I have selected host galaxy measurements over the redshift range below 1 to cover the redshift range down to the well-studied low redshift QSOs (Kotilainen et al. 1998, Kotilainen & Falomo 2000). The diagrams show the result, attempting to be comprehensive at higher redshifts, while selecting quantities derived with reliable methods and good data. They also do not impose any interpretation on the results, such as K-corrections, evolutionary assumptions, or extinction estimates. Given the different instrumental and observational limitations between the NIR and traditional optical bands, I have compiled (or applied corrections to obtain) the data for H-band and I-band. These are shown in the plots.

In addition to the studies referenced, a number of these are from ongoing and unpublished work of my own at the time of writing. They include NIR and optical data from CFHT with AO correction, and some *HST* WFPC2 data. The plots also show for comparison, selected points for low redshift QSOs, which I think are representative, but not at all complete, since there are hundreds of low redshift QSO hosts in the literature. The plots also distinguish between radio-loud and radio-quiet objects, since this is likely to be a fundamental and diagnostic difference. All points shown are derived for $H_0=50$ and $q_0=0$ cosmology. This is for easiest comparison rather than to represent a preferred choice.

While the comparison of results for the same objects from different investigators are encouraging in the few available instances, it is likely that both scatter and systematic differences arise from the different methods of removing PSFs, and overall image quality differences. I have omitted data that are less reliable - usually from poor resolution, smaller telescopes, or more noisy detectors. As a rough guide, the table shown compares the 'quality' of data from different sources. There are two important criteria in data for host galaxy studies: detection of faint extended flux, and image resolution to minimise the nuclear point source spread. Different telescopes have different strengths in these two areas, and there are additional criteria of detector noise and PSF complexity. Adding error bars to the diagrams to take all these into account would

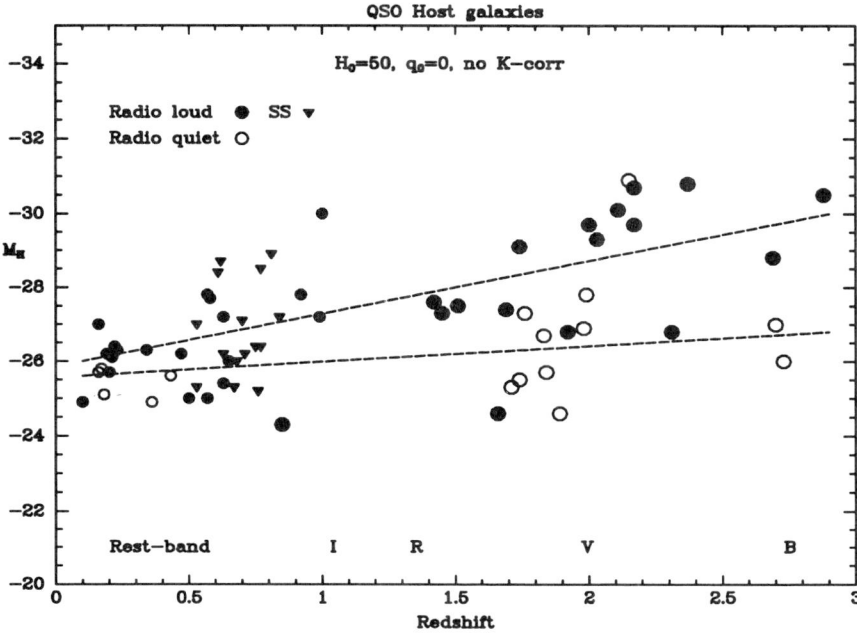

Figure 1. H-band absolute magnitudes of resolved host galaxies, as labelled. All values are for given cosmology and have no K-correction, and some have been derived from J or K band observations. $z < 0.3$ points refer to samples rather than individual objects, but are representative. The dotted lines are linear fits to the radio-loud and radio-quiet data as plotted. The rest wavelength bandpasses are given along the bottom.

be a large and unrewarding task, so I am assuming that the mix of different investigations in overlapping redshifts will show the overall trends without severe systematic errors.

Nothwithstanding these caveats, there are some clear points in the plots. Generally, the host galaxies are more luminous with increasing redshift. The increase is faster with redshift in visible wavelengths, and larger in the NIR in the radio-loud sources. However, the wavelength dependence is strongly affected by the redshift itself, as well as evolution in the SED. I have sketched in linear suggestions for the trends as observed. As noted by Falomo et al., at redshifts 1 and above, the radio-loud hosts are brighter than BCGs and L* galaxies, while radio-quiet ones are comparable with BCGs. All are unusually bright galaxies, and there a few enough null detections that this is unlikely to be an observational issue.

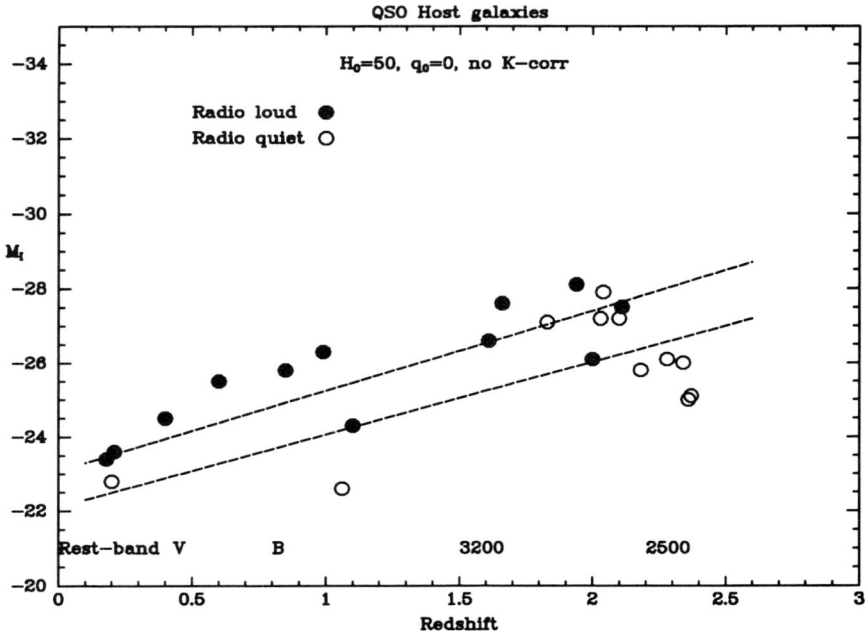

Figure 2. I-band absolute magnitudes as in Fig. 1.

Table 1. Comparative power of investigations

	NIR				CCD			
Telescope	Res(")[a]	Det[b]	PSF[c]	Telescope		Res(")	Det	PSF
VLT	0.5	230	50	CFHT	HRCAM	0.4	12	3
CFHT AO	0.15	91	60		AO	0.3	47	15
HST	0.15	100	40	HST	WFPC2	0.1	4-20	4-20
ESO	0.8	30	4		STIS	0.1	30	30
				KPNO		1.3	17	1
				WHT		0.7	88	12

[a] RES (") Image FWHM in arcsec
[b] Det = S/N achieved (Aperture x thruput x exp / noise)
[c] PSF = Goodness of PSF removal (Det x 0.1 / FWHM)

What can we infer from these results? Given that the observed rest wavelength changes with redshift as shown along the X-axes, the host galaxies are bluer as well as brighter, consistent with passive evolution or declining star-formation as time proceeds, and the hosts at redshifts 2 and higher are starburst objects. This has been noted by various workers. However, since QSO episodes are short compared with the lifetimes of galaxies, we are not necessarily seeing the evolution of a population of galaxies, but may also be looking at a changing population of hosts with time. We come back to this below.

As galaxy luminosity can reflect star-formation activity, increased content by merging, or stripping by tidal events, we can remove some parameters by looking at the average colours indicated by the lines through the H and I band plots. The colour plot shows this evolution for the two QSO types. We show for comparison the colour expected for constant star-formation, and passive evolution from starburst for galaxy formed at redshift about 4, with the adopted cosmology timescale. Dust and increased metallicity move the plots vertically up the diagram.

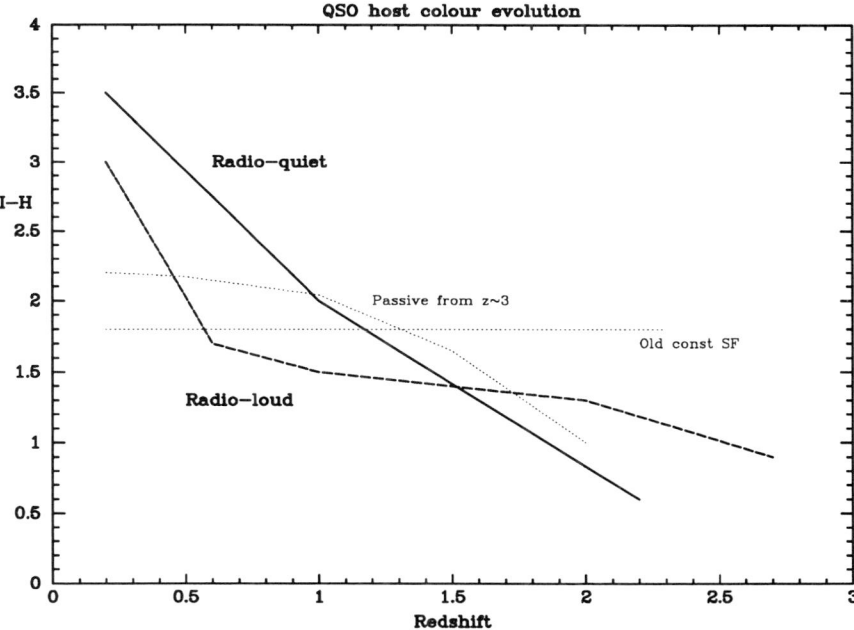

Figure 3. Locus of linear fits from Figs. 1 and 2 converted to I-H. The faint dotted lines are approximations from Gissel93 models for measured I-H as observed redshift changes. (The constant observed colour for constant star-formation arises because slow reddening is balanced by the moving bandpasses.) Radio-loud hosts appear to arise in a population that is actively forming stars over $z = 0.8$ to 2, while radio-quiet hosts have passively evolving populations. The observed redder colours at low redshift may arise from dust or increasing metal abundance.

The colour evolution is different for the two QSO types. At high redshift, both have colours of star-forming galaxies, but the radio-loud ones have an older population present. The radio-quiet hosts appear to evolve passively to redshift 1, and then are reddened by abundance changes and/or dust. The radio-loud hosts appear to have active star-formation at high levels down to redshift near 0.5, and then go the way of the radio-quiet.

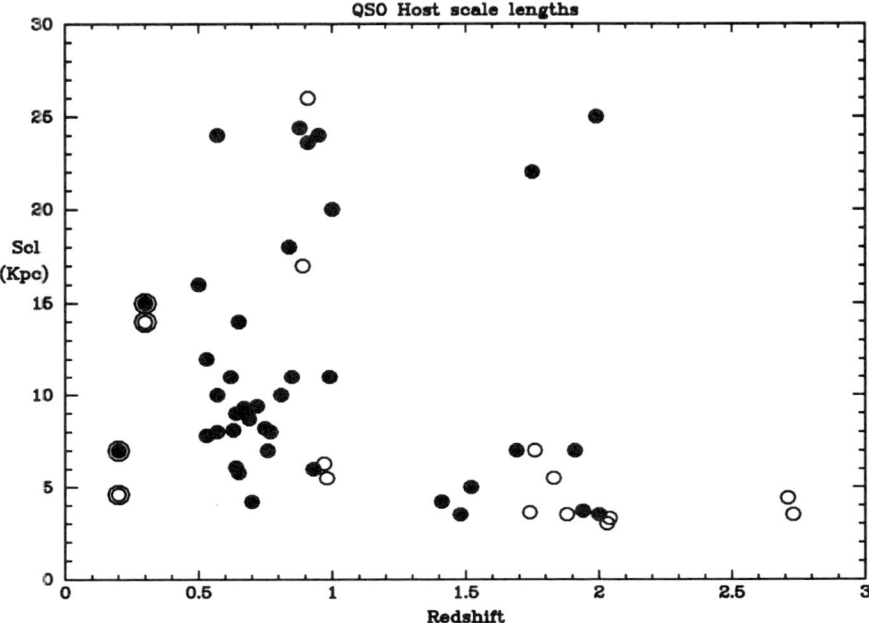

Figure 4. Host galaxy scale lengths for high redshift hosts, corrected for differences in published values of FWHM, enclosed flux fraction. The two low redshift points (circled) are for samples - the lower one being *HST* short exposure images, which do not detect faint outer parts of the galaxies.

The scale lengths are more subject to differences in PSF removal, but the plot shows what the best data indicate. Generally, the high redshift host galaxies are smaller, and become much smaller than present-day bright galaxies. There is a scatter of high values that may arise from tidal tails. If so, this is seen predominantly at redshifts 0.5 to 1. (when galaxy merging is generally known to be higher). This is consistent with hierarchical merging as galaxies evolve, as noted by other authors in the

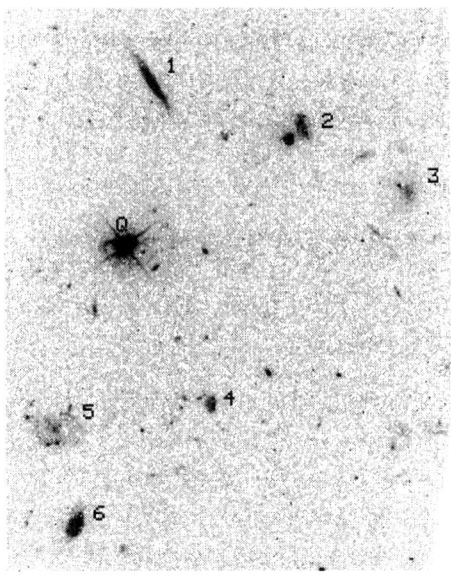

Figure 5. Radio-loud $z = 2$ QSO and companions imaged with HST/STIS (Hutchings 1998). The field is 40" x 30". QSO is extended, and is similar in size and blue colour to the compact group of irregular companion galaxies.

subject. It is also consistent with the evolution of non-active galaxies as seen by Lyman-break galaxy studies.

2. Cluster environment

There have been several studies of QSO galaxy companions. Many authors have noted the apparent connection between galaxy interactions and QSO activity, as well the signs of past tidal events in the profiles of QSO hosts at low redshift. These are much harder to see at high redshift, since tidal tails are old stars which are faint and close to the nuclear PSF. It is simpler to count and even get spectra for nearby companions, in order to characterise the galaxy environment of QSOs. However, such results are quite dependent on image quality, signal level, and detection threshold. The table again shows where we expect to do best in this regard. Fig. 5 shows a good example of some bright companions and a $z = 2$ QSO.

The compendium of high redshift QSOs is shown in the Fig. 6. The principal references for these are Hutchings et al. (1995), Hutchings (1995a), Teplitz et al. (1999), Wold et al. (2000, 2001), Haines et al. (2000), and some recent unpublished data of my own. The numbers plotted are estimated excess galaxies within 20 arcsec of the QSO in the sky, in some cases derived from different areas originally reported. The low-redshift values are means from many objects, and the vertical 'error bar' indicates the range among individual objects. A similar range is revealed at redshift 1.1, which is QSOs in the 'supercluster' region discussed by Hutchings et al. (1995). The dotted lines sketch in suggested increases in the RL and RQ environments above $z = 1$. Clearly these are very uncertain in view of the scatter where we have enough measures to know. However, it is generally accepted that QSOs do live in regions of enhanced galaxy populations, and that the number of galaxy companions probably increases at higher redshift. Several investigations show that the companion density on the sky falls quickly with radius, so that the plot is oversimplified in that respect.

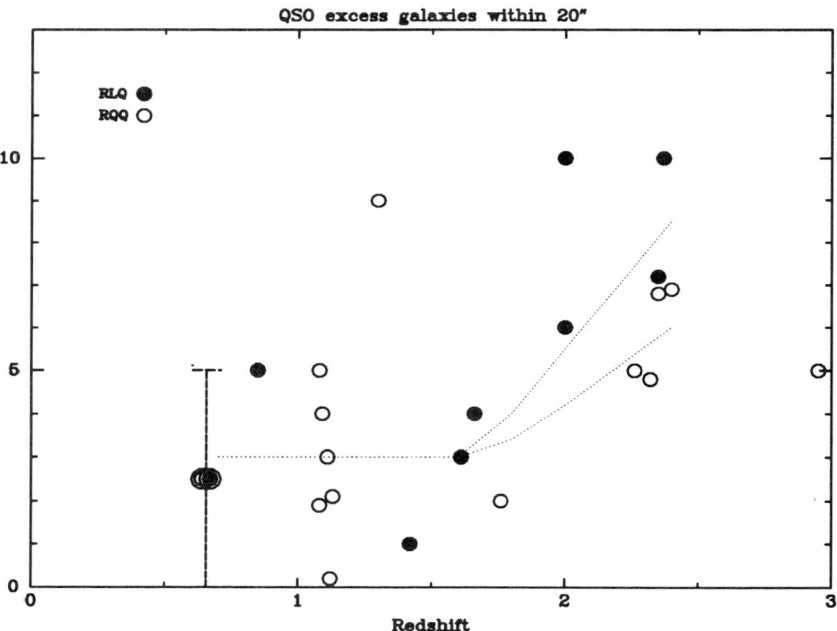

Figure 6. Counts of excess galaxies near QSOs. The values are corrected to refer to the same area of sky, and for differences in limiting magnitude. Dotted lines are suggested mean relationships for radio-loud and radio-quiet QSOs.

Another caveat is that at high redshift, the companions seen are compact, so that their detection depends on good signal and high resolution. The deepest detection of companions comes from long CCD exposures with *HST* . The small size of the faint galaxies means that *HST* can detect more of them, to a magnitude fainter than CFHT, for example. However, at magnitude fainter than 24, most galaxies are high redshift star-forming objects, so that their association with the QSO needs more than just blue colour from 2 filters. I have not plotted any of these very faint 'companions' in the diagram.

In the NIR, *HST* has the advantage of dark sky, but 8m class telescopes have so much more light-gathering power that they win - not only for companions but also the faint parts of host galaxies that indicate tidal events and evolved populations. However, no major investigations have been done yet.

The galaxy companions we know about, are different at high redshift, and suggestive of environments which may evolve into small groups (or a single large galaxy) at present time. The environment that triggers QSO activity almost certainly changes with cosmic time, ranging from initial merging of protogalaxies to tidal triggers involving gas at present time. Radio-loud sources are likely the more massive black hole (= galaxy bulge mass, age of galaxy, merging history,..?). It is also known in some cases that the companions are a foreground group, which may be connected with the QSO only by lensing action that makes the QSO more visible. We need more complete and careful investigations to get the required statistics on these issues.

3. Line emission

The classic paper on line emission from AGN at high redshift is some years ago, by McCarthy et al. (1991). This established the 'alignment effect' between radio and emission-line structure that seems to be accepted as the ionising effect of nuclear radiation and jet activity on a gas-rich environment. For low redshift QSOs, there is another famous paper by Stockton and MacKenty (1987) that investigates line emission, and finds extended and irregular emission line gas predominantly in extended radio-loud QSOs. This reveals the presence and complex dymanics of gas that lies well outside the stellar population and is presumably the result of tidal events that have fuelled the nucleus.

Work on radio-quiet high redshift QSOs has occured only recently, again with the help of larger telescopes, NIR detectors, and AO. Ohta et al. (2000) have detected [O II] emission in a $z = 4.7$ QSO, and Hutchings et al. (2000) find [O III] and Hα in a sample of radio-quiet or compact radio source QSOs. These all have line emission that lies within the host galaxy and thus corresponds to the NLR gas that has been mapped in low redshift Seyferts and QSOs. Where there is radio structure in these objects, the line emission is aligned with it. The emission line images indicate line equivalent widths typically 200 - 300Å in the redshifted frame, of which about half is typically from the nucleus.

More detailed work will extend this work to a more statistically significant sample and perhaps lead to an understanding of the internal dynamics of the high redshift host galaxies.

4. Summary

This review cannot do justice to the details of the different investigations. Other papers in this volume deal with individual programs, which are included in my summary plots.

Overall, the work of the past decade has established a credible set of investigations of high redshift QSO hosts and environments. This should become more detailed and extend to higher redshift with 8m class telescopes with AO, and advanced HST (and $NGST$) instrumentation. It seems well established that host galaxies at high redshift are luminous, with active star-formation, and generally are very compact for their luminosities. They also live in dense galaxy companion environments. The hosts and their companions must undergo significant merging and evolution, as the much less common present-epoch QSOs are found in larger hosts with less crowded environments. We are on the brink of studying the formation and evolution of the central black holes, which can be measured by the host spheroid morphology, and the nuclear BLR profiles. It is clear that the formation and triggering of QSOs is an integral part of the formation of galaxies, and that QSOs at high redshift offer a wealth of cosmological information for the years to come.

References

Aretxaga I., Terlevich, Boyle B.J., 1998, MNRAS, 296, 643
Falomo R., Kotilainen J.K., Treves A., 2001, ApJ, (astro-ph 0009181)
Hutchings J.B., 1995a, AJ, 109, 928
Hutchings J.B., 1995b, AJ, 110, 994
Hutchings J.B., 1998, AJ, 116, 20
Hutchings J.B., Crampton, D., Johnson A., 1995, AJ, 109, 73
Hutchings J.B., Crampton D., Morris S.L., Durand D., Steinbring E., 1999, AJ, 117, 1109
Hutchings J.B., Morris S.L., and Crampton D., 2001, AJ, (astro-ph 0012245)
Haines C.P., Clowes R.G., Campusano L.E., 2001, astro-ph 0012236
Heckman T.M., Lehnert M.D., van Breugel W., Miley G.K., 1992, ApJ, 370, 78
Kotilanen J.K., Falomo R., Scarpa R., 1998, A&A, 332, 503
Kotilainen J.K., Falomo R., 2000, A&A, 2000 (astro-ph 0008513)
Kukula M.J. et al., 2001, MNRAS (astro-ph 0010007)
Lehnert M.D., Heckman T.M., Chambers K.C., Miley G.K., 1992, ApJ, 393, 68

Lehnert M.D., van Breugel W., Heckman T.M., Miley G.K., 1999, ApJS, 123, 351
Lowenthal J.D., Heckman T.M., Lehnert M.D., Elias J.H., 1995, ApJ, 439, 588
McCarthy P.J., van Breugel W., Kapahi V.J., 1991, ApJ, 371, 478
Ohta K. et al., 2000, PASJ (astro-ph 0003107)
Ridgway S.E., Heckman T.M., Calzetti D., Lehnert M., 2001, ApJ (astro-ph 0011330)
Rix H-W. et al., 2000, astro-ph 9910190
Stockton A., MacKenty J.W., 1987, ApJ, 316, 584
Teplitz H.I., McLean I.S., Malkan M.A., 1999, ApJ (astro-ph 9902231)
Wold M., Lacy M., Lilje P.B., Serjeant S., 2000, MNRAS (astro-ph 9912070)
Wold M., Lacy M., Lilje P.B., Serjeant S., 2001, MNRAS (astro-ph 0011394)

Lutz Wisotzki

Knud Jahnke

THE LUMINOSITY FUNCTION OF QSO HOST GALAXIES

Lutz Wisotzki[1], Björn Kuhlbrodt[2] and Knud Jahnke[2]
[1] *Universität Potsdam*, [2] *Hamburger Sternwarte*
[1] lutz@astro.physik.uni-potsdam.de

Abstract We report on results from H band imaging observations of a complete sample of high-luminosity low-redshift QSOs. The luminosity function of QSO hosts is similar in shape to that of normal galaxies, although offset in normalisation by a factor of 10^{-4}. This supports the hypothesis that the parent population of quasars is identical to the general population of early-type field galaxies.

Introduction

Identifying the parent population of QSO host galaxies is one of the fundamental problems linking the QSO phenomenon to galaxy evolution in general. While the most widely adopted approach is to compare morphological (and increasingly also spectral) properties of normal and active galaxies, we investigate here the statistical distribution of galaxy luminosities. This short report highlights some results which will be elaborated in more detail in a series of forthcoming papers. We adopt $h = 0.5$ and $q_0 = 0.5$ throughout.

1. H band imaging of a complete QSO sample

In order to constrain the overall luminosity distribution of QSO hosts, the investigated targets must represent a *fair sample* of the QSO population. To obtain also the normalisation, the sample has furthermore to be *complete* (in the sense of comprising all QSOs within a given area that are brighter than a well-defined flux limit). Our sample has been selected from the Hamburg/ESO bright quasar survey (Wisotzki et al. 2000), with the following additional criteria: right ascension between 9^h and 16^h; nuclear absolute magnitudes $M_{B_J} < -23$; and redshifts $z < 0.3$. The resulting sample consisted of 30 targets distributed over

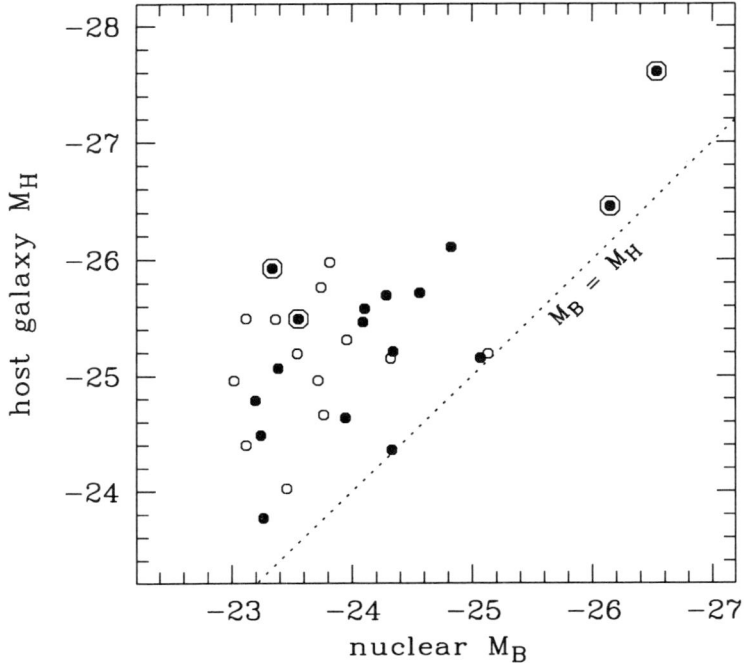

Figure 1. Host galaxy versus nuclear absolute magnitudes. Open symbols: disk-dominated hosts; filled dots: elliptical hosts. The four radio-loud QSOs are marked as circled dots.

2200 deg^2, all of which were subsequently observed. Because of the optical selection, the sample is predominantly radio-quiet, containing only four bona fide radio-loud quasars.

Observations were conducted in February 1999 with the ESO NTT and its near-infrared camera SOFI. With a seeing between 0$''$5 and 0$''$8, the host galaxy was clearly detected in all cases, and after PSF subtraction, unique morphological type assignment was possible for most objects. A summary of the results:

- A disk model is preferred in 11 objects (37%); there is often evidence for an additional substantial bulge component.

- A spheroidal model is preferred in 16 objects (53%); for the QSOs with $M_{B_J} < -24$, 10 out of 12 (83%) are spheroidal.

- Only three systems (10%) are strongly interacting or irregular.

- All hosts are very luminous, $M_H \lesssim M_H^\star$ with $M_H \simeq -24.5$ being a typical value for the general galaxy population (cf. below).

- There are no hosts with $M_{H,\text{gal}} > M_{B,\text{nuc}}$, thus the McLeod et al. (1999) diagonal boundary is confirmed and reinforced (in particular, there are no more upper limits).

- Except for this boundary, there is no convincing evidence for any correlation of $M_{H,\text{gal}}$ with $M_{B,\text{nuc}}$ (apart from the two extreme outliers 3C 273 and PKS 1302−102, which however are flat-spectrum radio sources and probably beamed).

2. Bivariate luminosity function

In order to fully describe a QSO host galaxy luminosity function (QHGLF), at least two independent variables are required: Host galaxy luminosity L_{gal} and nuclear luminosity L_{nuc}, leading to a *bivariate* LF $\Phi(L_{\text{gal}}, L_{\text{nuc}})$. Cosmological evolution of the QSO population demands Φ to depend additionaly on redshift z.

From the above results, we simplify the expression by tentatively assuming that QSO and host galaxy LFs are formally uncorrelated over the luminosity range of the sample under consideration.

$$\Phi = \phi(L_{\text{nuc}}, z)\, \psi(L_{\text{gal}}, z).$$

We adopt common parametric forms for both ϕ and ψ. The former can be well approximated, for low redshifts and not too low luminosities, by a single power law (Köhler et al. 1997, Wisotzki 2000), the latter is usually expressed by a Schechter function. For the evolution we assume a simple power-law form of pure number evolution, which is perfectly sufficient at low redshifts. The resulting expression for the QHGLF is

$$\Phi = \Phi_0 (1+z)^\kappa \left(\frac{L_{\text{nuc}}}{L^\star_{\text{nuc}}}\right)^\alpha \left(\frac{L_{\text{gal}}}{L^\star_{\text{gal}}}\right)^\beta e^{-L_{\text{gal}}/L^\star_{\text{gal}}}.$$

An important additional feature is the McLeod et al. (1999) boundary, which basically states that for a given host galaxy mass (and hence luminosity), there is a well-defined maximum nuclear luminosity, essentially given by the Eddington limit of the central black hole. Although the universal validity of this boundary has recently been questioned by Percival et al. (2000, also these proceedings), it is clearly present in our data. We therefore decided to incorporate it by multiplying the above expression by a factor $\mathcal{H}(L_{\text{gal}} - xL_{\text{nuc}})$ where $\mathcal{H}(t_1 - t_2) = 1$ for $t_1 > t_2$ and 0 elsewhere is the Heavyside step function, and x is an adjustable parameter. For L_{gal} measured in the H band and L_{nuc} measured in B, we adopt $x \approx 1$ (this corresponds to the diagonal dotted line in Fig. 1).

Numerical values of the QHGLF parameters have been estimated by maximum likelihood fitting of the above functional form to the observed

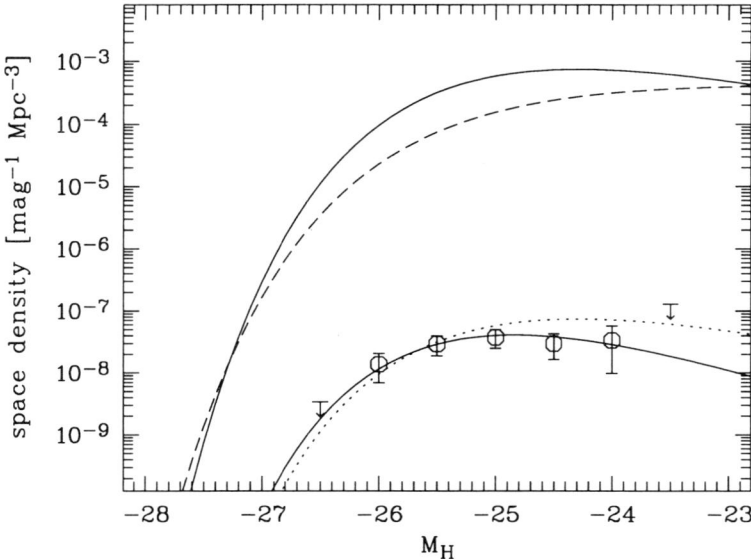

Figure 2. Estimates of the QHGLF. The points give the binned QHGLF with errors and upper limits from Poisson statistics. The lower solid curve is the best-fit maximum likelihood estimate. Also shown are some LFs of early-type field galaxies for comparison. Upper solid curve: Lin et al. (1999); dashed curve: Kochanek et al. (2001). The dotted line is the Lin et al. LF with rescaled normalisation to approximately match the QSO data points.

distribution of sources in $(M_{B,\mathrm{nuc}}, M_{H,\mathrm{gal}}, z)$ space. The resulting best-fit values are

$$\begin{aligned}
\Phi_0 &= 1.3 \times 10^{-7}\,\mathrm{Mpc}^{-3} \quad \text{for } M^\star_{B,Q} = -23 \\
\kappa &= 6.5 \\
\alpha &= -2.8 \\
M^\star_{H,\mathrm{gal}} &= -24.6 \pm 0.3 \\
\beta &= 0.5 \pm 0.5
\end{aligned}$$

We have also determined binned estimates of the QHGLF under the assumption that the QHGLF term $\phi(L_{\mathrm{nuc}}, z)$ is well-described by the above parameters. The resulting binned and maximum likelihood estimates are shown in Fig. 2.

The last two parameters are particularly interesting, as they describe the *shape* of the galaxy LF. In Fig. 3 we show confidence contours for these parameters and compare them with published values determined for the general field galaxy population (adapted to the H band by assuming typical galaxy colours). The characteristic luminosity $M^\star_{H,\mathrm{gal}}$ is

The luminosity function of QSO host galaxies 87

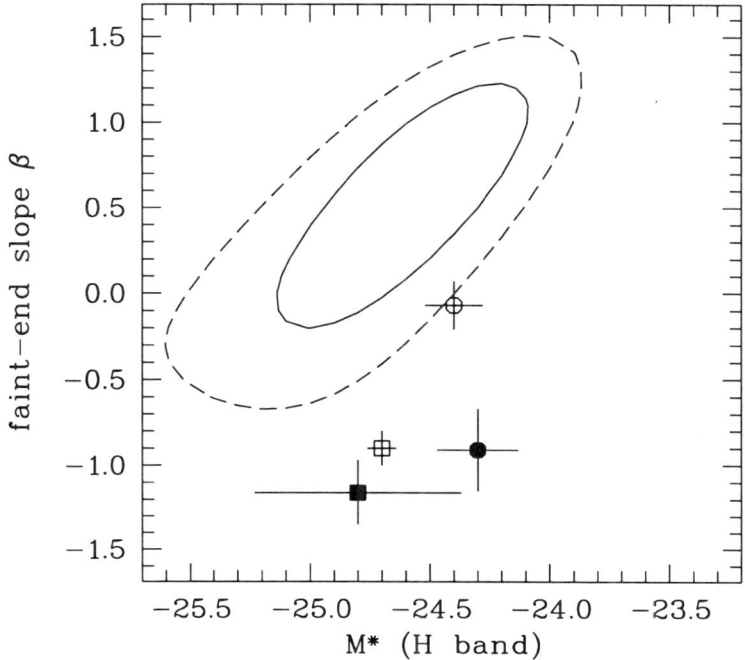

Figure 3. Likelihood contours for the HGLF parameters: The solid line corresponds to 68 %, the dashed line to 90 % confidence. Comparison values for the field galaxy LF are given by symbol markers. Open circle: Lin et al. (1999); filled circle: Gardner et al. (1997); open square: Kochanek et al. (2001); filled square: Loveday et al. (2000).

completely within the range found for field galaxies, while the faint-end slope β is only slightly flatter. The latter makes the QHGLF formally inconsistent with that of field galaxies, but Fig. 2 shows that the discrepancy even at the faint end is not large: The dotted line represents the Lin et al. (1999) LF of early-type field galaxies, rescaled to approximately match the normalisation of the QHGLF data points. The overall agreement is surprisingly good, and the slight mismatch at the faint end might even be explained by our limitation to only luminous QSOs.

To summarise, our results present no evidence that QSO host galaxies show a luminosity distribution widely different from that of inactive field galaxies. This lends considerable support to the hypothesis that the parent population of quasars consists of ordinary early-type galaxies, with very little if any bias towards higher luminosities or higher degree of morphological irregularities.

3. Implications for the QSO duty cycle

Dividing the space density of quasar hosts by that of the parent galaxies gives the ratio of active to inactive galaxies as a function of galaxy luminosity. This ratio has been interpreted in the past as the 'probability' of finding a QSO in a galaxy. By assuming that all major galaxies have massive black holes in their centres and using the argument of 'ensemble average equals time average', this is furthermore the *time fraction* that a typical galaxy spends in a 'quasar state' (also called the QSO duty cycle δ).

From our determination of the QHGLF, we found that the parent population can be probably identified with normal early-type field galaxies. Under these premises, we can estimate δ directly from the rescaling factor used in Fig. 2 to match the field galaxy LF to the QSO data points, giving $\delta \sim 10^{-4}$. Because of the similarity between the LF, there is virtually no trend of δ with luminosity. We can confidently exclude that the number density ratio between quasar hosts and luminous field galaxies approaches unity (implying $\delta \simeq 1$) as suggested recently by Hamilton et al. (2001).

Note that the above value for δ refers to zero redshift, since because of the explicit evolution term for the QSO space density in the analytic expression for Φ, the QHGLF normalisation is taken at $z = 0$. In this simple picture, QSO evolution would be interpreted as an increase in the duty cycle towards higher redshift with a rate of $\delta \propto (1+z)^{6.5}$.

With $0 < z < 0.3$ for our sample, this corresponds to a time span of $T \sim 5$ Gyrs. Multiplying this with the redshift-averaged value of $\delta(z)$, we can convert the duty cycle into a time scale of $<\Delta t> \simeq 10^6$ yrs, which is a robust lower limit to the total time a typical low-redshift quasar is switched on, and probably not too different from the actual time scale for quasar-type activity.

References

Gardner J.P., Sharples R.M., Frenk C.S., Carrasco B.E., 1997, ApJL 480, L99
Hamilton T.S., Casertano S., Turnshek D.A., 2001, astro-ph/0011255
Köhler T., Groote D., Reimers D., Wisotzki L., 1997, A&A, 325, 502
Kochanek C.S., Pahre M.A., Falco E.E. et al., 2001, astro-ph/0011456
Lin H., Yee H.K.C., Carlberg R.G. et al., 1999, ApJ 518, 533
Loveday J., 2000, MNRAS, 312, 557
McLeod K.K., Rieke G.H., Storrie-Lombardi L.J., 1999, ApJL, 511, L67
Percival W.J., Miller L., McLure R.J., Dunlop J.S., 2000, astro-ph/0002199
Wisotzki L., 2000, A&A, 353, 853
Wisotzki L., Christlieb N., Bade N. et al., 2000, A&A, 358, 77

QSO HOST GALAXY STAR FORMATION HISTORY FROM MULTICOLOUR DATA

Knud Jahnke[1], Björn Kuhlbrodt[2], Eva Örndahl[3] and Lutz Wisotzki[4]

[1,2] *Hamburger Sternwarte;* [3] *Astronomiska Observatoriet Uppsala;* [4] *Universität Potsdam*

[1] kjahnke@uni-hamburg.de, [2] bkuhlbrodt@uni-hamburg.de, [3] eva@astro.uu.se,

[4] lutz@astro.physik.uni-potsdam.de,

Abstract We investigate multicolour imaging data of a complete sample of low redshift ($z < 0.2$) QSO host galaxies. The sample was imaged in four optical ($BVRi$) and three near-infrared bands ($JHKs$), and in addition spectroscopic data is available for a majority of the objects.

We extract host luminosities for all bands by means of two-dimensional modeling of galaxy and nucleus. Optical and optical-to-NIR colours agree well with the average colours of inactive early type galaxies. The six independent colours are used to fit population synthesis models. We assess the presence of young populations in the hosts for which evidence shows to be very weak.

1. Goals

For an assessment of galaxy-formation timescales QSO hosts play a vital role, due to their obvious connection to black hole formation. Dating the nuclear activity and possibly connecting this to external events in the galaxy can help to decide on merger scenarios and the triggering mechanism for activity.

With this work we wanted to start an assessment of the stellar content of host galaxies. By decomposing the host into stellar components we will be able in the future to make detailed comparisons to inactive galaxies.

2. Why multicolour data?

In QSO host galaxy studies using single band or single optical–near infrared (NIR) colours is sufficient to characterise morphological properties like galaxy types, host and nuclear luminosities, apparent signs of interaction or to conduct environment studies (e.g. McLeod & Rieke

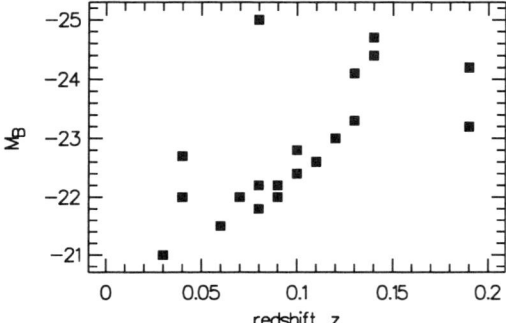

Figure 1. Sample properties. Absolute B band magnitude of QSOs as a function of redshift.

1995, Percival et al. 2001). Optical colours or spectra permit the characterisation of the dominant stellar population or allows assessment of black hole masses (McLure et al. 1999, Boisson et al. 2000). The NIR on the other hand yields, for low z, the best contrast of host against nucleus and allows to assess the mass-to-light ratio of a host.

For the separation of an SED into stellar populations of different ages, using only optical information becomes insufficient for a unique solution. In the NIR the emission of young populations rapidly decreases and old populations dominate. Thus for a study of the stellar components information about the entire SED from the optical to NIR wavelength range is needed.

For luminous AGN the spectral separation of nuclear and host components is very difficult and at the moment largely dependent on subjective or *ad hoc* criteria for the nuclear component. The S/N requirements limit studies to small redshifts and low nuclear luminosities, as the acquisition of spectra becomes very expensive, requiring 8m-class telescopes already at $z = 0.2$. While the quality of the spectral separation methods might change in the future the now available two-dimensional modeling software for QSO hosts allows a detailed and solid assessment of host galaxy fluxes and thus colour information with a high degree of reliability.

3. Sample and observations

We have compiled a sample of 20 objects with $z < 0.2$, drawn from the Hamburg/ESO survey (HES) for luminous QSOs (Wisotzki et al. 2000). The HES is a flux limited objective-prism survey, with a limiting nuclear magnitude $B_{\lim} \sim 17.5$ depending on the field, designed to detect QSOs solely on basis of their spectral properties. Thus unlike samples from many other QSO surveys, the sample is not biased against extended objects.

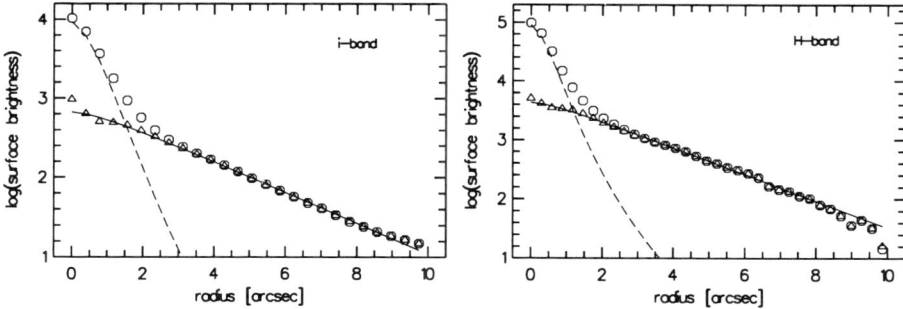

Figure 2. Radial profiles of HE 1310-1051 in *i* band (left) and *H* band (right). Circles: data points, dashed line: best fitting nucleus, solid line: best fitting disk model, triangles: residual galaxy light.

The sample used is a complete sample from a sky area of 611 deg^2, a low-z subsample of a sample defined by Köhler et al. (1997) to study the luminosity function of QSOs. Distribution in redshift and absolute magnitude are shown in Fig. 1, the sample represents moderately luminous QSOs when compared to the total population at all redshifts. The radio properties of most of the objects in the sample are not yet known, but as a subset of the QSO population most will be radio-quiet.

For all 20 objects we have acquired *BVRiJHKs* broadband photometry to evenly sample the SED over the optical–NIR wavelength interval. With three NIR bands we get some redundancy in the NIR to stabilise the stellar population fits. In addition we can make comparisons of sample properties to samples at higher redshift without the need for K-corrections. The *B* band images were integrated 30 s at the ESO 3.6m telescope (EFOSC2), *VRi* images 300–1200 s at ESO Danish 1.5m (DFOSC), and *JHKs* 160–900 s at ESO NTT (SOFI). In addition we have available optical spectra (3800–7500 Å) for 14 of the objects, taken with the ESO 3.6m telescope.

4. Fitting stellar populations

The nuclear contribution of the total QSO light has to be separated from the stellar light. We have developed a package for simultaneous two-dimensional modelling of a parametrised host model and the nuclear contribution. Luminosities for the hosts are determined from radial flux growth curves of the images, after subtracting the nuclear model resulting from the best fit (Fig. 2). More details about the modeling are given in the contribution by B. Kuhlbrodt et al. (these proceedings).

We could produce colours for 18 of the 20 objects. In the two remaining cases the separation was not yielding unique solutions due to highly disturbed morphological structure. We excluded these two from further analysis.

For the optical spectra we are currently developing a two-dimensional separation method similar to the imaging case. For some objects the current program already yields host spectra almost free of broad emission line components from the nucleus. Since this is not the case for all objects we use the optical spectra only for an independent cross-check of fit-results based on the broad-band colours derived from our imaging data.

To assess the primary stellar populations of the hosts, we fit stellar population synthesis model spectra to the multicolour data. For this we use single age, single metallicity population (SSP) spectra from the GISSEL96 library (Bruzual & Charlot 1996, Leitherer et al. 1996). We chose models with a Salpeter initial mass function and solar metallicity, ages 0.01–14 Gyr.

These specra were converted to $BVRiJHKs$ colours using ESO filter curves and fitted to the measured host galaxy colours via a least-χ^2 fit in two steps: 1) fitting only one SSP, age as free parameter, 2) fitting two SSPs, ages and mass-ratio of the two components free.

5. Results and Discussion

The general photometric properties of the sample comply well with values for inactive galaxies, but of course with a large object-to-object variation, $B-V=0.76$ (0.78 for an inactive Sab galaxy), $V-R=0.57$ (0.55 for intermediate type galaxies), and $V-K=3.2$ (3.2 for intermediate type galaxies), values taken from Fukugita et al. (1995) and Griersmith et al. (1982).

Of the 20 objects we could classify three as spheroidal, ten as disks from morphological analysis. Seven show signs of at least mild disturbance.

Fitting one or two SSPs is a strong simplification. At least for disks with a significant amount of onging continuous star formation this will surely oversimplify the picture.

Fitting one SSP in principle only compares general optical-to-NIR colours of host and SSP. Still the ages derived from the fits (Fig. 3) show a generally good agreement with ages expected from the morphological classification. The three classified spheroidals correspond to old populations of 7–17 Gyr while the majority of the disks have a clear tendency towards younger SSPs.

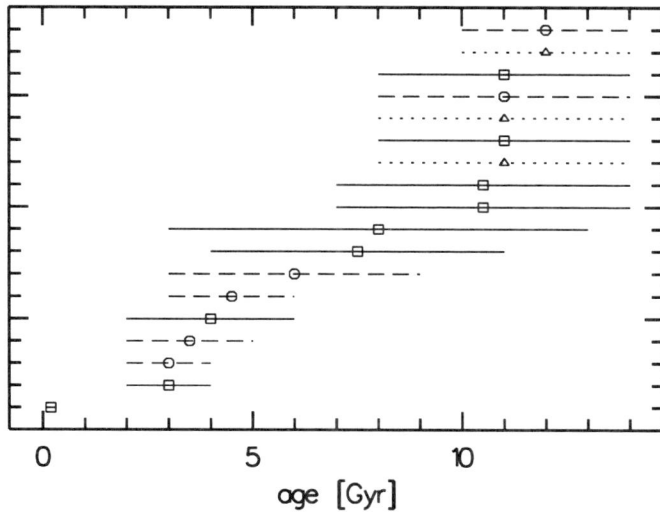

Figure 3. Fitting one single stellar population (SSP). Age-sequence of the best fitting SSP for all objects with range of ages from the fit. Solid lines/squares: morphological disk; dotted lines/triangles: spheroidal host; dashed lines/circles: undecided morphological classification.

When fitting two SSPs, we can distinguish contributions from old and young populations. For most objects though, contributions from a second population did not improve the fit by a great amount. If at all, the involved masses of a young population were small, only for two objects of the order of $\sim 2\,\%$. The resulting spectra for one of them are shown in Fig. 4. We see an excess of blue light in the data (points) compared to the dotted line (single SSP fit). Since both objects are morphologically classified as disks, models with continuous star formation will also be able to explain the blue component.

For the other objects no major second component was detected, and in fact all these objects are consistent with only one SSP and a uniform upper limit for the second, younger component of $\sim 0.5\,\%$ (by mass).

In total we find no signs for strong starburst activity, neither from the spectral fitting nor from general sample colours. The results are in favor of the idea that the parent population of QSO host galaxies is in fact the general field population of inactive galaxies.

In the future we will use spectral models representing continuous star formation for disk-type hosts and in addition combine spectral and colour information into one fitting criterium to make use of all information available. In order to do this, the contribution from the current

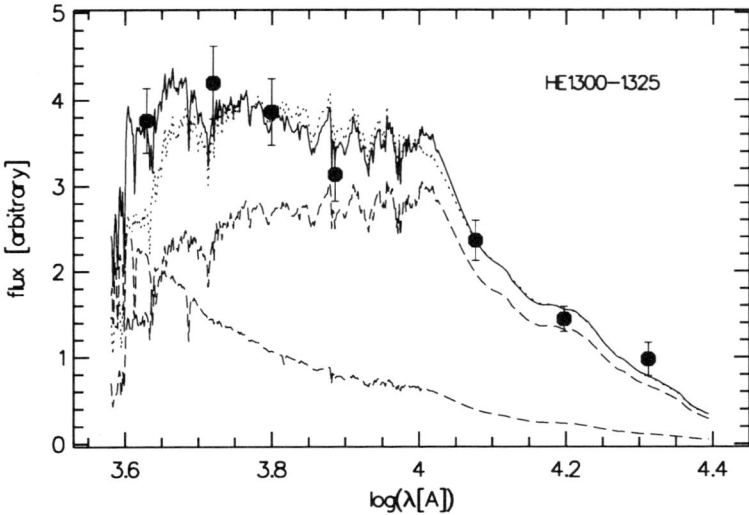

Figure 4. The dotted line respresents one fitted single stellar population (SSP), the solid line the best combination of two SSPs, compared to the data points for HE1300-1325. The two lowest dashed lines are the old and young SSPs contributing to the two-SSP spectrum. Wavelength scale is $\log(\lambda)$ in Å to better display the optical part.

main source of uncertainty, nucleus-galaxy separation, has to be reduced: we are currently improving our software for spectral separation to reach a confidence level comparable to that of the photometric separation.

References

C. Boisson, M. Joly, J. Moultaka, D. Pelat, M. Serote Roos, 2000, A&A, 357, 850B

G. Bruzual & S. Charlot, 1996, unpublished

M. Fukugita, K. Shimasaku, T. Ichikawa, 1995, PASP, 107, 945

D. Griersmith, A. R. Hyland, T. J. Jones, 1982, AJ, 87, 1106

C. Leitherer et al., 1996, PASP, 108, 996

T. Köhler, D. Groote, D. Reimers, L. Wisotzki, 1997, A&A, 325, 502

K. K. McLeod & G. H. Rieke, 1996, ApJ, 441, 96

R. J. McLure, M. J. Kukula, J. S. Dunlop, S. A. Baum, C. P. O'Dea, D. H. Hughes, 1999, MNRAS, 308, 377

W. J. Percival, L. Miller, R. J. McLure, J. S. Dunlop, 2001, MNRAS, in press, astro-ph/0002199

L. Wisotzki, N. Christlieb, N. Bade, V. Beckmann, T. Köhler, C. Vanelle, D. Reimers, 2000, A&A, 358, 77

NEAR-INFRARED IMAGING OF STEEP SPECTRUM RADIO QUASARS

Jary K. Kotilainen
Tuorla Observatory, University of Turku, Piikkiö, Finland; jarkot@astro.utu.fi

Renato Falomo
Osservatorio Astronomico di Padova, Padova, Italy; falomo@pd.astro.it

Abstract   We present near-infrared imaging of 19 steep spectrum radio quasars (SSRQ) at $0.5 < z < 1$. This sample is matched in redshift and optical and radio luminosity with our previously studied sample of 20 flat spectrum radio quasars (FSRQ). The host galaxy is clearly detected in 10, marginally in 6, and remains unresolved in 3 SSRQs. The SSRQ host galaxies are ~ 2 mag. brighter than L* galaxies, and ~ 1 mag. brighter than brightest cluster galaxies. The SSRQ hosts have similar luminosity to FSRQ hosts, and fall between the luminosities of lower and higher redshift radio-loud quasars. The nucleus-to-galaxy luminosity ratio of SSRQs is much smaller than that of FSRQs. We confirm for the most luminous SSRQs the trend noted for the FSRQs of a minimum host galaxy luminosity which increases linearly with the nuclear luminosity. Finally, FSRQs seem to reside in richer environments than SSRQs, as evidenced by a larger number of close companion galaxies.

Keywords: Galaxies:active, Galaxies:nuclei, Infrared:galaxies, Quasars:general

Introduction

Comparison of orientation-independent properties of quasars, e.g. host galaxies and environments, can crucially test orientation-based unified models (e.g. Urry & Padovani 1995) and shed light on the importance of interactions for triggering the quasar activity (e.g. Hutchings & Neff 1992) and the reciprocal effect of the quasar on the properties of the hosts. Possible cosmological evolution of the quasar population (e.g. Silk & Rees 1998) can be constrained by comparing quasar hosts and environments at different redshifts.

The host galaxies of low redshift ($z \leq 0.5$) quasars (e.g. McLeod & Rieke 1995, Taylor et al. 1996, Bahcall et al. 1997) are brighter than L* galaxies (Mobasher et al. 1993), and often as bright as brightest cluster galaxies (Thuan & Puschell 1989). While radio-loud quasars (RLQ) are found exclusively in giant elliptical galaxies, radio-quiet quasars (RQQ) reside in both ellipticals and spirals (Taylor et al. 1996).

While high redshift studies are important to characterize the evolution of host and nuclear properties, our knowledge of quasar host properties has been limited to low redshift, because of the increasing difficulty of resolving the quasar and the rapid cosmological dimming ($\alpha (1+z)^4$) of the host in contrast to the nucleus. Thus, few systematic and homogeneous studies of quasar hosts at $z > 0.5$ have been performed (e.g. Falomo et al. 2001, Kukula et al. 2001, Ridgway et al. 2001). Recently, we studied in the near-infrared (NIR) the host galaxies of 20 flat spectrum radio quasars (FSRQ) at intermediate redshift $0.5 < z < 1$ (Kotilainen et al. 1998, hereafter K98). While most FSRQs are characterized by rapid variability, high and variable polarization, core-dominated radio emission and superluminal motion (e.g. Padovani & Urry 1992), very few of these characteristics are shared by the more common steep spectrum radio quasars (SSRQ). This difference is explained in the unified models as synchrotron radiation relativistically beamed close to our line-of-sight in FSRQs, while SSRQs are viewed further away from the beaming axis.

We present here (and in Kotilainen & Falomo 2000) high spatial resolution NIR imaging of 19 SSRQs in the redshift range $0.5 < z < 1$. NIR offers many advantages over the optical, including larger host/nucleus contrast, negligible scattered light from the quasar, and small K-correction. The SSRQ sample was chosen to match the FSRQ sample (K98) in terms of redshift and optical and radio luminosity distribution, while the two samples are well separated in radio spectral index. Throughout this paper, $H_0 = 50$ km s^{-1} Mpc^{-1} and $q_0 = 0$ are used.

1. Observations, data reduction and modeling of the luminosity profiles

NIR images in the H-band (1.65 μm) were obtained of 19 SSRQs with IRAC2 (0.278 arcsec px^{-1}, field of view 1.2 arcmin^2) and SOFI (0.292 arcsec px^{-1}, field of view 5.0 arcmin^2) at ESO, La Silla. The quasars were shifted across the array between observations, thus keeping them always in the field, and using the other exposures as sky frames. Median seeing was ~1.0 arcsec FWHM during the observations. Data reduction consisted of correction for bad pixels by interpolating across neighboring pixels, sky subtraction using median averaged sky frames, flat-fielding, and combination of images, using field stars or the centroid of the light distribution of the quasar as reference.

After masking all companions around the quasars, radial luminosity profiles were extracted out to background noise level, typically $\mu(H) = $ 23-24 mag. arcsec^{-2}. The PSF characterization was performed using stars of various brightness in the quasar images, or using other frames obtained with similar seeing conditions. In the large field of view SOFI images, all quasars had suitable PSF stars in the field, while in the small field of view IRAC2 images only faint stars were usually available in the field. In these cases we used other fields containing sufficiently bright stars to derive the faint wing of the PSF.

The luminosity profiles were fitted into a point source (PSF) and bulge component by an iterative least-squares fit. At these redshifts, it was not possible to determine the morphology of the host galaxy, and the elliptical model was assumed. For unresolved quasars, we determined an upper limit to the brightness of the host galaxy by adding simulated host galaxies of various brightness to the observed profile until the simulated host became detectable within the errors of the luminosity profile.

2. Results and discussion

2.1. The host galaxies

We detect the host galaxy clearly for 10 (53 %) SSRQs and marginally for 6 (32 %) SSRQs, while the host remains unresolved for 3 (16 %) SSRQs. In Fig. 1 we show the location of the SSRQ and FSRQ hosts in the H-band absolute magnitude vs. redshift diagram. The average H-band absolute magnitude of the resolved SSRQ hosts is $M(H) = -27.2 \pm 1.1$. The SSRQ hosts are therefore very luminous, much brighter than an L* galaxy ($M(H) = -25.0 \pm 0.3$; Mobasher et al. 1993). It is evident that the SSRQ hosts are preferentially selected from the high-luminosity tail of the galaxy luminosity function. Indeed, we find no SSRQ host with $M(H) > -25$, indicating that these quasars cannot be hosted by a galaxy with L < L*, similarly to what was found by Taylor et al. (1996) for low redshift RLQs. The SSRQ hosts are on average of similar luminosity to those of FSRQ hosts (K98).

There is a positive trend of host luminosity with redshift (Fig. 1) as the FSRQ and SSRQ hosts fall between the luminosities of lower redshift ($M(H) \sim -26$) and higher redshift ($M(H) \sim -29$) RLQs. This is consistent with passive stellar evolution in ellipticals (Bressan et al. 1994), but inconsistent with hierarchical galaxy formation (e.g. Kauffmann & Hähnelt 2000), which predicts $\sim$L* hosts at z = 2 - 3 that undergo

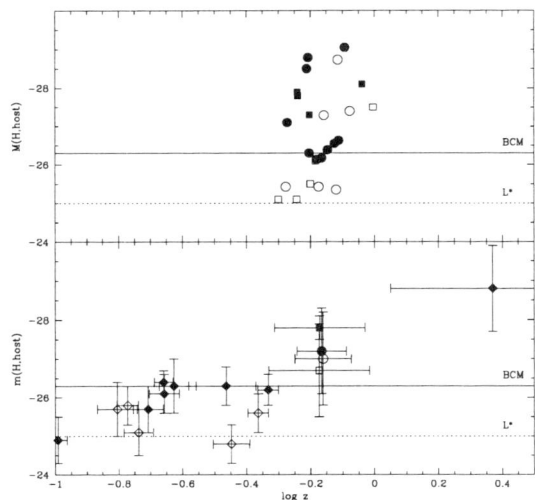

Figure 1. **Upper panel:** The absolute H-band magnitude of SSRQs and FSRQ hosts vs. redshift. Resolved and marginally resolved SSRQs are marked as filled and open circles, respectively. FSRQs (K98) are marked as filled (resolved) and open (marginally resolved) squares. The luminosities of L* galaxies (Mobasher et al. 1993) and brightest cluster galaxies (Thuan & Puschell 1989) are indicated as dashed and solid lines, respectively. **Lower panel:** As in the upper panel, except for the mean values of the SSRQ and FSRQ hosts in comparison with RLQ (filled diamonds) and RQQ (open diamonds) samples from literature.

mergers to become present-day giant ellipticals. The trend suggests that major merging has already happened at $z = 2 - 3$.

2.2. The nuclear component

The average absolute magnitude of the SSRQ nuclei (M(H) = -28.2 ± 1.2) is ~ 1.5 mag. fainter than that of the nuclei of FSRQs (K98; M(H) = -29.7 ± 0.8). This difference is even more evident when considering the nucleus/galaxy (LN/LG) luminosity ratio, which is 3.8 ± 3.2 and 21 ± 11 for the SSRQs and FSRQs, respectively. Whereas the majority of the FSRQs have LN/LG > 10, only one SSRQ is above this limit, and the distribution of the LN/LG ratio for the SSRQs is similar to that of low redshift RLQs (e.g. Taylor et al. 1996). SSRQs clearly exhibit a nuclear component that is systematically fainter than that of FSRQs. This is consistent with the beaming model with larger Doppler amplification factor for FSRQs than SSRQs.

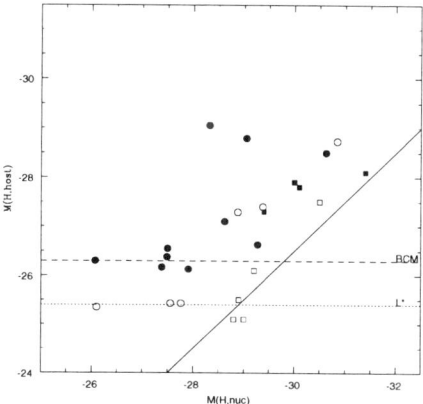

Figure 2. The H-band nuclear vs. host luminosity. For symbols, see Fig. 1. The solid line is the limiting mass-luminosity envelope from McLeod & Rieke (1995).

In Fig. 2, we show the relation between the luminosities of the nucleus and the host galaxy for the SSRQs and FSRQs. We confirm for the SSRQs the trend noted for FSRQs (K98) for the more powerful quasars to reside in more luminous hosts, in the sense that there is a lower limit to the host luminosity, which increases with nuclear luminosity. This trend is in agreement with the relationship found by Magorrian et al. (1998) between the mass (luminosity) of the black hole and the mass (luminosity) of the bulge in nearby galaxies. Our results for SSRQs and FSRQs therefore suggest that the Magorrian et al. relationship extends to host galaxy masses at cosmological distances. On the other hand, the much weaker correlation found for more nearby, lower luminosity active galaxies (e.g. McLeod & Rieke 1995) may indicate that the onset of the correlation occurs only after a certain level in nuclear and/or galaxy luminosity has been reached.

2.3. The close environment of SSRQs and FSRQs

Close companions have been found around many quasars (e.g. Hutchings 1995, Bahcall et al. 1997), but the physical association has been confirmed only in a few cases through spectroscopy (e.g. Heckman et al. 1984, Canalizo & Stockton 1997). These observations have sustained the idea that nuclear activity can be triggered by strong tidal interactions and/or galaxy mergers. RLQs occur preferentially in dense groups of galaxies, and only rarely in rich galaxy clusters. There have been suggestions (Stockton 1978, Hutchings & Neff 1990) that FSRQs (and in general, core-dominated RLQs) have a smaller frequency of companions with respect to the more common SSRQs. This difference, if confirmed, is important for understanding the connection between the characteristics of the nuclear activity and the environment.

We have compared the frequency of companion galaxies around the SSRQs and FSRQs, within a projected distance of 100 kpc, and brighter than $M^*(H) = -23$ at the redshift of each quasar. The average number of companions around the FSRQs and SSRQs is 1.19 ± 1.01 and 0.53 ± 0.68, respectively. The fraction of FSRQs and SSRQs with no close companion is 25% and 58%, respectively. Contrary to the previous suggestions (see above), the close environment of FSRQs is at least comparably rich than that of SSRQs. This argues against a tight connection between the formation/evolution of the radio structure and the quasar environment.

References

Bahcall J.N., Kirhakos S., Saxe D.H., Schneider D.P., 1997, ApJ, 479, 642
Bressan A., Chiosi C., Fagotto F., 1994, ApJS, 94, 63
Canalizo G., Stockton A., 1997, ApJ, 480, L5
Falomo R., Kotilainen J., Treves A., 2001, ApJ, 547, 124
Heckman T.M., Bothun G.D., Balick B., Smith E.P., 1984, AJ, 89, 958
Hutchings J.B., 1995, AJ, 110, 994
Hutchings J.B., Neff S.G., 1990, AJ, 99, 1715
Hutchings J.B., Neff S.G., 1992, AJ, 104, 1
Kauffmann G., Hähnelt M., 2000, MNRAS, 311, 576
Kotilainen J.K., Falomo R., 2000, A&A, 364, 70
Kotilainen J.K., Falomo R., Scarpa R., 1998, A&A, 332, 503 (K98)
Kukula M.J., Dunlop J.S., McLure R.J. et al., 2001, MNRAS, in press
Magorrian J., Tremaine S., Richstone D. et al., 1998, AJ, 115, 2285
McLeod K.K., Rieke G.H., 1995, ApJ, 454, L77
Mobasher B., Sharples R.M., Ellis R.S., 1993, MNRAS, 263, 560
Padovani P., Urry C.M., 1992, ApJ, 387, 449
Ridgway S.E., Heckman T.M., Calzetti D., Lehnert M., 2001, ApJ, in press
Silk J., Rees M.J., 1998, A&A, 331, L1
Stockton A., 1978, ApJ, 223, 747
Taylor G.L., Dunlop J.S., Hughes D.H., Robson E.I., 1996, MNRAS, 283, 930
Thuan T.X., Puschell J.J., 1989, ApJ, 346, 34
Urry C.M., Padovani P., 1995, PASP, 107, 803

ADAPTIVE-OPTICS IMAGING OF LOW AND INTERMEDIATE REDSHIFT QUASARS

Isabel Márquez[1], Patrick Petitjean[2,3], Bertrand Theodore[4,5], Malcolm Bremer[6], Guy Monnet[7] and Jean-Luc Beuzit[8]

[1] *Instituto de Astrofísica de Andalucía (CSIC), Apdo. 3004, 18080 Granada (Spain)*
[2] *Institut d'Astrophysique de Paris, CNRS, 98bis Bd Arago, F-75014 Paris, France*
[3] *UA CNRS 173 – DAEC, Observatoire de Paris-Meudon, F-92195 Meudon Cedex, France*
[4] *Service d'Aéronomie du CNRS, BP 3, F-91371 Verrière le Buisson, France*
[5] *ACRI, 260 route du Pin Montard, BP 234, F-06904 Sophia-Antipolis, France*
[6] *Bristol Univ. (Dept. of Physics) H H Wills Physics Laboratory, Tyndall Av, Bristol BS8 1TL, United Kingdom*
[7] *European Southern Observatory, Karl Schwarzschild Straße 2, D-85748 Garching-bei-München, Germany*
[8] *Canada-France-Hawaii Telescope Corporation, 65-1238 Mamaloha Highway, Kamuela, HI 96743, USA*

Abstract We present the results of AO imaging (H and K bands) of 12 low and intermediate redshift ($z < 0.6$) quasars using the PUEO system mounted on the Canada-France-Hawaii telescope. Five quasars are radio-quiet and seven are radio-loud. The images, obtained under poor seeing conditions, and with the QSOs ($m_V > 15.0$) themselves as reference for the correction, have typical spatial resolution of FWHM ~ 0.3 arcsec before deconvolution. The deconvolved H-band image of PG 1700+514 has a spatial resolution of 0.15 arcsec and reveals a wealth of details on the companion and the host-galaxy.

Four out of the twelve quasars have close companions and obvious signs of interactions. The two-dimensional images of three of the host galaxies unambiguously reveal bars and spiral arms. The morphology of the other objects are difficult to determine from one dimensional fit to the host galaxies; deeper images are needed.

Analysis of mocked data shows that elliptical galaxies are always recognized as such, whereas disk hosts can be missed for small disk scale lengths and large QSO contributions.

Keywords: Galaxies: active – Galaxies: quasars – Galaxies: fundamental parameters – Galaxies: photometry – Infrared: galaxies

1. Sample selection and observations

The quasars were selected such that the nuclei were bright enough to be used as the wavefront reference point source. The sample of radio-quiet quasars were all PG quasars with $m_b < 16.5$ and with redshift less than 0.6. The radio-loud objects were selected from 3C, 4C, B2 and PKS catalogues with the same magnitude and z criteria. The final objects observed (see Table 1) were selected based upon the suitability for the observing conditions on the observing runs. We used the CFHT adaptive optics bonnette (PUEO) and the IR camera KIR. The weather conditions were poor during both runs and the FWHM of the seeing PSF was never better than 0.8 arcsec. The adaptative-optics correction was performed on the QSOs themselves. The final images have a typical resolution of FWHM $\sim$ 0.3 arcsec. After each science observation an image of a star with similar magnitude as the QSO was taken in order to determine the PSF and use it to deconvolve the images. Due to rapid variations in the wheather conditions however, it was not always possible to follow this predefined procedure.

A synthetic PSF function, derived from the stellar images was used to deconvolve each of the images. As it was not always possible to apply a standard procedure due to fluctuating seeing conditions, a careful although, somewhat arbitrary choice of the PSF had to be done. In Fig. 2 we show the images of PKS 1700+514 obtained using, for the deconvolution, three different PSFs from stars observed during the same night. These have respectively, FWHM = 0.30, 0.42 and 0.48 arcseconds. The initial image of the object has a resolution of FWHM = 0.26 arcsec and the star observed just after the science exposure has FWHM = 0.48 arcsec. It is apparent that the best result is obtained using the star with the FWHM closest to that of the science exposure. Here, we were guided in the exercice by the existence of the *HST* image by Hines et al. (1999). In general, this illustrates the crucial role played by a careful PSF determination in AO observations.

Here we only repport the results for two objects, together with the analysis of simulated images. The results for the whole sample are fully described in a forthcoming paper (Márquez et al. 2001).

2. PKS 1302-102

The image obtained at CFHT under moderately good seeing conditions is of similar quality to that obtained with HST by Bahcall et al. (1995; see for comparison Hutchings et al. 1994) with FWHM = 0.24 arcsec after deconvolution (see Fig. 1). The two objects at 1 and 2 arcsec from the quasar are well-detached, and are more clearly seen when both the PSF and a model for the host galaxy (obtained by masking the companions and fitting ellipses to the isophotes) are subtracted. It is unlikely that these companions are intervening objects as strong associated metal line absorption would be expected at such small impact parameter when no such absorption is detected in the HST spectrum down to $W_{obs} \sim 0.2$ Å (Jannuzi et al. 1998). The host-galaxy of this quasar has been detected by HST and fitted with a $r^{1/4}$ profile (Disney et al. 1995). Márquez et al. (1999) derive that the galaxy contributes 40% the total flux in the J-band when McLeod & Rieke (1994) measure this contribution to be 31% of the total flux in the H-band after fitting an exponential profile to the host galaxy. We have performed a similar fit on the present data and found that the contribution of the galaxy amounts to 39% in H and 18% in K. However, we find that an $r^{1/4}$ profile is a better fit. In that case, the contribution of the host-galaxy to the total light is of 70% (60%) in H (K), in good agreement with the values derived by subtracting an scaled version of the PSF and directly integrating the residual flux.

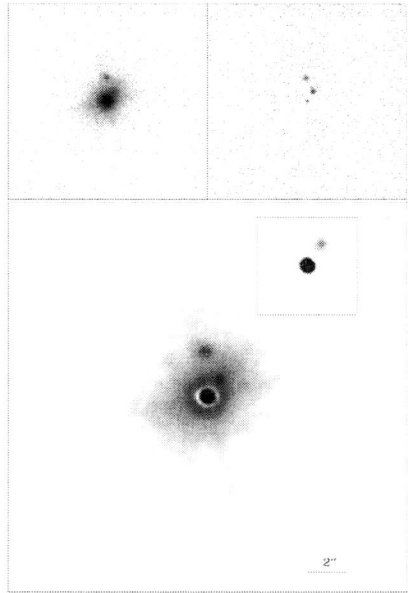

Fig. 1: Images of PKS 1302-102 in the H-band. The top-left panel and bottom panel correspond to the image, respectively, before and after deconvolution. The spatial resolutions are respectively FWHM $\sim$ 0.32 and 0.24 arcsec. The top right panel shows the two companions after subtraction of the PSF and a model for the host galaxy. The inset in the bottom pannel corresponds to a higher contrast version of the inner 4×4 arcsec^2.

3. PG1700+514

Ground-based imaging revealed an extension about 2 arcsec northeast of the quasar (Stickel et al. 1995) which was shown by adaptive-optics imaging and follow-up spectroscopy to be a companion with a redshift 140 km s^{-1} blueward of the quasar (Stockton et al. 1998). NICMOS observations lead Hines et al. (1999) to argue that the companion is a collisionally induced ring galaxy. The image obtained at CFHT is shown in Fig. 2. We confirm the findings by Stockton et al. (1998) that the companion has the appearance of an arc with several condensations. We used different PSF to deconvolve the image. The best deconvolution is obtained using the star with the FWHM closest to that of the AGN (0.30 arcsec). The image has a final resolution of 0.16 arcsec and is probably the best image obtained yet on this object. The companion is seen as a highly disturbed system with a bright nucleus and a ring-like structure; the nucleus beeing decentered with respect to the ring. The host-galaxy is clearly seen around the quasar with a bright extension to the south-west. In addition, we detect a bright knot to the south-east which is not seen in the NICMOS data probably because of the presence of residuals in the PSF subtraction. The comparison between the HST and CFHT images of PG 1700+514 shows how powerful AO can be, and bodes well for the use of the technique on 10 m-class telescopes.

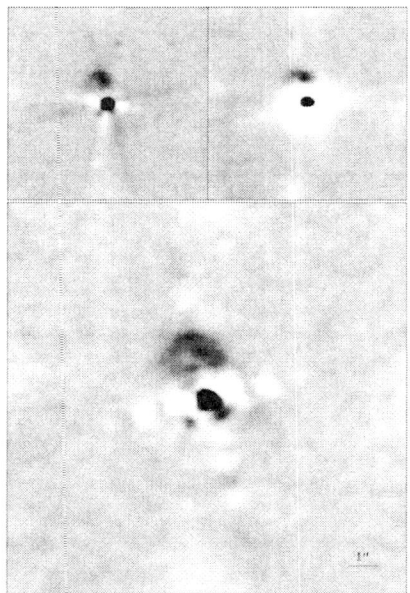

Fig. 2: Images of PG 1700+514 after deconvolution using the PSF given by three stars observed during the same night. The best deconvolution is obtained using the star with the FWHM closest to that of the quasar. The resulting image (bottom) has a final resolution of FWHM = 0.16 arcsec.

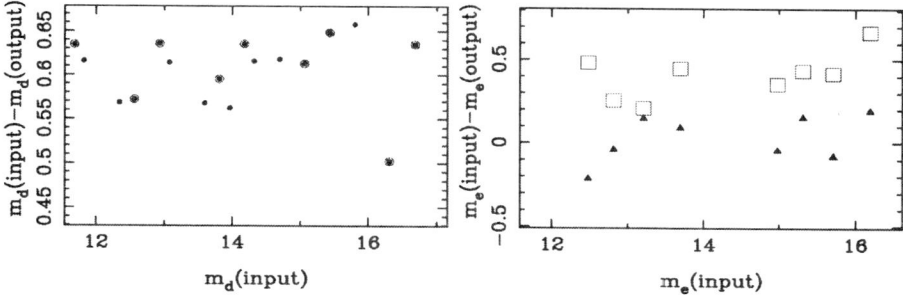

Fig. 3: Test of the fitting procedure on disk (left) and elliptical (right) galaxies. We plot here the difference between the input and output magnitudes in the case the host-galaxy has the same magnitude as the quasar versus the input magnitude. Left: When a dot is surrounded by a circle this means that the fit by an $r^{1/4}$ law is at least as good as the fit by an exponential disk. Right: Triangles correspond to the systems where the QSO contributes to the total luminosity as the host galaxy does, squares represent those cases in which the host luminosity is half that of the QSO.

4. Analysis

In order to test our fitting procedure (see Márquez et al. 2001), we have generated images of model elliptical (E) and disk (S) galaxies with scale-lengths and effective surface brightness within the range derived from the data. A point source is added in the center of the galaxy to mimic the quasar. An appropiate amount of noise is added, and then the images are convolved with a typical observed PSF. The mocked images are analyzed in the same way as real data.

We first note that an E galaxy is always recognized as such by the fitting procedure, whereas a S galaxy is better fitted by a $r^{1/4}$ law when the unresolved point-source contributes more than half the total light. This means that, at least with data of similar quality to those presented here, the fraction of E galaxies in the sample may be overpredicted. Going deeper, at least 0.5 to 1 magnitude, should help solve this problem as it is apparent that the distinction between spiral and E profiles is easier when the galaxy is detected at larger distances from the central point-source.

It is interesting to note that the output magnitudes are brighter than the input in both cases, E or S galaxies (see Fig. 3). The reason for this is probably the difficulty in determining the extension of the PSF wings which, if not subtracted properly, will artificially increase the flux of the host galaxy. In the case of Ss, the difference is as large as 0.6 magnitudes when the contribution of the point-source is the same as the contribution of the host-galaxy (see left side in Fig. 3). For Es, the difference is less but still important when the QSO dominates the total flux.

5. Conclusions

Adaptive-optics imaging in the H and K bands has been used to study the morphology of QSO host-galaxies at low and intermediate redshifts ($z < 0.6$). We detect the host-galaxies in 11 out of 12 quasars, of which 5 are radio-quiet and 7 are radio-loud quasars. The images, obtained under poor seeing conditions, and with the QSOs themselves as reference for the correction, have typical spatial resolution of FWHM ~ 0.3 arcsec before deconvolution. In the best case, the deconvolved H-band image of PG 1700+514 (with a spatial resolution of 0.16 arcsec) reveals a wealth of detail on the companion and the host-galaxy, and is probably the best-quality image of this object thus far.

Four of the quasars in our sample have close companions and show obvious signs of interactions. The two-dimensional images of three of the host galaxies unambiguously reveal bars and spiral arms. For the other objects, it is difficult to determine the host galaxy morphology on the basis of one dimensional surface brightness fits alone.

We simulated mocked images of host-galaxies, both spirals and ellipticals, applying the same analysis as to the data. Disk hosts can be missed for small disk scale-lengths and large QSO contributions. In this case, the host galaxy can be misidentified as an elliptical galaxy. Elliptical galaxies are always recognized as such, but with a luminosity which can be overestimated by up to 0.5 magnitudes. The reason for this is that the method used here tends to attribute some of the QSO light to the host. This is also the case for disk galaxies with a strong contribution of the unresolved component.

References

Bahcall J.N., Kirkhakos S., Schneider D.P., 1995, ApJ, 450, 486
Disney M.J., Boyce P.J., Blades J.C. et al., 1995, Nature 376, 150
Hines D. C., Low F. J., Thompson R. I., Weymann R. J., Storrie-Lombardi L. J., 1999, ApJ, 512, 140
Hutchings J.B., Morris S.C., Gower A.C., Lister M.L., 1994, PASP, 106, 642
Jannuzi B.T., Bahcall J.N., Bergeron J. et al., 1998, ApJS, 118, 1
Kukula M.J., Dunlop J.S., McLure R.J. et al., 2000, astro-ph/0010007
Márquez I., Durret F., Petitjean P., 1999, A&AS 135, 83
Márquez I., Petitjean P., Theodore, B. et al. 2001, A&A, in press
McLeod K.K., Rieke G.H., 1994, ApJ, 431, 137
Stickel M., Fried J.W., McLeod K.K., Rieke G.H., 1995, AJ, 109, 1979
Stockton A., Canalizo G., Close L.M., 1998, ApJ, 500, 121

SUBARU OBSERVATIONS OF THE HOST GALAXIES AND THE ENVIRONMENTS OF THE RADIO GALAXY 3C324 AT $Z = 1.1$

Toru Yamada
National Astronomical Observatory of Japan
yamada@optik.mtk.nao.ac.jp

Introduction

3C 324 is a proto-typical powerful radio galaxy with strong UV-optical alignment effect. The object is known to sit in the dense environment of the galaxies at the same distance. In this contribution, we present the results of the NIR imaging observations made during the telescope comissioning period of the Subaru 8.2m telescope. The data was analyzed with the very deep *HST* WFPC2 optical images. We discuss the nature of the 3C 324 host galaxy as well as the properties of the galaxies in the surrounding cluster. The results presented here has been published in Yamada et al. (2000), Kajisawa et al. (2000a,b) and to be published in Nakata et al. (2001).

1. Old host of 3C 324: the nuclear activity as secondary event ?

We observed the field of the 3C 324 and the associated clusters in K' band with the Subaru 8.2m telescope equipped with the CISCO, NIR camerawith $\sim 2 \times 2$ arcmin^2 f.o.v. We could collect the 800 sec integration with the excellent seeing with FWHM $\sim 0.37''$ and in total 3000 sec with $\sim 0.8''$. The data with high resolution is used for the study of the host galaxy morphology (Yamada et al. 2000) and all the data is used for the analysis on the cluster galaxies (Kajisawa et al. 2000a,b).

The host galaxy of 3C 324 is clearly resolved and seen to be a spheroidal galaxy well approximated by a de Vaucouleurs profile in the Subaru NIR images (Fig. 1). The effective (half-light) radius evaluated from profile fitting is $1''.3$ (11.2 kpc), which is about half the value previously published in the literature. After subtraction of the model galaxies, we clearly detect the 'aligned component' in the K'-band image (Fig. 2).

The peak of the K'-band light coincides with the position of the radio core, which strongly implies that the engine of the powerful radio sources is indeed hosted at the nucleus of the giant elliptical galaxy. The NIR peak also corresponds to the gap in the rest-frame UV emission, which may be due to a dust lane. It is very likely that we see the obscuring structure from an almost edge-on view. The host galaxy has a very red $R_{F702W} - K$ color and the near-infrared light of the galaxy is likely to be dominated by an old stellar population, while the relatively blue $B_{F450W} - R_{F702W}$ color suggests that there may be some small amount of star-formation activity.

These mohological and color information imply that the host galaxy of 3C 324 is an old well developed elliptical galaxy. Considering the (possibly) relatively short time-scale of the nuclear activity, we argue that the nuclear activity appers a posteriori as some secondary event in the history of the galaxy. Signiture of recent star formations (dust, clue colors) do suggest some recent minor merger events with a gaseous system(s).

The colors of the 'aligned' components located inside the host galaxy, which are obtained after subtracting the host component, may be explained by nebular continuum emission with a small amount of a dust while those outside the host galaxy are better modeled by optically-thin dust scattering of the nuclear light.

2. The cluster around 3C324

We also investigated the optical and near-infrared colors of the K'-selected galaxies in the clusters at $z \sim 1.2$ near 3C 324 (Kajisawa et al. 2000a,b). The distribution of the colors of the galaxies in the cluster region is found to be fairly broad and it is very likely that the scatter is intrinsic one and not due to the contamination by the fourground galaxies. Although the 'red finger' of the relatively bright galaxies whose $R - K$ color is consistent with those of passive evolution models for old galaxies is recognized, the sequence is found to be truncated at $K' \sim 20$ mag. and there are few galaxies with similar red color in the cluster region at the fainter magnitude range. The wide color distribution of the possible cluster member galaxies may imply that there is significant scatter in the star-formation history among them provided that they are really associated at the same redshift.

We found that the bulge-dominated galaxies selected by the quantitative morphological classification form a broad sequence in the color-

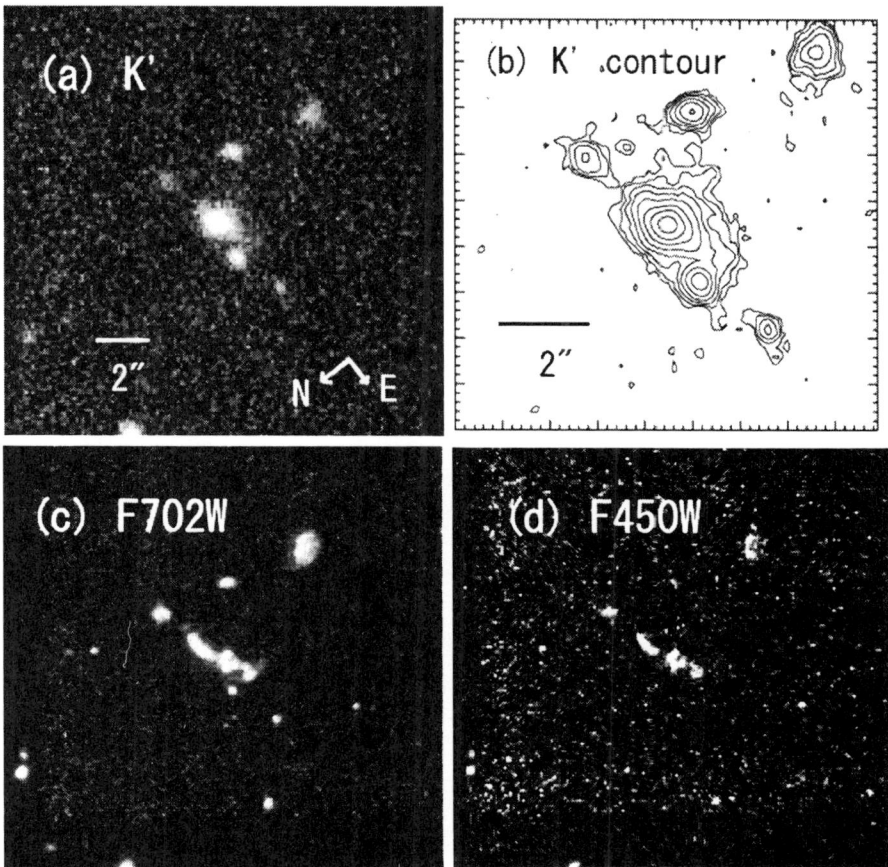

Figure 1. K' images of 3C 324 taken with the Subaru telescope (panel a) and the isophotal contour map for the image after boxcar smoothing with 3×3 pixels (panel b). The lowest contour in panel (b) corresponds to the 1σ noise level of the sky before the smoothing; the contour interval is 0.5 mag. arcsec^{-2}. The images HST/WFPC2 taken in the $F702W$ (panel c) and the $F450W$ filters (panel d) are also shown. The box spans 17″ in panels (a), (c), and (d) and 10″ in the panel (b).

magnitude diagram and the slope of this sequence is much steeper than that expected in a metallicity sequence model of color-magnitude relation with coeval galaxy formation followed by passive evolution.

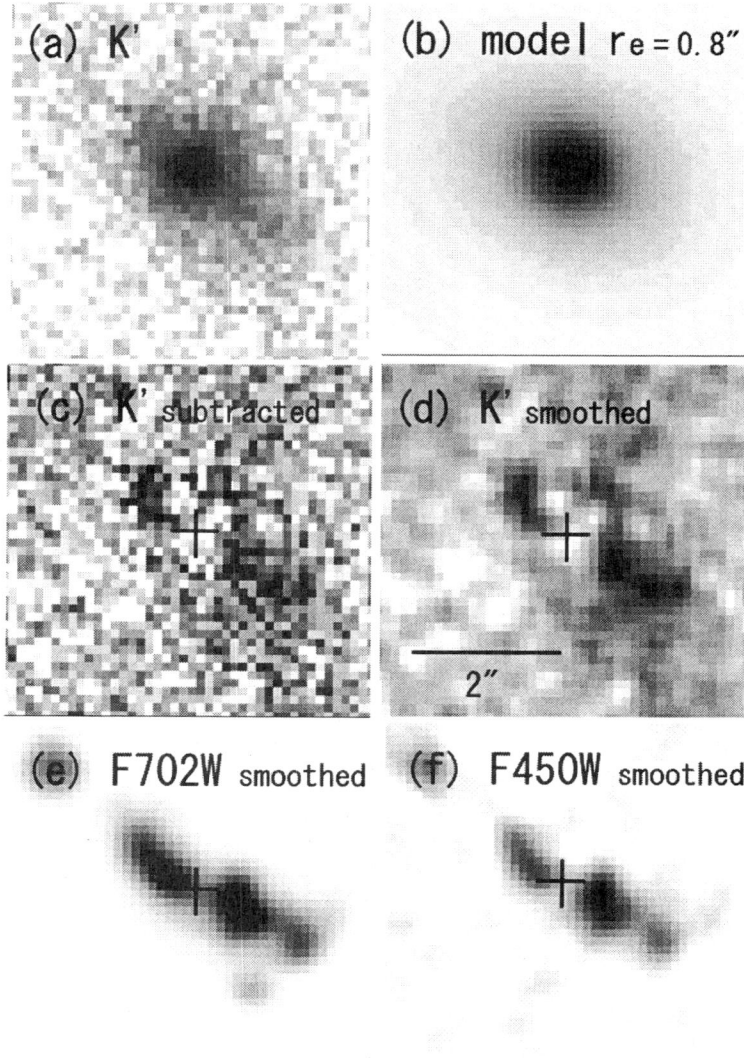

Figure 2. Observed K'-band image after removing of the close companions (panel (a)) and the image of the adopted model galaxy with an effective semi-major axis of $0''.8$ is shown in (b). The resultant image after subtraction of the model host in shown in (c) and a boxcar-smoothed version of this is shown in (d). A NIR alignment effect is clearly detected. For comparison, we show the Gaussian-smoothed HST images in panels (e) and (f). The crosses show the position of the light peak in the K'-band image.

3. Sample of cluster galaxies selected by photometric redshift

Combined with the CISCO and HST data with the newly analyzed data from the optical image obtained using the Suprime-Cam that was equipped at the Cassegrain focus during the telescope comissioning period, we constructed a $BVRIK$ photometric-redshift selected sample of the candidate cluster galaxies (Nakata et al. 2001, in preparation). With this new sample, we re-examin the color and luminosity distribution and basically confirmed the qualitative arguments made in Kajisawa et al. (2000b).

Strong spatial segregation of the color and K'-band luminosity is seen in the sky distribution of the cluster galaxies selected by photometric redshift; the redder and the brighter objects tend to be located near 3C 324 while the bluer or fainter galaxies show more diffuse distribution. We consider that the 3C 324 cluster is still at the dynamically young stage and the formation of the massive clusters of galaxies may have been spatially biased to the very central region of the cluster. Finally, we note that the spatioal distribution of the cluster galaxy show very elongated structure. Interestingly, the axis is very similar to that of the radio axis.

I thank all my collaborators in the papers Yamada et al. (2000), Kajisawa et al. (2000a,b), and Nakata et al. (2001)

References

Kajisawa, M. et al., 2000a, PASJ, 52, 61
Kajisawa, M. et al., 2000b, PASJ, 52, 53
Yamada, T. et al., 2000, PASJ, 52, 43

Toru Yamada

From left to right, James Dunlop, William Percival, Matt Jarvis and Chris Willot.

EXTREMELY RED RADIO GALAXIES

Chris J. Willott, Steve Rawlings and Katherine M. Blundell
Astrophysics, University of Oxford, UK

Abstract At least half the radio galaxies at $z > 1$ in the 7C Redshift Survey have extremely red colours ($R - K > 5$), consistent with stellar populations which formed at high redshift ($z \gtrsim 5$). We discuss the implications of this for the evolution of massive galaxies in general and for the fraction of near-IR-selected EROs which host AGN, a result which is now being tested by deep, hard X-ray surveys. The conclusion is that many massive galaxies undergo at least two active phases: one at $z \sim 5$ when the black hole and stellar bulge formed and another at $z \sim 1 - 2$ when activity is triggered by an event such as an interaction or merger.

Introduction

Radio sources are known to reside in massive, luminous host galaxies. Therefore, they are an excellent way of selecting such galaxies out to high redshifts and tracing their evolution. The obscuring torus, which forms the basis of the unified schemes, blocks the non-stellar nuclear emission from our line-of-sight providing a much clearer view of the host galaxy properties than is the case for quasars. 3C radio galaxies at $z \gtrsim 0.6$ have complex optical continuum and emission line structures aligned along their radio axes, which can generally be interpreted as due to recent (jet-induced) star-formation or non-thermal processes associated with the active nucleus (e.g. Best et al. 1998). Lacy et al. (1999a) have shown that the strength of the alignment effect decreases with decreasing radio luminosity such that fainter samples do not suffer from this problem.

The 7C Redshift Survey (7CRS) is a low-radio frequency (151 MHz) flux-selected sample in three small patches of sky covering a total sky area of 0.022 sr. The flux-limit is 0.5 Jy – a factor of 25 times lower than the revised 3CR sample (Laing et al. 1983). We now have complete optical/near-infrared identifications for all the radio sources and > 90% spectroscopic redshifts. Further details of the survey are given in Willott et al. (2001a) and Lacy et al. (1999b). Only 7 out of 76 sources in the

7C-I and 7C-II regions lack spectroscopic redshifts. For these sources we have obtained optical and near-IR photometry in order to constrain their redshifts using photometric redshift techniques. We find that most of these seven galaxies have very red colours. A full account of this work is given in Willott et al. (2001b). A flat cosmology with parameters $\Omega_M = 0.3$, $\Omega_\Lambda = 0.7$ and $H_0 = 70$ km s^{-1}Mpc^{-1} is assumed throughout.

1. Results of photometric redshift fitting

$RIJHK$ photometry was obtained for the seven sources without spectroscopic redshifts. Six of these objects have extremely red colours ($R - K > 5.5$), similar to those expected from evolved stellar populations at $z > 1$. These data were fit with a range of instantaneous burst model galaxies (Bruzual & Charlot, in prep.) with redshift, age and reddening as free parameters. Best-fit models were found by searching for the minimum in the χ^2 distribution. Fits with reduced $\chi^2 < 1$ were found for five objects and the remaining two had best-fit reduced $\chi^2 < 2$. An example of the SED-fitting is shown in Fig. 1.

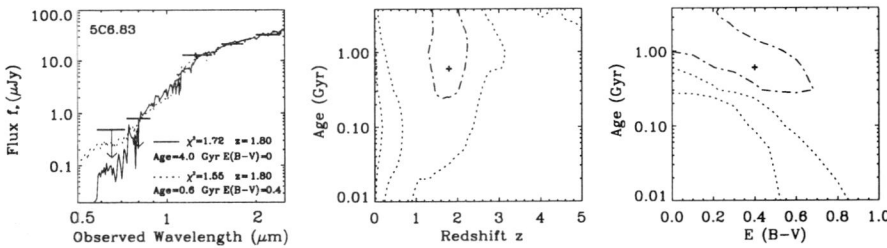

Figure 1. Example of model-fitting to the optical to near-IR SEDs of 7CRS radio galaxies. The left panel shows the broad-band photometric data of 5C6.83 (7CB021111.3+303948; $R > 24.5$; $K = 18.3$) with best-fit model galaxies. The best fit unreddened model is shown as a solid line and the best-fit reddened model as a dotted line. Details of these fits are given in the bottom-right corner. The centre and right panels show how the χ^2 of the fit depends upon redshift, age of the stellar population and reddening. The minimum in reduced χ^2 (1.55) is shown as a cross.

The best-fit redshifts for these radio galaxies ranges from $z = 1.05$ to $z = 2.35$ with typical uncertainties $\Delta z = \pm 0.3$. Considering other factors such as the $K - z$ relation and the lack of emission lines in their optical spectra, we believe that all the galaxies are likely to fall within the redshift range $1 < z < 2$. The flatness of the SEDs from J to K strongly argues against $z > 2.5$. We know from near-IR spectroscopy that the colours are not strongly affected by emission line contamination, but in two cases there is a marginal emission line in the near-IR which could be Hα at $z \approx 1.5$, consistent with the SED-fitting. Quite a wide range of parameters provide acceptable fits for many of the galaxies. The degeneracy of age and reddening appears to be the strongest cause of this as shown by the diagonal shape of contours in the age-reddening plane in Fig. 1 (from old and dust-free at top-left to young and dusty at bottom-right). In contrast there is little correlation between redshift and age in the contours and the best-fit redshifts are in most cases similar for both the reddened and unreddened cases.

2. The colours of 7CRS radio galaxies

Since all members of the 7CRS (regions I and II) have been securely identified, have spectroscopic or photometric redshifts and have R and K-band imaging (Willott et al., in prep), we can investigate the colour evolution of the radio galaxies. In Fig. 2 we plot the observed $R - K$ colour against redshift for the 49 narrow-line radio galaxies and the 2 broad-line radio galaxies (the 23 quasars are not plotted because their magnitudes are clearly dominated by non-stellar emission). The solid curve is a model featuring an instantaneous starburst at redshift $z = 5$. At redshifts $z \leq 1$ the colours of most of the radio galaxies are close to the model curve suggesting that these colours result from very old galaxy populations with little (unobscured) current star-formation. Deviations to bluer colours can be caused by a small amount of more recent star-formation or AGN-related processes like scattering of quasar light.

Moving to $z \gtrsim 1$ we find a marked increase in the scatter of the $R - K$ colours of the radio galaxies. This is probably due to a combination of two effects. First, at redshifts higher than $z \sim 0.8$, the observed R-band samples the rest-frame light below the 4000 Å break. The ratio of fluxes below and above 4000 Å is a very strong function of the amount of current or recent star-formation. Therefore small differences in the amount of recent star-formation will have a more dramatic effect on the observed colour at $z > 0.8$. Lilly & Longair (1984) showed that the optical/near-IR colours of 3CR galaxies at $z > 1$ are inconsistent with a no-evolution model. Their observed $R - K \approx 4$ colours are similar to those of the bluest 7CRS radio galaxies. For the 3C objects these colours are best explained by recent star-formation in a few cases (Chambers & McCarthy 1990), but in most cases, as further evidenced by large optical polarizations (e.g. di Serego Alighieri et al. 1989), they are probably caused by an extra non-stellar rest-frame UV component, typically scattered light from the quasar nucleus (see Best et al. 1998).

It is clear from Fig. 2 that, at $z > 1$, the bluer 7CRS galaxies are much more likely to have redshifts measured from optical spectroscopy. This suggests that the bluer sources have stronger emission lines and are

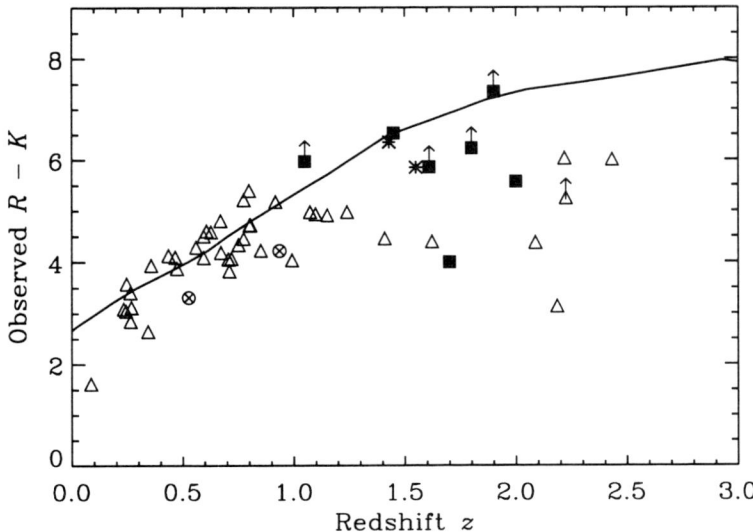

Figure 2. Observed $R - K$ colour as a function of redshift for radio galaxies in the 7C-I and 7C-II regions of the 7CRS. Triangles are narrow-line radio galaxies, crossed circles are broad-lined radio galaxies and filled squares are the seven 7CRS objects with photometric redshift estimates. The asterisks are the two red radio galaxies from the LBDS survey (Dunlop 1999). The solid line shows the evolution of the expected observed colour of a galaxy which formed in an instantaneous starburst at redshift $z = 5$ using the models of Bruzual & Charlot (in prep.). This model provides a good fit to the upper envelope of the galaxy colours up to at least redshift $z = 2$.

therefore good candidates for having non-stellar (AGN) and/or starburst components in the rest-frame UV. At $z > 1$, radio galaxies from the 6CE sample, which has a flux-limit in between those of 3CRR and the 7CRS, also tend to have strong emission lines and blue colours (Rawlings et al. 2001). Therefore the reason that such a high fraction of extremely red radio galaxies at high-redshift has not been seen in previous complete samples is due to their higher radio flux-limits selecting only higher luminosity $z > 1$ radio galaxies.

3. Do the red colours indicate old galaxies?

We have shown that at least half the $z > 1$ radio galaxies in the 7CRS have red colours ($R - K > 5$), consistent with those expected from evolved stellar populations at these redshifts. In addition, the fact that the bluer high-redshift radio galaxies have stronger emission lines, suggests that they too may have underlying host galaxies which are very red. But can we be certain that these colours are an indicator of age, rather than the effects of reddening by dust? The answer is that we cannot be completely certain in the absence of high signal-to-noise spectroscopy of stellar absorption features. Due to the faintness of these galaxies in the optical ($I \gtrsim 24$), this will have to be performed in the near-infrared utilizing the low-background available with OH-suppression spectrographs on large telescopes. Instead, we have to look at circumstantial evidence, such as that attained for similar objects. Dunlop et al. (1999 and references therein) have studied two red radio galaxies at $z \sim 1.5$ drawn from a faint radio sample. Deep Keck spectroscopy shows that the red colours of these galaxies are due to an old stellar population and not due to reddening by dust. The ages inferred are still controversial, although ages of ~ 3 Gyr remain the best estimate (Nolan et al. 2001). For these ages observed at $z = 1.5$, the star-formation must have ceased at $z \sim 5$.

4. Implications for AGN activity in elliptical galaxies

Although powerful radio galaxies are very rare objects, we can use our findings to predict the relationship between other AGN (both radio-loud and radio-quiet) and the near-infrared selected ERO population. Extrapolating down the radio luminosity function (Willott et al. 2001a) to less powerful radio galaxies, we find that if a similar fraction of these lower power radio galaxies have similar colours, they imply a space density of red radio galaxies which is about 3% of the space density of near-IR selected EROs (Daddi et al. 2000). Hence we find that a small, but significant, fraction (~ 3 %) of field EROs are likely to be hosting radio-loud AGN.

However, it is well-known that luminous radio sources have limited lifetimes ($\sim 10^8$ years) which are much smaller than the Hubble time. The time elapsed between $z = 2$ and $z = 1$ is approximately 3.5 Gyr. Therefore if individual radio sources have lifetimes of only $\sim 10^8$ years, then the number of galaxies undergoing radio activity during this period would be a factor of 30 greater than that observed. Hence all of the near-IR selected EROs could plausibly undergo such a period of radio activity. A caveat to this is that the typical lifetimes of weak radio sources such as those which would dominate the ERO population are not well-constrained and could have longer lifetimes of ~ 1 Gyr. In such a case, only $\sim 10\%$ of high-z EROs would undergo a period of radio activity at some point (see Willott et al. 2001b for more details).

The hardness of the X-ray background requires that the space density of optically-obscured quasars exceeds that of optically-luminous quasars (e.g. Comastri et al. 1995), which in turn are well known to outnumber radio-loud quasars by at least an order of magnitude. Many of the hard X-ray sources discovered in Chandra surveys have very red galaxy counterparts with weak or absent emission lines (Crawford et al. 2001, Cowie et al. 2001). These objects are likely to be the radio-quiet analogues of the 7CRS EROs discussed here. The hard X-ray properties of the ERO population will be investigated with XMM-Newton and Chandra surveys of ERO fields. In the HDF-North Caltech area, 4 out of 33 EROs ($R - K > 5$) have hard X-ray detections (Hornschemeier et al 2001). Assuming an elliptical galaxy fraction in the field ERO population of 70% (Moriondo et al. 2001, Stiavelli & Treu 2000) this corresponds to $\sim 20\%$ of these ellipticals being observed to undergo a phase of AGN activity. Better statistics will accurately determine the duty cycle of AGN-activity in massive galaxies and provide constraints on quasar lifetimes (which are currently not well-constrained for radio-quiet quasars). It seems likely that rather than being an oddity, AGN activity is common in massive galaxies, not only during the high-redshift formation epoch, but also at a later stage once the major episode of star (and black hole) formation has long since ceased.

References

Best P.N., Longair M.S., Röttgering H.J.A., 1998, MNRAS, 295, 549
Chambers K.C., McCarthy P.J., 1990, ApJL, 354, L9
Comastri A., Setti G., Zamorani G., Hasinger G., 1995, A&A, 296, 1
Cowie L.L. et al., 2001, ApJL, in press, astro-ph/0102306
Crawford C.S., Fabian A.C., Gandhi P., Wilman R.J., Johnstone R.M., 2001, MNRAS, submitted, astro-ph/0005242
Daddi E. et al., 2000, A&A, 361, 535
di Serego Alighieri S., Fosbury R.A.E., Tadhunter C.N., Quinn P.J., 1989, Nature, 341, 307
Dunlop J.S., 1999, in The Most Distant Radio Galaxies, ed. P.N. Best, H.J.A. Röttgering, M.D. Lehnert, (KNAW Colloq.; Dordrecht: Kluwer), p. 14
Hornschemeier A.E. et al., 2001, ApJ, in press, astro-ph/0101494
Lacy M., Ridgway S.E., Wold M., Lilje P.B., Rawlings S., 1999a, MNRAS, 307, 420
Lacy M. et al., 1999b, MNRAS, 308, 1096
Laing R.A., Riley J.M., Longair M.S., 1983, MNRAS, 204, 151
Lilly S.J., Longair M.S., 1984, MNRAS, 211, 833
Moriondo G., Cimatti A., Daddi E., 2001, A&A, in press, astro-ph/0010335
Nolan L.A., Dunlop J.S., Jimenez R., Heavens A.F., 2001, MNRAS, submitted, astro-ph/0103450
Rawlings S., Eales S.A., Lacy M., 2001, MNRAS, 322, 523
Stiavelli M., Treu T., 2000, To appear in the proceedings of the conference "Galaxy Disks and Disk Galaxies", ASP Conf. series, eds. Funes and Corsini, astro-ph/0010100
Willott C.J., Rawlings S., Blundell K.M., Lacy M., Eales S.A., 2001a, MNRAS, 322, 536
Willott C.J., Rawlings S., Blundell K.M., 2001b, MNRAS, in press, astro-ph/0011082

THE ENVIRONMENTS OF RADIO-LOUD QUASARS

J.M. Barr, M.N. Bremer
Physics Department, Bristol University, UK
j.barr@bristol.ac.uk,

J.C. Baker
Department of Astronomy, Berkeley, USA

Abstract We have obtained multi-colour imaging of a representative, statistically complete sample of low-frequency selected ($S_{408MHz} > 0.95$ Jy) radio loud quasars at intermediate ($0.6 < z < 1.1$) redshifts. These sources are found in a variety of environments, from the field through to rich clusters. We show that statistical measures of environmental richness, based upon single-band observations are inadequate at these redshifts for a variety of reasons. Environmental richness seems correlated with the size and morphology of the radio source, as expected if the energy density in the radio lobes is approximately the equipartition value and the lobes are in pressure equilibrium with a surrounding intragroup/cluster medium. Selecting on radio size therefore efficiently selects dense galactic sytems at these redshifts.

Introduction

Our sample of quasars were a subset of the statistically complete Molonglo Quasar Sample (MQS) (Kapahi et al. 1998), with redshifts $0.65 < z < 1.10$ and with $0 < RA < 14$. Several sources were randomly excluded due to observing-time limitations. Radio luminosities of the sources ranged between $27.4 < \text{Log}(P_{408MHz}) < 28.2$, a factor 5 lower than the most luminous (3CR) sources at comparable redshifts.

Most fields were imaged in the optical using either EFOSC or EFOSC2 on the ESO 3.6m telescope with a field of view of 5'x 5', with some observed with the CTIO and AAT 4m telescopes. Filter bands were chosen to straddle the 4000Å break giving the greatest contrast for early

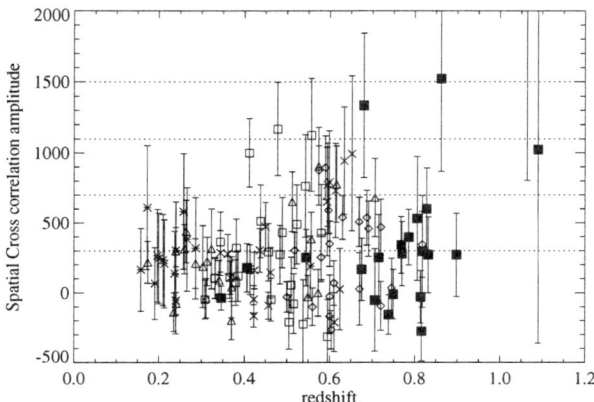

Figure 1. Values of the spatial cross correlation amplitude as a function of redshift for Radio Loud quasar fields. Filled squares, this work. Diamonds, Wold et al. (2000). Triangles, Yee & Ellingson (1993). Squares, Ellingson et al. (1991). Crosses, Yee & Green (1987). Asterisks, McClure & Dunlop (2000). The horizontal lines represent Abell classes 0,1,2,3.

type galaxies at the appropriate redshift (typical images being complete to $R \sim 24$). A subset of these fields were imaged in J and K using IRAC2 on the ESO/MPI 2.2m. with a smaller (2'x 2') field-of-view (typically complete to $K \sim 18.5$).

1. Single filter clustering statistics

We analysed clustering of faint galaxies around the quasars in two ways. Firstly, we carried out "traditional" statistical tests of clustering on single-band data (or at least not colour-selecting the galaxies), as practiced by others (e.g. see Wold et al., these proceedings). The spatial cross-correlation amplitude B_{gq} and the magnitude-limited overdensity within 0.5 Mpc, $N_{0.5}$ (Hill & Lilly, 1991) were computed for our fields in either the R or I band. The results for B_{gq} are shown in Fig. 1. They show broad agreement with those of Wold et al. (2000) and McClure & Dunlop (2000) to a redshift of 0.9. Above this, our data are not deep enough to detect a significant portion of cluster galaxies. Normalising to the expected cluster luminosity function then causes spurious results with large errors. Our observations support the notion that RLQs inhabit a wide variety of environments as found in other studies (*e.g.* Yee & Green 1987).

Additionally we find several cases where B_{gq} is misleading. This may be for several reasons. Firstly, interpretation of the the statistical tests in terms of environmental richness assumes that the quasar is at the centre of any (roughly circular) overdensity. At high redshift, clusters are generally not relaxed spheroidal systems, and may have several bright ellipticals within them, rather than one central dominant galaxy. We find that our quasars are not always at the centres of overdensities, and in some cases there are other galaxies that have colours and magnitudes of first-ranked cluster ellipticals within the overdensity.

Secondly, at the magnitude levels of interest, the number density of background objects, and the significant variation in that quantity on the scale-lengths of distant clusters make it very difficult to determine the background in one band to subtract from any overdensity. Thirdly, even when an obvious overdensity is detected, it can sometimes be made up of galaxies with the wrong optical-IR colours to be at the redshift of the quasar. In at least two cases we find an overdensity which gives a large value of B_{gq} but is a foreground agglomeration, found to be so by reference to optical and infrared colours. Detections such as these can lead to the misidentification of clusters and an increase in the average value of the cross-correlation amplitude.

2. Multicolour analysis

Because of the problems with single-band statistical measures of clustering, we used the colours of galaxies in the quasar fields to estimate clustering. Specifically, we determine the colours expected for passively evolving ellipticals at the centres of clusters at the redshift of each quasar and compare these to the colours of the observed galaxies.

Optical $V - I, R - I$ colours can be used to isolate the 4000Å break for ellipticals at $z > 0.6$. Near-infrared J and K filters are also useful for identifying galaxies above a redshift of 0.5. Specifically, objects with a $J - K$ colour of ~ 1.7 and above are galaxies at redshift 0.5 and higher (Pozzetti & Manucci 2000). Spirals at these redshifts can also be isolated due to their red $J - K$ (due to dust reddening), but relatively blue optical colours. The $J - K$ colour sets unresolved high redshift galaxies completely apart from stars and brown dwarfs which have $J - K < 1.5$.

For each object we determine the distribution of a magnitude limited sample of objects in the field as a function of their colour. Fig. 2 shows an example of this. Each greyscale square is the smoothed surface density of objects with $21 < I < 23$ with colours ranging between $0.5 < V - I < 3.5$ in bins of 0.5 mag. in colour in the 5'x 5' field of a $z = 0.9$ quasar. The grey levels are normalised to give the same object density

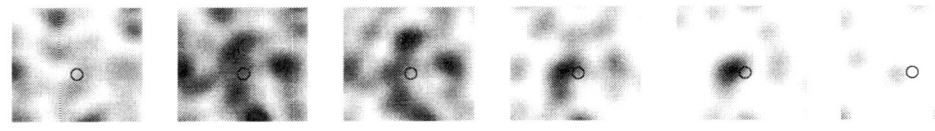

Figure 2. The density of $21 < I < 23$ objects in the 5'x5' field of MRC B0450-221 ($z = 0.9$). The figure shows progressively redder objects in 0.5 mag. bins from $0.5 < V - I < 1.0$ on the left through to $3.0 < V - I < 3.5$ on the right. The position of the quasar is indicated by the circle; it is on the edge of an overdensity of $2.5 < V - I < 3.0$ objects.

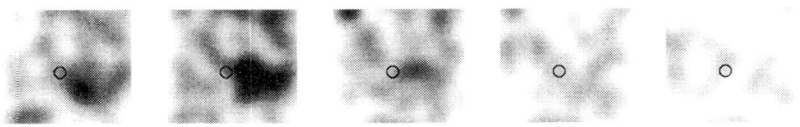

Figure 3. The density of objects in the field of MRC B0346-279. The figure shows the density of objects with $0 < R - Z' < 2.5$ in 0.5 mag. bins. It can clearly be seen that the large value of $N_{0.5}$ is caused by clustering of blue ($0.5 < R - Z' < 1.0$) objects to the West of the quasar.

for the same grey level in each colour bin. Most objects are bluer than $V - I < 2$. An overdensity of red objects close to the quasar can be seen in the $2.5 < V - I < 3$ bin, the colour expected for ellipticals at $z = 0.9$. J and K imaging of the central region confirm that these red objects have $J - K$ and $I - K$ colours of ellipticals at z=0.9. Narrow band imaging in redshifted [OII] and follow-up spectroscopy confirm faint star forming galaxies in the field at the quasar's redshift, but not in the region of the red overdensity. This is a clear case of a cluster around a $z = 0.9$ quasar and is discussed in detail in Baker et al. (2001).

In contrast Fig. 3 shows the distribution of faint objects around a $z = 1$ quasar. The R-band image clearly shows a strong overdensity of $22 < R < 24$ close to the quasar, leading to a strong $N_{0.5}$ signal, indicating at first sight that the quasar inhabits a rich cluster. However, the $R - Z'$ colour distribution peaks at $0.5 < R - Z' < 1.0$, far too blue for ellipticals at $z = 1$. Similarly the $R - K$ colours of these objects are also too blue to be ellipticals at the quasars redshift. Thus we have a case where the colour information rules out this overdensity being a cluster at the quasar redshift (unless the morphological mix is extremely unusual).

The environments of radio-loud quasars

 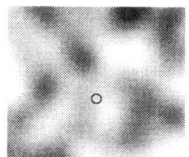

Figure 4. Distribution of the density of $20 < I < 22$ objects with $1.4 < R - I < 1.6$ for two 5'x5' fields centred on MRC quasars at $z = 0.77$. Both greyscales are normalised to the same surface densities.

Fig. 4 compares the spatial distribution of objects with $20 < I < 22$ and $1.4 < R - I < 1.6$ in the fields of two quasars with almost identical redshifts ($z = 0.77$). The first cluster shows a clear overdensity of objects with this colour, consistent with the colour ellipticals at $z = 0.77$, the second does not (and shows no obvious clustering of objects of any other colour). The $R - K$ colours of the objects around the first quasar are again consistent with being in a cluster associated with the quasar.

Similar analysis of other objects in the sample lead to detections of several other systems of galaxies, from compact groups to clusters along with other systems that at first sight appear to be groups or clusters at the quasar redshift, but have colours consistent with lower (or different) redshift objects.

3. A possible correlation with radio size

The range of radio luminosities for our sample covers less than an order of magnitude, and the lookback time difference between the lowest and highest redshift members is less than 2Gyr, so we cannot probe if these quantities correlate with environmental richness. However, we may expect a correlation with the extent of the radio lobes in low-frequency selected radio sources. If the radio lobes are in pressure equilibrium with a surrounding medium, then if their internal energy densities/pressures are close to (or scale roughly as) the equipartition value, there should be relationship between true (unprojected) angular size and the pressure of the external medium (*e.g.* see Miley 1980, Bremer et al. 1992). For sources that evolve self similarly, $P_{equip} \propto (\frac{F}{\theta^3})^{4/7}$ where F is the flux of the source (essentially constant in a flux-limited sample) and θ is the angular size of the source. The smaller the source, the higher the equipartition pressure, and therefore the higher the external pressure (assuming source pressures scale with the equipartition value, at least statistically). Sources surrounded by a dense ICM should therefore be smaller than those in the field, or lower mass groups.

Excluding flat spectrum point sources, there are seven sources which have a radio extent of less than 20". We find four obvious clusters, of which three are confirmed by IR data. The fourth has an obvious overdensity with the correct optical colours, but requires IR data to confirm the nature of the constituent galaxies.

Of the remaining nine larger objects, only three show signs of overdensities with the right colours. Two of these are asymmetric sources which appear to have groups associated with their disturbed radio morphology (*e.g.* Bremer, Baker & Lehnert 2001).

4. Summary

Powerful low-frequency selected radio-loud quasars at redshifts $0.6 < z < 1.1$ exist in a wide variety of environments, from the field through compact groups to rich clusters.

The quasar is not always directly centred on any overdensity we find, nor is any overdensity confined to within 0.5 Mpc of the active galaxy. This has the effect of making clustering statistics like B_{gq} and $N_{0.5}$ rather blunt tools for analysing individual clusters. We find that colours must be used when trying to distinguish clusters from their backgrounds. In particular near infrared colours can be used to accurately extract the high redshift galaxy population.

The smallest extended sources are more likely to be classified as being in clustered environments (though not necessarily in clusters). This is to be expected from equipartition arguments. Factoring in an estimate of true source size and excluding core-dominated sources results in an extremely efficient way of selecting fields containing high redshift groups and clusters. The richness of these systems vary from field to field, but (given the results of Martin Hardcastle in these proceedings) we could expect comparable richnesses to many X-ray selected clusters at similar redshifts. Blind, wide-field optical searches for clusters find systems of comparable richness (but with far lower efficiency). Despite the obvious effectiveness of colour selection in detecting these systems, follow-up 8m imaging and spectroscopy is still required to determine the parameters of the systems.

References

Baker J.C., Hunstead R.W., Brinkmann W., 1995, MNRAS, 277, 553
Baker J.C., Hunstead R.W., Bremer M.N., Bland-Hawthorn J., Athreya R. M. Barr J.M., 2001, AJ, 121, 1821.

Bremer M.N., Crawford C.S., Fabian A.C., Johnstone R.M., 1992, MNRAS, 254, 614
Bremer, M.N., Baker, J.C., Lehnert, M., 2001, MNRAS, submitted.
Ellingson E., Yee H.K.C., Green R.F., 1991, ApJ, 371, 49
Hill G.J., Lilly S.J., 1991, ApJ, 367, 1
Kapahi V.K., Athreya R.M., Subrahmanya C.R., Baker J.C., Hunstead R.W., McCarthy P.J., van Breugel W., 1998, ApJS, 118, 327
McClure R.J., Dunlop J.S., 2000, MNRAS, submitted (astro-ph/0007219)
Miley G., 1980, ARA&A, 18, 165
Pozzetti L., Mannucci F., 2000, MNRAS, 317, L17
Wold M., Lacy M., Lilje P.B., Serjeant S., 2000, MNRAS, 316, 267
Yee H.K.C., Green R.F., 1987, ApJ, 319, 28
Yee H.K.C., Ellingson E., 1993, ApJ, 411, 43

Malcolm Bremer and Joanne Baker

Martin Hardcastle

EXTENDED X-RAY EMISSION AROUND RADIO-LOUD QUASARS

Martin J. Hardcastle
Department of Physics, University of Bristol
Tyndall Avenue, Bristol BS8 1TL, UK
m.hardcastle@bristol.ac.uk

Abstract   It has only recently become possible to separate extended from nuclear X-ray emission in intermediate-redshift radio-loud quasars. I discuss the observations that exist so far and their implications for the quasars' cluster environments; several quasars at $0.3 < z < 0.7$ have been found to lie in reasonably rich cluster environments. I briefly comment on the improvements to be expected from the new generation of X-ray satellites, and show, as an illustration, a new *Chandra* detection of extended emission around a $z = 0.66$ quasar.

Introduction

There are several reasons why we should expect to be able to see luminous extended thermal X-ray emission around radio-loud, steep-spectrum quasars at intermediate redshifts ($0.3 < z < 1.0$). Firstly, there is optical evidence for overdensities of galaxies around such objects (see for example Bremer, this volume, Wold et al., this volume). Secondly, all models of extragalactic double radio sources require that the pressure in the external medium should be comparable to the pressure inside the radio lobes, which implies the presence of a reasonably luminous distribution of hot, X-ray-emitting gas around them. Thirdly, hot gas around the lobes is required to produce effects such as Faraday rotation and the Laing-Garrington effect. And finally, we have known for many years that some powerful radio galaxies lie in luminous X-ray clusters, and so unified models for radio galaxies and quasars require that at least some radio-loud quasars must do the same.

Detecting an X-ray-emitting cluster environment around a distant quasar, and measuring its luminosity and temperature, is valuable for several reasons. Most fundamentally, it gives us quite direct information about the mass of the host system, in contrast to the indirect approach needed to infer this information from optical observations. Detection of high-redshift X-ray clusters can be used to constrain cosmological models (e.g. Donahue et al. 1998, Fabian et al. 2001). For all quasars, we would like to understand whether there is any link between the presence of a cooling flow and the fuelling of the AGN. And for radio-loud quasars the detection of an extended X-ray environment allows us to test unified models and to constrain the dynamics of radio sources.

The problem with any attempt to detect these environments has of course historically been the same as the problem of imaging host galaxies around quasars in the optical: quasars are bright point X-ray sources with non-thermal spectra and luminosities comparable to that of a rich cluster at energies of a few keV. The quasar emission is therefore likely to dominate the measured X-ray flux. The problem is compounded by the (comparatively) low spatial and spectral resolution available with the past generation of X-ray observatories.

1. The past: *ROSAT* searches for quasar clusters

ROSAT was the best observatory for searching for host cluster emission until 2000, but the best resolution available with the *ROSAT* HRI was 5 arcsec, while the PSPC had a resolution ~ 20 arcsec. Moreover, seeing a cluster under the quasar emission depended on knowing the shape of the point-spread function, which was affected by aspect uncertainties on board the satellite and by processing errors on the ground. (Since bright point X-ray sources are rare, it was not usually possible to observe a calibration source in the same field as the quasar). Early work with *ROSAT* was done by Hall et al. (1995, 1997); they looked at one radio-quiet and three radio-loud objects with $0.2 < z < 0.7$, selecting their targets because they were known to lie in cluster environments. They found extended X-ray emission around the radio-quiet quasar but were only able to set limits on the cluster luminosity of the radio-loud objects.

More recently two groups (Crawford et al. 1999, Hardcastle & Worrall 1999) have detected extended X-ray emission around a number of 3CR quasars. These two studies were carried out independently but the techniques used were very similar. The data were corrected for the effects of aspect error using the centroiding and restacking technique of Harris et al. (1998) — for this process, the bright quasar nucleus is an advantage, since it allows accurate centroids to be found. Once this had been done, models consisting of a point source and β model were fitted to the radial profiles of the quasars (this involves the assumption that

Extended X-ray emission around radio-loud quasars

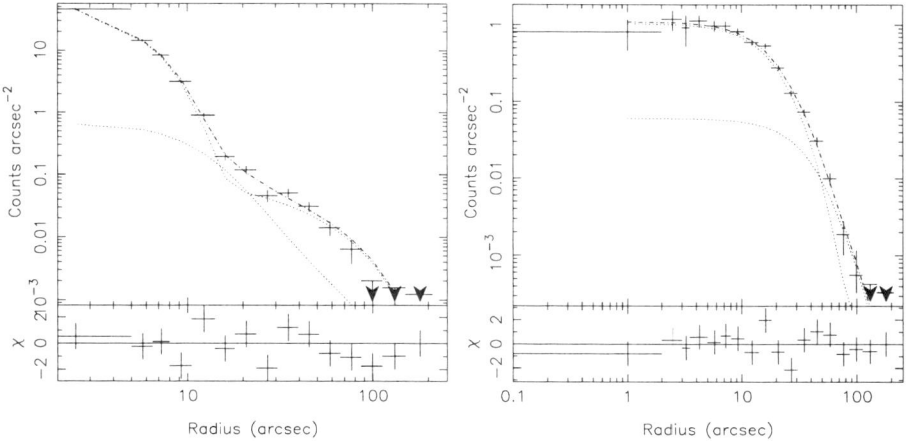

Figure 1. The ROSAT radial profiles, with best-fitting point and β models, of 3C 48 (left: HRI data) and 3C 254 (right: PSPC data).

the quasar is at the centre of the cluster emission). Examples of some of the fits can be seen in Fig. 1.

The 7 target sources of Crawford et al. were chosen because they were thought to lie in cluster environments, though in some cases (e.g. 3C 48) the evidence was only based on extended emission-line diagnostics. They detected all 7 objects. By contrast, in Hardcastle & Worrall (1999) we looked at all 3CRR radio sources that had been observed with *ROSAT*. We detected similar environments, with similar luminosities, around the five objects we had in common with Crawford et al.; these detected objects had extended emission which made up $\sim 10-20\%$ of the total soft emission from the quasar. However, we also found a large number (22) of sources with *no* detected extended environment; we used the observations to determine upper limits on the cluster luminosities. In many cases these non-detections were from short *ROSAT* exposures, and the upper limits are high.

What conclusions can we draw about the environments of radio-loud quasars? In Fig. 2, I plot the inferred 2–10 keV luminosities for 3CRR objects as a function of redshift. It will be seen that most of the high-redshift objects, including all the quasars, have extended X-ray luminosities at or consistent with the level of a moderately rich cluster (Abell class 0–1). At lower redshifts (and hence lower luminosities of the radio source), the detected environments are substantially less luminous. The detected quasars occupy much the same region on this plot as radio galaxies of comparable redshift luminosity, as expected from unified

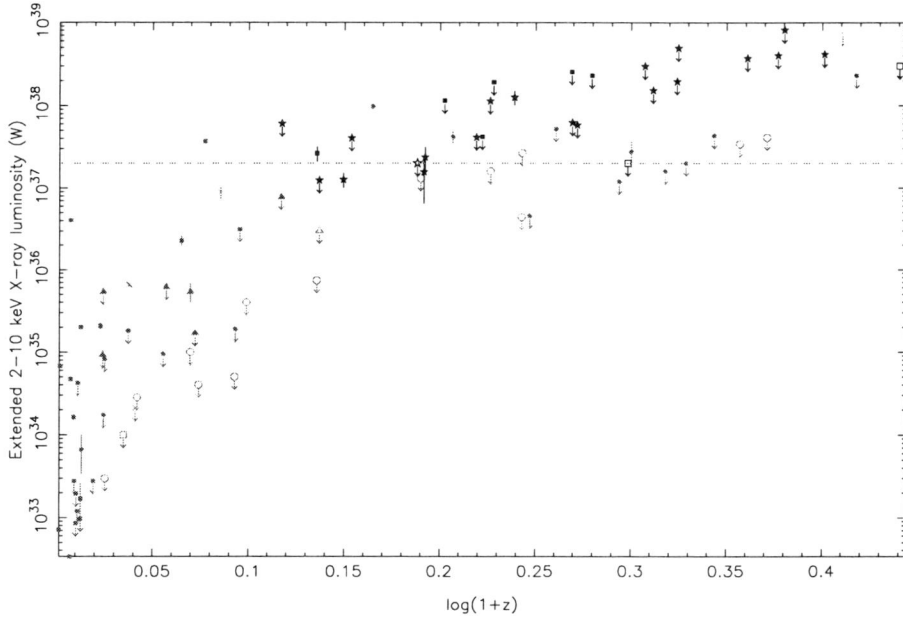

Figure 2. The extended X-ray luminosity of 3CRR radio sources as a function of redshift. Quasars are in bold; other objects are radio galaxies. The dotted line shows a representative moderately rich cluster luminosity of 2×10^{44} ergs s^{-1}.

models; this is consistent with the picture emerging from optical observations. Since the 3CRR objects represent at any given redshift the most radio-luminous objects in the universe, we would predict from this that more typical radio-loud objects, with lower radio luminosities (perhaps selected from radio surveys with a lower flux limit) should lie in cluster environments which are no richer than these.

We have also used these observations together with the radio data on the sources to compare the inferred internal pressure (the minimum pressure in the radio lobes) with the external pressure due to the X-ray-emitting medium (or an upper limit if no detection has been made). We find that the minimum pressures in both radio galaxies and quasars are systematically lower, by up to an order of magnitude, than the external thermal pressures, confirming an earlier result by Leahy & Gizani (1999) based on a smaller sample. Since radio sources cannot in fact have pressures much lower than the external pressure, the inference is that minimum pressure estimates are wrong by at least an order of magnitude (we explore the consequences of this result in more detail elsewhere;

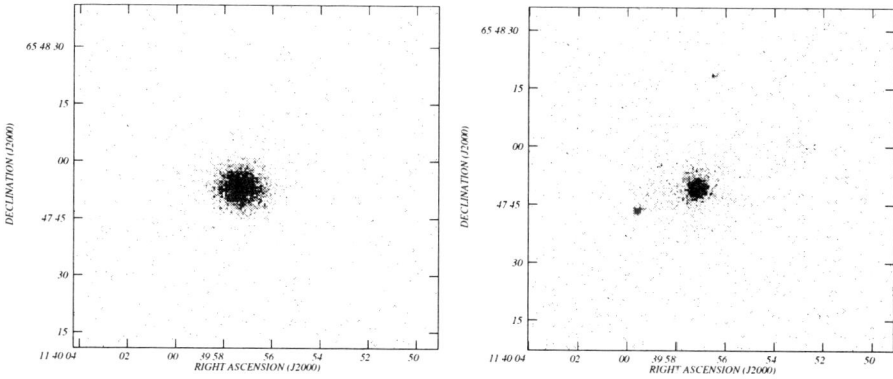

Figure 3. Observations of 3C 263 with ROSAT (left) and Chandra (right). Black is 5 counts per 0.5-arcsec pixel in both cases.

Hardcastle & Worrall 2000). Thus, at least some progress has been made with *ROSAT* towards answering the questions posed in section 4.

2. The future: *Chandra* and *XMM-Newton*

The new imaging X-ray observatories open up new horizons for this kind of study. *Chandra* is likely to be most important, because its 0.5-arcsec resolution makes spatial decomposition of the quasar nucleus from the extended emission a comparatively trivial task. We have begun a programme of *Chandra* observations of quasars and radio galaxies at intermediate redshifts, with the intention of characterising their hot gas environments much more accurately than was possible with *ROSAT*. The improvement provided by *Chandra* is striking. Fig. 3 shows images of the $z = 0.66$ quasar 3C 263 taken with *ROSAT* (the data analysed by Hall et al.) and *Chandra* (Hardcastle, Birkinshaw, Cameron, Harris & Worrall, in preparation). Whereas no extension was visible either to Hall et al. or to us after detailed analysis of the *ROSAT* data, extended emission (and a detection of the radio hotspot, probably inverse-Compton emission) is clearly visible by eye in the *Chandra* image; the quasar lies in a moderately luminous X-ray environment with $L \sim 10^{44}$ ergs s^{-1}, consistent with the upper limits from the *ROSAT* data. *Chandra* will also be important in comparisons between radio galaxies and radio-loud quasars, since we find that at least some radio galaxies have substantial nuclear soft X-ray emission which requires high resolution for its separation from the extended component (Worrall et al. 2001). If these early results are representative, we can expect to see substantial advances in this area in the coming years.

References

Crawford C.S., Lehmann I., Fabian A.C., Bremer M.N., Hasinger G., 1999, MNRAS, 308, 1159

Donahue M., Voit G.M., Gioia I., Luppino G., Hughes J.P., Stocke J.T., 1998, ApJ, 502, 550

Hall P.B., Ellingson E., Green R.F., Yee H.K.C., 1995, AJ, 110, 513

Hall P.B., Ellingson E., Green R.F., 1997, AJ, 113, 1179

Hardcastle M.J., Worrall D.M., 1999, MNRAS, 309, 696

Hardcastle M.J., Worrall D.M., 2000, MNRAS, 319, 562

Fabian A.C., Crawford C.S., Ettori S., Sanders J.S., 2001, MNRAS, in press (astro-ph/0101478)

Leahy J.P., Gizani N.A.B., 1999, in 'Life Cycles of Radio Galaxies', ed. J. Biretta et al., New Astronomy Reviews, astro-ph/9909121

Worrall D.M., Birkinshaw M., Hardcastle M.J., 2001, MNRAS, submitted

WFPC2 IMAGING OF QUASAR ENVIRONMENTS: A COMPARISON OF LBQS AND *HST* ARCHIVE QUASARS

Rose A. Finn
Chris D. Impey
Eric J. Hooper

Introduction

Studies of the large-scale environments of quasars provide insight into the role of environment in triggering quasar activity and can be used to corroborate theories of radio power. This paper addresses two main questions. First, have previous quasar environment studies been biased by preferentially studying very strong radio sources or unusual optically selected quasars? Second, are radio-loud quasars located in different environments than radio-quiet quasars? We use HST WFPC2 data to address these questions. To investigate possible biases in quasar sample selection, we compare 16 quasars from the Large Bright Quasar Survey (LBQS) to a sample of quasars drawn from the Hubble Space Telescope (HST) Archive, consisting of mainly PG and PKS quasars. Quasars from the LBQS are representative of the radio-quiet quasar population as a whole and are therefore more likely to reflect the incidence of clustering around quasars in general. To adress the second question, we compare the environmental properties of the radio-loud and radio-quiet quasars within each sample.

1. Quasar Samples

The LBQS (Hewett, Foltz & Chaffee 1995) sample contains 16 quasars in the redshift range $0.39 < z < 0.504$, of which six are radio-loud. The 16 LBQS quasar fields were observed in the $F675W$ filter.

A comparison sample of 27 quasars was selected from the HST Archive. The 19 quasars observed in $F606W$ have lower redshifts, with $0.15 < z < 0.29$, and six of the 19 are radio-loud. The remaining 8 quasars of the Archive sample were observed in $F702W$. The redshifts range from $0.223 \leq z \leq 0.514$, and 4 out of 8 are radio-loud.

2. Results

Our two main results are that 1) the LBQS quasars lie in less dense environments than the more luminous Archive quasars, and 2) radio-loud and radio-quiet quasars are found in similar environments. Comparison with previous environment studies puts these findings in a different light. First, the relatively sparse environments of the radio-quiet LBQS quasars are consistent with previous ground-based studies, but the radio-loud LBQS quasars are in unusually sparse environments when compared to other radio-loud quasars. Second, the richer environments of the Archive radio-loud quasars are consistent with previous ground-based studies, but the Archive radio-quiet quasars are in unusually dense environments compared with the radio-quiet quasars studied by Smith et al. (1995) and Ellingson et al. (1991). Furthermore, by comparing our values of the amplitude of the spatial correlation function, B_{gq}, to the Longair & Seldner (1979) value of 90 $(h_{100}^{-1}$ Mpc$)^{1.77}$ for Abell 0 clusters, we find that the Archive quasars are located in galaxy environments slightly less rich than Abell 0 clusters. When compared with the galaxy-galaxy covariance amplitude of 20 $(h_{100}^{-1}$ Mpc$)^{1.77}$ from Davis & Peebles (1983), the LBQS quasars are in environments comparable to the typical galaxy. This is not surprising for the radio-quiet LBQS subsample, but one might expect the radio-loud LBQS quasars to be in denser environments.

How do we make sense of these results? Fisher et al. (1996) note that the $F606W$ quasars are among the most luminous, and this might explain why the $F606W$ radio-quiet quasars have denser environments than average. Furthermore, the LBQS quasars are less luminous on average than the Archive quasars, so maybe this explains why the LBQS are in relatively sparse environments. However, a Spearman rank test indicates a 60% probability that B_{gq} and M_V are uncorrelated. Therefore, it appears that optical luminosity can not explain why the Archive radio-quiet are in dense environments nor why the LBQS radio-loud quasars are in sparse environments.

Radio luminosity might help explain the sparse environments of the radio-loud LBQS quasars because the radio-loud LBQS quasars have lower radio luminosities than the Archive radio-loud quasars. We find no correlation between radio power and environment when considering both radio-loud and quiet quasars. We note that almost all the LBQS quasars are a factor of 10-100 less powerful that the radio-loud Archive quasars, and it is not clear that the same emission mechanism holds across this large range in radio power. However, when considering the radio-loud quasars only, a Spearman rank test indicates a modest correlation (95% probability) between radio power and environment, which is dependent on the one point at extreme values of radio luminosity and B_{gq} (85% propability of a correlation if this point is removed). Most of the radio data for the radio-quiet quasars are upper limits, so we cannot test for an independent correlation between radio power and environment for radio-quiet quasars. The fact that the correlation in the radio-loud data does not apply to both radio-loud and quiet quasars implies that two emission mechanisms may be at work (e.g. Stocke et al. 1992, Hooper et al. 1996), even though there appears to be a continuum of radio power among quasars.

3. Conclusion

Finally, what is the true incidence of clustering around quasars? As discussed in the Introduction, the LBQS is among the most representative surveys of the currently known radio-quiet quasar population, and the radio-quiet quasars make up $\approx$90% of the whole. Therefore, most quasars, like the LBQS sample presented here, lie in environments comparable to the typical galaxy. The Archive radio-quiet quasars have unusually dense environments and are thus not a representative sample of radio-quiet quasars. The clustering associated with radio-loud quasars correlates with radio power, and the environments of the radio-loud quasars presented here range from that of a typical galaxy to Abell 0 clusters. We present data for only 16 LBQS quasars, and a larger sample is needed to strengthen these results. The LBQS also makes an excellent sample for extending quasar environment studies to higher redshift to look for evolution.

References

Davis, M., Peebles, P. J. E., 1983, ApJ, 267, 465
Ellingson, E., Yee, H. K. C., Green, R. F., 1991, ApJ, 371, 49
Fisher, K., Bahcall, J., Kirhakos, S., Schneider, D., 1996, ApJ, 468, 469
Hewett, P. C., Foltz, C. B., Chaffee, F. H., 1995, AJ, 109, 1498
Hooper, E. J., Impey, C. D., Foltz, C. B., Hewett, P. C., 1996, ApJ, 473, 746
Longair, M. S., Seldner, M., 1979, MNRAS, 189, 433
Smith, E. P., Heckman, T. M., Bothun, G. D., Romanishin, W., Balick, B., 1986, ApJ, 306, 64
Stocke, J. T., Morris, S. L., Weymann, R. J., Foltz, C. B., 1992, ApJ, 396, 487

Stanislaw Rys

Matteo Guainazzi

DECELERATION AND ASYMMETRY IN QSO RADIO MAP

Stanisław Ryś
Astronomical Observatory, Jagiellonian University
PL – 30244 KRAKÓW, ul. Orla 171, POLAND
strys@oa.uj.edu.pl

Abstract   The plasma elements in the jets move with relativistic velocities decreasing with constant deceleration. Using such an approximation for both the jet and counter jet of a double–lobed radio source, we are able to derive a new formula for the asymmetry observed in radio structures. The model was applied to the description of the asymmetries in radio maps of 10 QSO's and 12 galaxies. For selected sources the bulk velocity of plasma elements exceeds that found from a kinematic model (i.e. with constant velocity), and the sources will be larger in the future than they are in their present day maps.

1.   Description of the radio structure

Our reasoning is as follows: A considerable number of sources show double structure. Such structures allow us to investigate the jet by comparison of events appearing on both sides of the AGN. Since both jets are produced by the same central object, they start with the same values of a number of parameters and begin to evolve in a similar environment. If a radio-source is intrinsically symmetrical, we can describe the temporal properties of its central engine and events in the surrounding medium, neglecting lot of the assumptions which concern particular phenomena (such as those in hydrodynamic simulations).

The basic assumption of the model is that the jet consists of a plasma flow from the central source which is described by a sequence of elementary infinitesimally-small plasma elements (a Lagrangian description of the fluid). The plasma element moves at relativistic speed with constant deceleration along a straight line and they have the same properties in their own reference frame. Asymmetries are produced entirely due to the fact that the opposite parts (with respect to the central engine) of the structure are seen at different epochs of their evolution and possess different velocities in the observer's rest frame.

We can write an equation describing the motion (cf. Rybicki & Lightman 1975) as:

$$t = \frac{c}{g}(\beta_0 \gamma_0 - \beta \gamma) \qquad r = \frac{c^2}{g}(\gamma_0 - \gamma) \qquad (1)$$

where $\beta_0 = \beta(t=0)$ and $\gamma_0 = (1-\beta_0^2)^{-1/2}$. These equations (1) give us information on $r \sim t$ dependencies but do not take into account the position of the observer. From the geometry of the problem (Ryś 2000) we obtain an equation connecting the Lorentz factor of a plasma element (γ) and its twin ($\bar{\gamma}$) - located on opposite sides of the central engine.

$$\sqrt{\gamma^2 - 1} + (\gamma_0 - \gamma)\cos(\theta) = \sqrt{\bar{\gamma}^2 - 1} - (\gamma_0 - \bar{\gamma})\cos(\theta) \qquad (2)$$

The bar above the γ denotes that the symbol concerns the counter jet. We can easily solve this equation in two ways. If the value of γ is known we calculate $\bar{\gamma}$, or having $\bar{\gamma}$ we calculate γ. The essence of my method consists in interchanging the positions of the twin plasma elements between two opposite region of radio emission. While performing this, we make suitable modifications of the positions and fluxes in agreement with the particular model adopted (for more detail see Ryś 2000).

When interchanging the positions of the twinned elements, we must take into account the changes in brightness due to their different ages and corrected with respect to Doppler boosting effects. I adopt a temporal evolution model in the rest frame of a plasma element in the form of an exponential dependence on time $S(t) = S_0 \cdot \exp(-T/\tau)$, where τ - is a characteristic time scale, and S_0 denotes the initial value of the flux. We obtain the transformation rule for the time between the observer (t) and plasma element (T) rest frames starting from the formula for infinitesimal time intervals ($dT = dt/\gamma(t) = invariant$). Hence, the rule which should be used during transformation of the map is:

$$\frac{\bar{S}}{S} = \left[\frac{\bar{\gamma}(1 + \bar{\beta}\cos(\theta))}{\gamma(1 - \beta\cos(\theta))}\right]^{-(3+\alpha)} \left(\frac{\bar{\gamma}^2(1+\bar{\beta})}{\gamma(1+\beta)}\right)^{+(c/g\tau)} \qquad (3)$$

where α- is the spectral index. Now in contrast to the kinematical model, the Doppler and Lorentz factors depend on the plasma element positions inside of the investigated structure. We should notice that we obtain power-law dependencies between radio fluxes received by the observer from the twins, because of the logarithmic solution for the transformation of time $T(t)$. But it is most important that, in the case of the decelerated movement, the formula for transformation of the time possesses a logarithmic shape and the time interval of order $10^6 \div 10^8$ in the observer's frame gives only about $(6 \sim 8)/g$ years in the plasma element frame. Such a time is much shorter than the life-time of electrons frequently calculated for extended radio sources.

The model was applied to a description of the asymmetry of 23 structures (10 QSO's and 13 galaxies) selected from the GB/GB2 and DRAGN (Leahy et al. 1996) surveys. The results of the fittings of this simple dynamical approximation of evolution of radio structures are:
- for selected sources the median value of the bulk velocity of plasma elements is $<\beta_0^Q> = 0.65 \pm 0.17$, $<\beta_0^G> = 0.83 \pm 0.15$ (cf. Hardcastle et al. 1999), and in the future the sources will be larger than they are seen in their present day maps.
- the model allows us to find the value of velocity and its component directed towards the observer. This is the first method (as far as I know), which provides information about the value and the direction of velocity vectors from a single frequency radio map.

The deceleration seems to be the important parameter in the description of evolution of radio structure. We know that the most important influence on the observed asymmetry is induced by the Doppler effect (Rees 1967). This fact leads to a simple unification scheme described in the review by Urry and Padovani (1995). At present the idea of unifying all varieties of radio structures using only one parameter (i.e. orientation dependent Doppler correction) is hopeless, but we think that by including deceleration we will be able to restore the idea of a unified scheme with a relativistic correction which should be done when we try to distinguish physically different types of AGN.

Acknowledgments I thank LOC for hospitality and I am grateful to my colleagues K. Maślanka and G. Stachowski for our fruitful discussion.

References

Hardcastle M.J., Alexander P., Pooley G.G., Riley J.M., 1999, MNRAS, 304, 135
Leahy J.P., Bridle A.H., Strom R.G., 1996, DRAGN Atlas,
    (http://www.jb.man.ac.uk/atlas)
Machalski J., Condon J.J., 1983, AJ, 88 ,143.
Rees M.J., 1967, MNRAS, 135, 345
Rybicki G.B., Lightman A.P., 1979, in *Radiative Processes in Astrophysics*,
    John Wiley & Sons, Inc.
Ryś S., 2000, A&A, 355, 79
Urry C.M., Padovani P., 1995, PASP, 107, 803

From left to right, Xavier Barcons, J.M. Rodríguez Espinosa, Andrew Sheinis, Heino Falche and Jochen Heidt

Joseph Rhee

SPATIALLY RESOLVED SPECTROSCOPY OF EMISSION-LINE GAS IN QSO HOST GALAXIES

Andrew I. Sheinis
UCO/Lick Observatory
University of California at Santa Cruz 95064
sheinis@ucolick.org

Abstract We present off-nuclear spectra of 3 radio loud QSO's, 3C249.1, 3C273 and 3C323.1, taken with the echellette spectrograph and imager (ESI) at Keck observatory. From these spectra we have extracted the spatial profile along the slit of the [OIII]λ5007Åline. Fitted Gaussian distributions to each of these profiles show emission-line gas out to several tens of kiloparsecs from the galaxy nucleus. Most observations show several gas components at distinct velocities and velocity dispersions, much of which is above the escape velocity for any resonable mass galaxy. In addition, we show slitless spectroscopy images for one other object, 3C48. From the slitless spectroscopy images we can extract 2-dimensional spatial as well as velocity information on the emission line gas.

1. Spatially resolved Off-Nuclear spectroscopy

We have taken off-nuclear spectra of 3 radio loud QSO's, 3C249.1, 3C273 and 3C323.1 (Boroson and Oke 1984), using the Echellette spectrograph and imager (ESI) (Sheinis et. al 2000) at Keck observatory. Each object was observed at several different position angles and offsets from the nucleus. From these spectra we have extracted the spatial profile along the slit of the [OIII]λ5007Åline.

Each image is the median of four 15 minute exposures. The data were processed using the Information Data Language (IDL). They were first rectified to remove the instrument distortion, bias subtracted then a two dimensional sky model was subtracted. After this extraction we fit a linear combination of two Gaussian distributions to each of these profiles, using IDL.

Fig. 1 shows the two-dimensional extraction image for each object. For 3C249.1, three slit position are shown, 3 seconds east, 3.5 seconds north and 3 seconds west. One position is shown for each of the remaining objects, 3C273 and 3C323.1. Those positions are 4 seconds east and 3 seconds east. The lower panels of Fig. 1 show the results of the Gaussian fit for each object, namely the mean, dispersion and magnitude of the Gaussian fit for each gas component.

The plots show bright extended emission at tens of kiloparsecs from the galaxy nucleus. This emission is observed to have peak velocity FWHM $=2.35 \times \sigma$ of 540, 940 and 535 km/sec for the three slit positions of 3C249.1, 500 km/sec for 3C273 and 470 km/sec for 3C323.1. These velocities are most likely well beyond escape velocity for these galaxies.

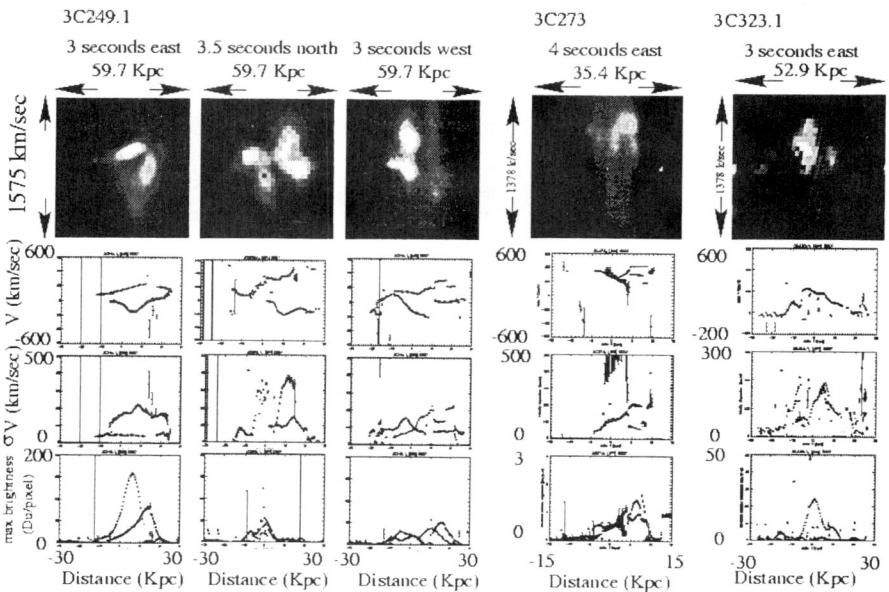

Figure 1. 3C249.1, 3C273 and 3C323.1 seen in [OIII], spatially resolved.

2. Slitless Spectroscopy

We have obtained slitless spectroscopy images of several objects. The four panels of Fig. 2 show ten minute exposures of one well-studied object, 3C48 (Canalizo 2000) taken at four different position angles, through a six arcsecond wide slit. They have been processed identically to the above images. Additionally, a modelled QSO continuum was carefully subtracted to reveal details within a half arcsecond of the nucleus.

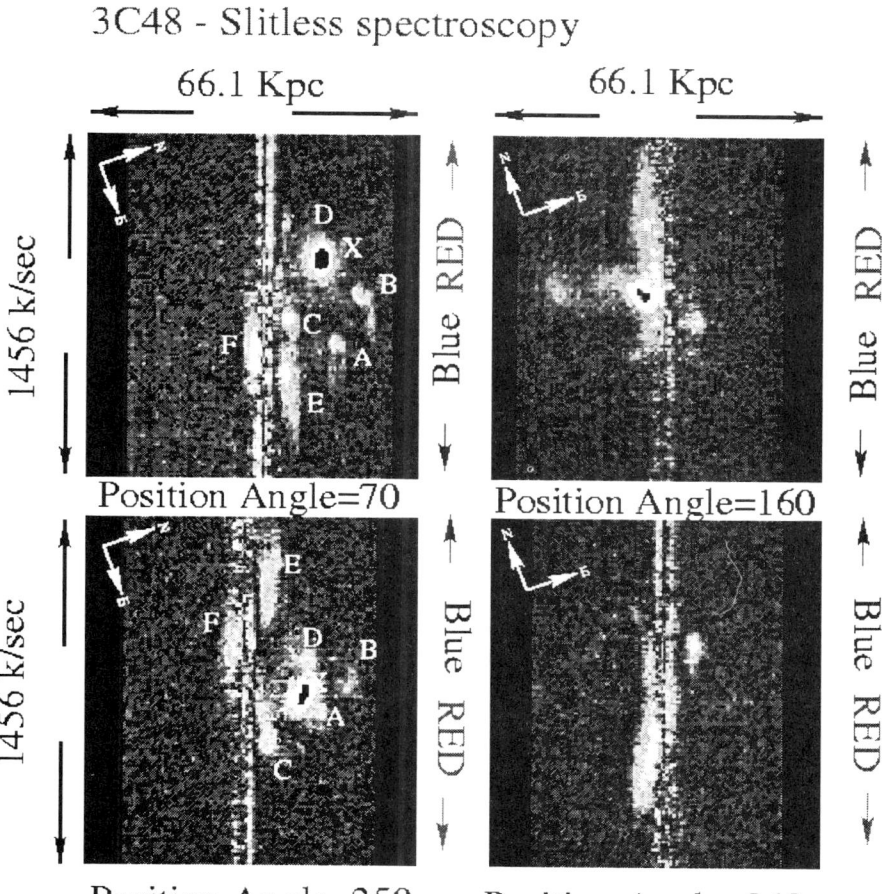

Figure 2. Slitless spectroscopy of 3C48, seen in [OIII].

In order to deconvolve the velocity and spatial information in the y axis of each image, we have produced pairs of images that have the dispersion axis rotated relative each other, by 180 degrees. In objects (like 3C48) that contain well defined clumps of gas moving at a definite velocity, position and velocity can be inferred directly by comparing the two images. In objects with a more complex spatial and velocity structure this method becomes less effective.

Fig. 2 shows the emission-line gas of 3C48 as viewed through a 6 arcsecond wide by 20 arcsecond long slit. The two images show two different position angles separated by 180 degrees. In first pair of images of 3C48 (upper and lower left) we see one bright gas knot (X) redshifted by $\approx$ 200 km/sec relative to four smaller knots (A-D) that appear to be at similar velocities. All 5 knots are located to the north of the nucleus. Closer to the nucleus we see one higher dispersion ($\approx$ 600 km/sec), higher velocity knot ($\approx$ 500km/sec) (E) to the north and another (F) to the south. A similar evaluation can be done to the rightmost pair of images.

This method has shown promise to produce spatial and velocity information in emission-line gas.

References

Sheinis, A.I., Miller, J. S., Bolte, M., Sutin, B.M., 2000, SPIE., 4008, 522
Boroson, T.A., Oke, J.B., 1984, ApJ, 281, 535
Canalizo, G., Stockton, A., 2000, ApJ, 528, 201

HOST GALAXIES AND THE SPECTRAL VARIABILITY OF QUASARS

Fausto Vagnetti
Dipartimento di Fisica, Università di Roma "Tor Vergata"
Via della Ricerca Scientifica 1, I-00133 Roma, Italy
fausto.vagnetti@roma2.infn.it

Dario Trèvese
Dipartimento di Fisica, Università di Roma "La Sapienza"
Piazzale A. Moro 2, I-00185 Roma, Italy
dario.trevese@roma1.infn.it

Introduction

Variability of the spectral energy distribution (SED) of active galactic nuclei (AGNs) can provide clues for understanding both the main emission processes and the origin of their variations. So far, multi-wavelength monitoring has been possible for a limited number of bright objects, thanks to large international cooperations (see Ulrich, Maraschi & Urry 1997, and refs. therein). The "ensemble" analysis of the light curves was obtained in the past only from single-band observations of statistical samples of QSOs (Giallongo, Trèvese & Vagnetti 1991, Hook et al. 1994, Cristiani et al. 1996, Di Clemente et al. 1996). Cutri et al. (1985), Kinney et al. (1991), and Edelson et al. (1990), found a hardening of the spectrum in the bright phase in small QSO samples. A positive correlation of the spectral index α ($f_\nu \propto \nu^\alpha$) with brightness variations $\Delta \log f_\nu$ has been found by Trèvese et al. (2000). The recent publication, by Giveon et al. 1999, of two-band monitoring of 42 PG QSOs, allows a statistical study of the spectral variability of the entire set of light curves (Trèvese & Vagnetti 2001). They find a positive correlation of the $B - R$ color with the brightness variations δB. These results confirm the interpretation of the variability-redshift (v-z) correlation suggested by Giallongo, Trèvese & Vagnetti (1991) and Di Clemente et al. (1996). An independent analysis of the same light curves, performed by Cid

Fernandes et al. (2000) implies the existence of an underlying spectral component, redder than the variable one, which could be identified either with the host galaxy, considered by Romano & Peterson (1998), or with a non-flaring part of the QSO spectrum. We evaluate the effect of the host galaxy constant SED on the variable QSO+host SED, through numerical simulation based on templates of the QSO and host SEDs provided by Elvis et al. (1994). We also compare the results with the spectral variations deduced from Giveon et al. (1999), concluding that the nuclear spectral variability is dominant.

1. Discussion and Conclusions

From an atlas of 47 normal QSO continuum spectra, Elvis et al. (1994) derive the average QSO template spectrum corrected for the contribution of a template host galaxy spectrum. On the basis of these templates we produce a synthetic (QSO+host) SED. We characterize the relative contribution of the nucleus and the host by the parameter: $\eta \equiv \log(L_H^Q/L_H^g)$, where L_H^Q and L_H^g are the total H band luminosities of the QSO and the host galaxy respectively. To test (disprove) the hypothesis that the spectral energy distribution of the active nucleus maintains its shape while changing brightness, we represent variability by small changes of $\Delta\eta(t)$ around each adopted η value. For each synthetic spectrum we compute the instantaneous spectral slope as a function of frequency and time: $\alpha(\nu,t) \equiv \partial \log L_\nu / \partial \log \nu$. We characterize the continuum SED variability by the $\beta(\nu) \equiv \partial \alpha(\nu)/\partial \log L_\nu$ parameter representing the variation of the local spectral slope per unit change of $\log L_\nu$.

Variability of QSO spectra has been analyzed on the basis of the data of Giveon et al. (1999), consisting of the light curves of a sample of 42 nearby and bright QSOs ($z < 0.4, B < 16$ mag), belonging to the Palomar-Green (PG) sample, monitored for 7 years with median sampling interval of 39 days, in B and R, with the 1 m telescope of the Wise Observatory. For each time interval τ between two observations, it is possible to define $\beta(\tau) \equiv [\alpha(t+\tau) - \alpha(t)]/[\log L_{\bar{\nu}}(t+\tau) - \log L_{\bar{\nu}}(t)]$, $\bar{\nu} = \sqrt{\nu_B \nu_R}$. We consider, for each object, the spectral variability parameter β_m defined as the average of $\beta(\tau)$ values for all $\tau \leq 1000$ days and the average value of the spectral slope α_m. Both these quantities can be compared with the values (β,α) obtained from the synthetic spectrum. In Fig. 1 the continuous curve represents β versus α for η ranging from -3 (dominant host galaxy, in B and R) to 3 ($\sim$ pure QSO). To evaluate the effect of redshift we also show the dashed curve computed for $z = 0.4$, the maximum redshift of the Giveon et al. (1999) sample. On average, the distribution of points is clearly shifted towards top right of Fig. 1. This means that the relative weight of nuclear and stellar component cannot account for the observed spectral variability, which therefore must be intrinsic of the active nucleus.

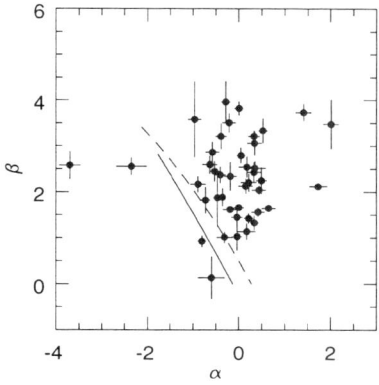

Figure 1 The observed β versus α values for PG QSOs. The two curves ($z = 0$ continuous, $z = 0.4$ dashed), represent $\beta(\alpha)$ of the composite spectrum, for $-3 < \eta < +3$.

We have computed synthetic QSO+host SEDs on the basis of templates of the nuclear and the star light spectra. We have computed the spectral slope α and the spectral variability β for the sample of PG QSOs monitored in the B and R bands by the Wise Observatory group. A comparison of the observed distribution of points in the $\beta - \alpha$ plane with the synthetic $\beta(\alpha)$ allows to disprove the hypothesis that the observed variations of the QSO spectral shape are solely due to a change of the relative weights of the nuclear and stellar component. In the framework of the analysis of Cid Fernandes et al. (2000), our result implies that the constant spectral component should be identified with a non flaring part of the accretion disk.

References

Cristiani, S., Trentini, S., La Franca et al., 1996, A&A, 306 395
Cutri, R.M., Wisniewski W.Z., Rieke, G.H., Lebofsky, H.J., 1985, ApJ, 296, 423
Cid Fernandes, R., Sodré, L. Jr., Vieira da Silva, L. Jr., 2000, ApJ, 544, 123
Di Clemente A., Giallongo, E., Natali, G., Trèvese, D., Vagnetti, F., 1996, ApJ, 463, 466
Edelson, R.A., Krolik, J. H., Pike, G. F., 1990, ApJ, 359, 86
Elvis, M. et al., 1994, ApJS, 95, 1
Giallongo E., Trèvese D., Vagnetti F., 1991, ApJ, 377, 345
Giveon, U., Maoz, D., Kaspi, S., Netzer, H., Smith, P.S., 1999, MNRAS, 306, 637
Hook, I.M., Mc Mahon, R.G., Boyle, B.J., Irwin, M.J., 1994, MNRAS, 268, 305
Kinney, A. L., Bohlin, R.C., Blades, J.C., York, D.G., 1991, ApJS, 75, 645
Romano, P., Peterson, B.M., 1999, ASP Conf. Ser., 175, 55
Trèvese, D., Kron R. G., Bunone, A., 2000, ApJ (in press), astro-ph/0012408

Trèvese, D., Vagnetti, F. 2001, (in preparation)
Ulrich, M.H. , Maraschi, L., Urry, C. M., 1997, ARA&A, 35, 445

III

TIDAL INTERACTIONS/MERGERS. ULIRGS.

Michael Rowan-Robinson

Dave Sanders

THE AGN-STARBURST CONNECTION IN ULTRALUMINOUS INFRARED GALAXIES

Michael Rowan-Robinson

Astrophysics Group, Blackett Laboratory, Imperial College of Science Technology and Medicine, Prince Consort Rd, London SW7 2BW

mrr@ic.ac.uk

Abstract Models for the SEDs of infrared galaxies are reviewed, with particular application to hyperluminous infrared galaxies.

Keywords: Infrared: galaxies – Galaxies: starburst – Galaxies: active

Introduction

Starbursts and AGN dust tori have very distinct spectral energy distribution signatures and this allows us to estimate the relative contribution of these two components to the spectra of infrared galaxies.

Here I review briefly the ingredients that go into modelling infrared galaxy spectral energy distributions and discuss the application of these to the most luminous objects, quasars and hyperluminous infrared galaxies.

1.   Ingredients for model SEDs

The key ingredients for models for the spectral energy distributions (SEDs) of infrared galaxies are **(1) A model for interstellar dust grains.** Detailed models have been provided by Mathis et al. (1977), Draine and Lee (1984), Rowan-Robinson (1986, 1992), Desert et al (1990), Siebenmorgen & Krugel (1992), Dwek (1998). **(2) An assumed density distribution for the dust.** Detailed physical density profiles have been developed by Yorke (1977) for HII regions, by Efstathiou & Rowan-Robinson (1991) for protostars and by Eftathiou et al. (2000) for starbursts. However most authors have followed the approach of Rowan-Robinson (1980), who explored models with $N(r) \propto r^{-\beta}$. Silva et al. (1998) have investigated exponential density profiles for the cirrus component. **(3) The geometry of the dust.** The most common assumption is spherical symmetry. Axi-symmetric models have been investigated by Efstathiou & Rowan-Robinson (1990, 1991, 1995), Pier & Krolik (1992), Granato et al. (1994, 1997), Silva et al. (1998). Rowan-Robinson (1995) gave an approach to modelling emission from an ensemble of discrete compact clouds. **(4) A radiative transfer code.** The first accurate spherically symmetric radiative transfer code was by Rowan-Robinson (1980) and the first accurate axisymmetric code by Efstathiou et al. (1990). Pier and Krolik (1992), Krugel and Siebenmorgen (1994), Granato et al. (1997) and Silva et al. (1998) have also developed axisymmetric codes. **(5) Source components** Typically, SEDs can be understood as a superposition of several distinct source components. Rowan-Robinson & Crawford (1989) were able to fit the IRAS colours and spectral energy distributions of galaxies detected in all 4 IRAS bands with a mixture of 3 components, emission from interstellar dust ('cirrus'), a starburst and an AGN dust torus. Recently Xu et al. (1998) have shown that the same 3-component approach can be used to fit the ISO-SWS spectra of a large sample of galaxies. To accomodate the Condon et al. (1991) and Rowan-Robinson and Efstathiou (1993) evidence for higher optical depth starbursts, Ferrigno et al. (2000, in prep.) have extended the Rowan-Robinson & Crawford (1989) analysis to include a fourth component, an Arp220-like, high optical depth starburst, for galaxies with $\log L_{60} > 12$. Efstathiou et al. (2000) have given improved radiative transfer models for starbursts as a function of the age of the starburst, for a range of initial dust optical depths, and similar analysis has been carried out by Silva et al. (1998) and Siebenmorgen (1999, 2000).

Sanders et al. (1988) proposed, on the basis of spectroscopic arguments for a sample of 10 objects, that all ultraluminous infrared galaxies contain an AGN and that the far infrared emission is powered by this. Sanders et al. (1989) proposed a specific model, in the context of a discussion of the infrared emission from PG quasars, that the far infrared emission comes from the outer parts of a warped disk surrounding the AGN. This is a difficult hypothesis to disprove, because if an arbitrary density distribution of dust is allowed at large distances from the AGN, then any far infrared spectral energy distribution could in fact be generated. Rowan-Robinson (2000a) discussed whether the AGN dust torus model of Rowan-Robinson (1995) can be extended naturally to explain the far infrared and submillimetre emission from hyperluminous infrared galaxies, but concluded that in many cases this does not give a satisfactory explanation. Rigopoulou et al. (1996) observed a sample of ultraluminous infrared galaxies from the IRAS 5 Jy sample at submillimetre wavelengths, with the JCMT, and at X-ray wavelengths, with ROSAT.

They found that most of the far infrared and submillimetre spectra were fitted well with the starburst model of Rowan-Robinson & Efstathiou (1993). The ratio of bolometric luminosities at 1 keV and 60 μm lie in the range $10^{-5} - 10^{-4}$ and are consistent with a starburst interpretation of the X-ray emission in almost all cases. Even more conclusively, Genzel et al. (1998) have used ISO-SWS spectroscopy to show that the majority of ultraluminous IR galaxies are powered by a starburst rather than an AGN.

2. Applications to quasars and hyperluminous infrared galaxies

Quasars and Seyfert galaxies, on the other hand, tend to show a characteristic mid infrared continuum, broadly flat in νS_ν from $3 - 30\mu m$. This component was modelled by Rowan-Robinson & Crawford (1989) as dust in the narrow-line region of the AGN with a density distribution $n(r) \alpha\, r^{-1}$. More realistic models of this component based on a toroidal geometry are given by Pier & Krolik (1992), Granato & Danese (1994), Efstathiou & Rowan-Robinson (1995). Rowan-Robinson (1995) suggested that most quasars contain both (far IR) starbursts and (mid IR) components due to (toroidal) dust in the narrow line region.

One of the major discoveries of the IRAS mission was the existence of ultraluminous infrared galaxies, galaxies with $L_{fir} > 10^{12} h_{50}^{-2} L_\odot (h_{50} = H_o/50)$. The peculiar Seyfert 2 galaxy Arp 220 was recognised as having an exceptional far infrared luminosity early in the mission (Soifer et al. 1984).

The conversion from far infrared luminosity to star formation rate has been discussed by many authors (e.g. Scoville & Young 1983, Rowan-Robinson et al. 1997). Rowan-Robinson (2000b, in prep.) has given an updated estimate of how the star-formation rate can be derived from the far infrared luminosity, finding

$\dot{M}_{*,all}/[L_{60}/L_\odot] = 2.2\ \phi/\epsilon\ \mathrm{x} 10^{-10}$

where ϕ takes account of the uncertainty in the IMF (= 1, for a standard Salpeter function) and ϵ is the fraction of UV light absorbed by dust, estimated to by 2/3 for starburst galaxies (Calzetti 1998). We see that the star-formation rates in ultraluminous galaxies are $> 10^2 M_\odot yr^{-1}$. However the time-scale of luminous starbursts may be in the range $10^7 - 10^8$ yrs (Goldader et al. 1997), so the total mass of stars formed in the episode may typically be only 10% of the mass of a galaxy.

Here I discuss an even more extreme class of infrared galaxy, hyperluminous infrared galaxies, which I define to be those with rest-frame infrared (1-1000 μm) luminosities, $L_{bol,ir}$, in excess of $10^{13.22} h_{50}^{-2} L_\odot$

($=10^{13.0}h_{65}^{-2}L_\odot$). For a galaxy with an M82-like starburst spectrum this corresponds to $L_{60} \geq 10^{13}h_{50}^{-2}L_\odot$, since the bolometric correction at 60 μm is 1.63. While the emission at rest-frame wavelengths 3-30 μm in these galaxies is often due to an AGN dust torus (see below), I argue that their emission at rest-frame wavelengths $\geq 50\mu$m is primarily due to extreme starbursts, implying star formation rates in excess of 1000 $M_\odot/yr$. These then are excellent candidates for being primeval galaxies, galaxies undergoing a very major episode of star formation. More details are given in Rowan-Robinson (2000a).

For a small number of these galaxies we have reasonably detailed continuum spectra from radio to UV wavelengths. Figs. 1-6 show the infrared continua of some of these hyperluminous galaxies, with fits using radiative transfer models (specifically the standard M82-like starburst model and an Arp220-like high optical depth starburst model from Efstathiou et al. (2000) and the standard AGN dust torus model of Rowan-Robinson (1995).

I now discuss the individual objects in turn:

F10214 + 4724

The continuum emission from F10214 was the subject of a detailed discussion by Rowan-Robinson et al. (1993). Green & Rowan-Robinson (1995) have discussed starburst and AGN dust tori models for F10214. Fig. 1 shows M82-like and Arp 220-like starburst models fitted to the submillimetre data for this galaxy. The former gives a good fit to the latest data. The 60 μm flux requires an AGN dust torus component. To accomodate the upper limits at 10 and 20 μm, it is necessary to modify the Rowan-Robinson (1995) AGN dust torus model so that the maximum temperature of the dust is 1000 K rather than 1600 K. I have also shown the effect of allowing the dust torus to extend a further factor 3.5 in radius. This still does not account for the amplitude of the submm emission. The implied extent of the narrow-line region for this extended AGN dust torus model, which we use for several other objects, would be 326 $(L_{bol}/10^{13}L_\odot)^{1/2}$ pc consistent with 60-600 $(L_{bol}/10^{13}L_\odot)^{1/2}$ pc quoted by Netzer (1990). Evidence for a strong starburst contribution to the IR emission from F10214 is given by Kroker et al. (1996) and is supported by the high gas mass detected via CO lines. Granato et al. (1996) attempt to model the whole SED of F10214 with an AGN dust torus model, but still do not appear to be able to account for the 60 μm emission.

SMMJ02399 − 0136

A starburst model fits the submm data well and the ISO detection at 15 μm gives a very severe constraint on any AGN dust torus component.

The starburst interpretation of the submm emission is supported by the gas mass estimated from CO detections (Frayer et al. 1998).

F08279 + 5255

An M82-like starburst is a good fit to the submm data and an AGN dust torus model is a good fit to the 12-100 μm data. The high gas mass detected via CO lines (Downes et al. 1999) supports a starburst interpretation, though the ratio of L_{sb}/M_{gas} is on the high side. The submm data can also be modelled by an extension of the outer radius of the AGN dust torus.

H1413 + 117

The submm data is well fitted by an M82-like starburst and the gas mass implied by the CO detections (Barvainis et al. 1994) supports this interpretation. The extended AGN dust torus model discussed above does not account for the submm emission. However Granato et al. (1996) model the whole SED of H1413 in terms of an AGN dust torus model.

15307 + 325

A starburst model gives a natural explanation for the 60-180 μm excess compared to the AGN dust torus model required for the 6.7 and 14.3 μm emission (Verma et al. 2000), but the non-detection of CO poses a problem for a starburst intepretation.

PG1634 + 706

The IRAS 12-100 μm data and the ISO 150-200 μm data (Haas et al 1998) are well-fitted by the extended AGN dust torus model and an upper limit can be placed on any starburst component. The non-detection of CO is consistent with this upper limit.

In Fig. 7 we show the far infrared luminosity derived for an assumed starburst component, versus redshift, for hyperluminous galaxies, with lines indicating observational constraints at 60, 800 and 1250 μm. Three of the sources with (uncorrected) total bolometric luminosities above $10^{14} h_{50}^{-2} L_\odot$ are strongly gravitationally lensed. IRAS F10214+4724 was found to be lensed with a magnification which ranges from 100 at optical wavelengths to 10 at far infrared wavelengths (Eisenhardt et al. 1996, Green & Rowan-Robinson 1996). The 'clover-leaf' lensed system H1413+117 has been found to have a magnification of 10 (Yun et al. 1997). Downes et al. (1999) report a magnification of 14 for F08279+5255. Also, Ivison et al. (2000) estimate a magnification of 2.5 for SMMJ02399-0136 and Frayer et al. (1999) quote a magnification of 2.75 ± 0.25 for SMMJ14011+0252. These magnifications have to be corrected for in estimating luminosities (and dust and gas masses) and these corrections are indicated in Fig. 7. Farrah et al. (2001) have used *HST* to image a further 9 hyperluminous infrared galaxies and have

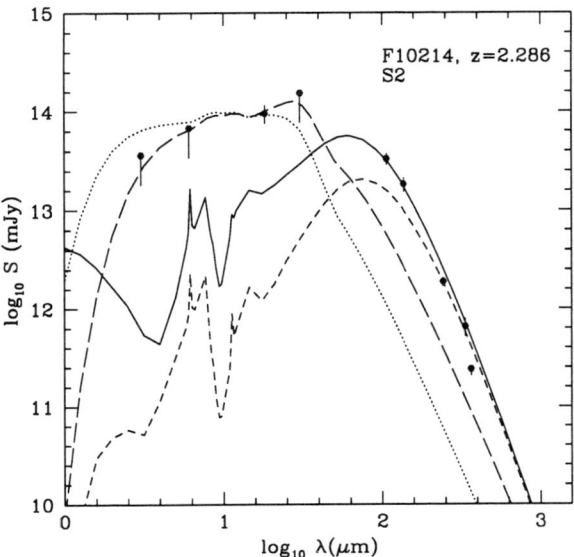

Figure 1. Observed (rest-frame) spectral energy distribution for F10214, modelled with M82-type starburst (solid curve), Arp 220-type starburst (broken curve), AGN dust torus (dotted curve, modified AGN dust torus model - long-dashed curve).

found no new cases of gravitational lensing. They have also studied the host galaxies in samples of ULIRGs and HLIRGs containing quasars and found that the proportion with an elliptical galaxy host increases towards the very highest luminosities.

Although lensing may affect some of the most luminous objects, there is strong evidence for a population of galaxies with far ir luminosities (after correction) in the range $1-3 \times 10^{13} h_{50}^{-2} L_\odot$. I have argued that in most cases the rest-frame radiation longward of 50 μm comes from a starburst component. The luminosities are such as to require star formation rates in the range $3-10 \times 10^3 h_{50}^{-2} M_\odot$yr, which would in turn generate most of the heavy elements in a $10^{11} M_\odot$ galaxy in $10^7 - 10^8$ yrs. Most of these galaxies can therefore be considered to be undergoing their most significant episode of star formation, i.e. to be in the process of 'formation'.

It appears to be significant that a large fraction of hyperluminous IR galaxies are Seyferts, radio-galaxies or QSOs. This is in part a selection effect, in that luminous AGN have been selected for submm photometric follow-up. For the 12 objects found from direct follow-up of IRAS samples or 850 μm surveys (Rowan-Robinson 2000a), 50% are QSOs or Seyferts and 50 % are narrow-line objects, similar to the proportions

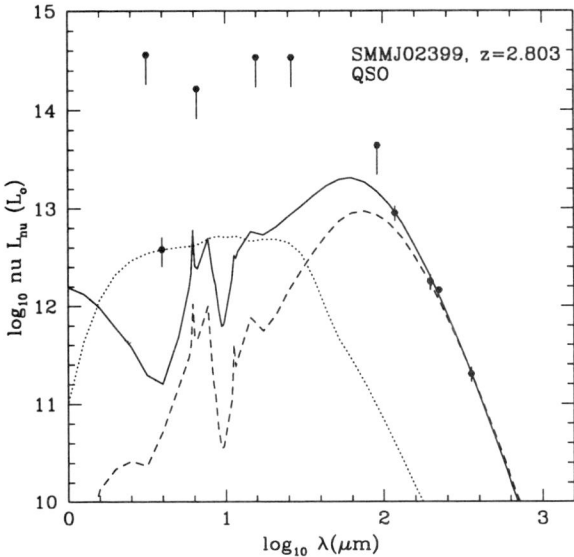

Figure 2. Observed spectral energy distribution for SMMJ02399, notation as for Fig. 1.

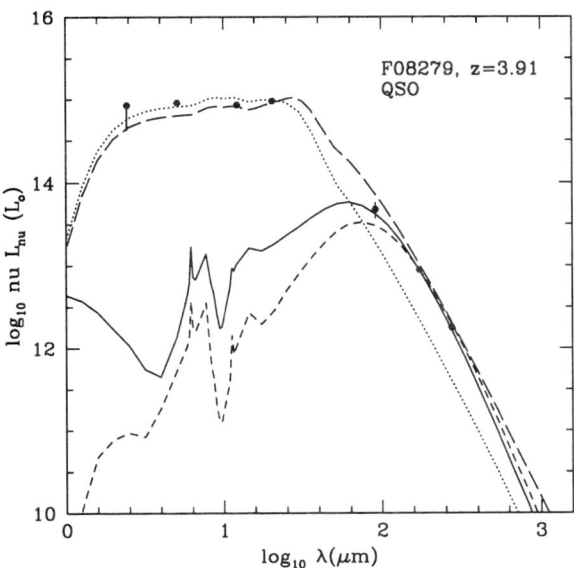

Figure 3. Observed spectral energy distribution for F08279, notation as for Fig. 1.

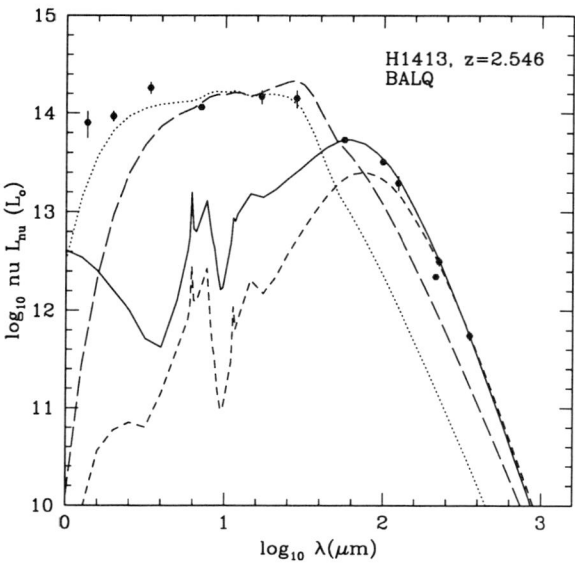

Figure 4. Observed spectral energy distribution for H1413, notation as for Fig. 1.

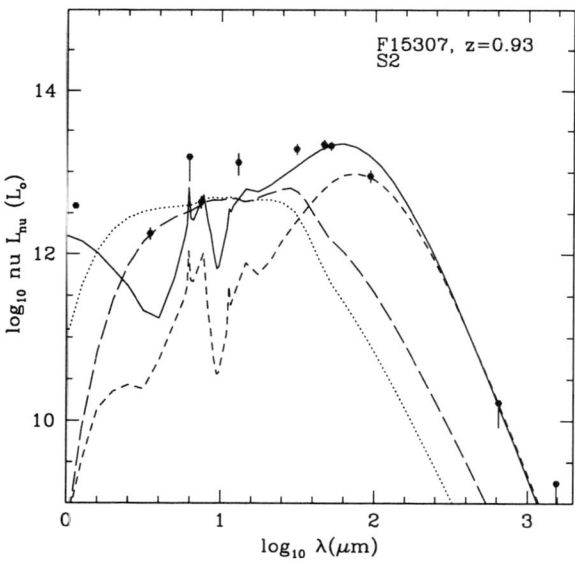

Figure 5. Observed spectral energy distribution for F15307, notation as for Fig. 1.

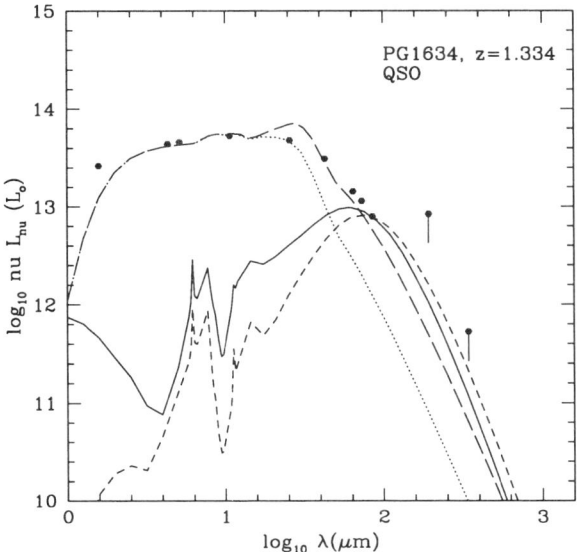

Figure 6. Observed spectral energy distribution for PG1634, notation as for Fig. 1.

found for ultraluminous IR galaxies. However despite the high proportion of ultraluminous and hyperluminous galaxies which contain AGN, this does not prove that an AGN is the source of the rest-frame far infrared radiation. The ISO-LWS mid-infrared spectroscopic programme of Genzel et al. (1998) has shown that the far infrared radiation of most ultraluminous galaxies is powered by a starburst, despite the presence of an AGN in many cases. Wilman et al. (1999) have shown that the X-ray emission from several hyperluminous galaxies is too weak for them to be powered by a typical AGN.

In the Sanders et al. (1989) picture, the far infrared and submillimetre emission would simply come from the outer regions of a warped disk surrounding the AGN. Some weaknesses of this picture as an explanation of the far infrared emission from PG quasars have been highlighted by Rowan-Robinson (1995). A picture in which both a strong starburst and the AGN activity are triggered by the same interaction or merger event is far more likely to be capable of understanding all phenomena (cf Yamada 1994, Taniguchi et al. 1999).

Where hyperluminous galaxies are detected at rest-frame wavelengths in the range 3–30 μm (and this can correspond to observed wavelengths up to 150 μm), the infrared spectrum is often found to correspond well to emission from a dust torus surrounding an AGN (e.g. Figs. 1, 3, 4,

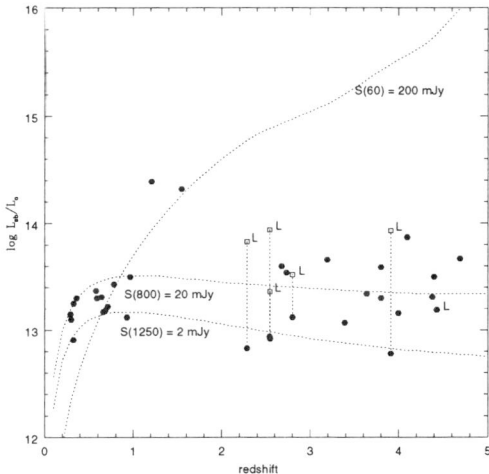

Figure 7. Bolometric luminosity in starburst component for galaxies with luminosities $> 10^{13} L_\odot$ (Tables 1-4). The galaxies labelled L are known to be lensed. Loci corresponding to the limits set by S(60) = 200 mJy, S(800) = 20 mJy, and S(1250) = 2 mJy are shown.

6). This emission often contributes a substantial fraction of the total infrared (1-1000 μm) bolometric luminosity. The advocacy of this paper for luminous starbursts relate only to the rest-frame emission at wavelengths $\geq 50\mu$m. Fig. 8 shows the correlation between the luminosity in the starburst component, L_{sb}, and the AGN dust torus component, L_{tor}, for hyperluminous infrared galaxies, PG quasars (Rowan-Robinson 1995), and IRAS galaxies detected in all 4 bands (Rowan-Robinson & Crawford 1989) (this extends Fig. 8 of Rowan-Robinson 1995). The range of the ratio between these quantities, with $0.1 \leq L_{sb}/L_{tor} \leq 10$, is similar over a very wide range of infrared luminosity (5 orders of magnitude), showing that the proposed separation into these two components for hyperluminous IR galaxies is not at all implausible.

The radiative transfer models can be used to derive dust masses and hence, via an assumed gas-to-dust ratio, gas masses. For the M82-like starburst model used here the appropriate conversion is $M_{dust} = 10^{-4.6} L_{sb}$, in solar units (Green & Rowan-Robinson 1996). Rowan-Robinson (2000a) has compared these estimates with gas mass estimates derived from CO observations, when available, and found good agreement (within a factor of 2).

The range of ratios of L_{sb}/M_{gas} for hyperluminous galaxies is consistent with that derived for ultraluminous starbursts. There is a tendency

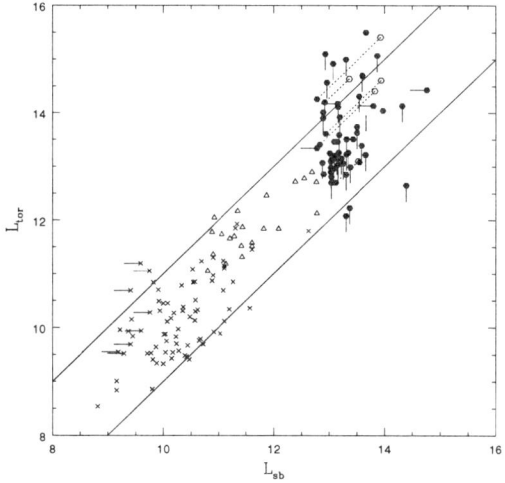

Figure 8. Bolometric luminosity in AGN dust torus component versus bolometric luminosity in starburst component: filled circles, hyperluminous IR galaxies (this paper); open triangles, PG quasars (Rowan-Robinson 1995); crosses, IRAS galaxies detected in 4 bands (Rowan-Robinson & Crawford 1989, galaxies with only upper limits on L_{tor} omitted).

for the time-scale for gas-consumption, assuming a star formation rate given by Eq. (1), to be shorter for the more luminous objects, in the range $10^7 - 10^8$ yrs (alternatively this could indicate a higher value for the low-mass cutoff in the IMF). The cases where a strong limit can be set on M_{gas} are also, generally, those where the SEDs do not support the presence of a starburst component. After correction for the effects of gravitational lensing, gas masses ranging up to 1-3 x $10^{11} M_\odot$ are seen in most hyperluminous galaxies, comparable with the total stellar mass of an L_* galaxy ($10^{11.2}(M/4L)h_{50}^{-2}$). Hughes et al. (1997) argue that a star-forming galaxy can not be considered primeval unless it contains a total gas mass of $10^{12} M_\odot$, but this seems to neglect the fact that 90 % of the mass of galaxies resides in the dark (probably non-baryonic) halo.

3. Conclusions and discussion

(1) Where there is significant emission at restframe wavelengths > 50 µm in luminous infrared galaxies, it is generally due to a starburst component.

(2) The bolometric luminosity of hyperluminous infrared galaxies is dominated by an AGN dust torus in about 50 % of cases, and by a starburst in the rest.

(3) The broad correlation found by Rowan-Robinson (1995) between the luminosities in starburst and AGN dust tori components extends to luminosities of $10^{14} L_\odot$, indicative of a link between the feeding of an AGN and the fuelling of a starburst.

(4) Some hyperluminous galaxies appear to be forming stars at $> 10^3 M_\odot$ per year and so are likely to be forming a very significant fraction of their final stellar mass.

(5) There is a good correlation between the gas-mass deduced, using normal gas-to-dust ratios, from the far infrared emission, and that detected directly from molecular line observations. Several objects have gas masses of order of the baryonic mass in an L_* galaxy, consistent with their being essentially primeval galaxies.

References

Barvainis R. et al., 1994, Nature, 371, 586
Desert F.-X. et al., 1990, A&A, 237, 215
Downes D. et al., 1999, ApJ, 513, L1
Draine B.T., Lee H.M., 1984, ApJ, 285, 89
Dwek E., 1998, ApJ, 501, 643
Efstathiou A., Rowan-Robinson M., 1995, MNRAS, 273, 649
Efstathiou A. et al., 2000, MNRAS, 313, 734
Eisenhardt P.R. et al., 1996, ApJ, 461, 72
Frayer D.T. et al., 1999, ApJ, 514, L13
Frayer D.T. et al., 1998, ApJ, 506, L27
Genzel R. et al., 1998, ApJ, 498, 579
Goldader J.D., Joseph R.D., Doyon R., Sanders D.A., 1997, ApJ, 474, 104
Granato G.L., Danese L., 1994, MNRAS, 268, 235
Granato G.L., Danese L., Franceschini A., 1997, ApJ, 460, L11
Granato G.L., Danese L., Franceschini A., 1996, ApJ, 460, L11
Green S., Rowan-Robinson M., 1996, MNRAS, 279, 884
Hughes D.H., Dunlop J.S., Rawlings S., 1997, MNRAS, 289, 766
Ivison R. et al., 2000, MNRAS, in press (astro-ph/9911069)
Kroker H. et al., 1996, ApJ, 463, L55
Krugel E., Siebenmorgen R., 1994, A&A, 282, 407
Mathis J.S., Rumpl W., Nordsieck K.H., 1977, ApJ, 217, 425
Pier A., Krolik J., 1992, ApJ, 401, 99
Rigopoulou D. et al., 1996, MNRAS, 278, 1049
Rowan-Robinson M., 1980, ApJS, 44, 403
Rowan-Robinson M., 1986, MNRAS, 219, 737
Rowan-Robinson M., Crawford J., 1989, MNRAS, 238, 523
Rowan-Robinson, M., 1992, MNRAS, 258, 787
Rowan-Robinson M. et al., 1993, MNRAS, 261, 513
Rowan-Robinson M., 1995, MNRAS, 272, 737

Rowan-Robinson, M., Efstathiou, A., 1993, MNRAS, 263, 675
Rowan-Robinson M. et al., 1997, MNRAS, 289, 490
Rowan-Robinson M., 2000a, MNRAS, 316, 885
Sanders D.B. et al., 1988, ApJ, 325, 74
Sanders D.B. et al., 1989, ApJ, 347, 29
Scoville, N.Z., Young, J.S., 1983, ApJ, 265, 148
Siebenmorgen R., Krugel E., 1992, A&A, 259, 614
Silva L., Granato G.L., Bressan A., Danese L., 1998, ApJ, 509, 103
Taniguchi Y., Ikeuchi S., Shioya Y., 1999, ApJ, 514, L9
Wilman R.J. et al., 1998, 300, L7
Yamada T., 1994, ApJ, 423, L27
Xu C. et al., 1998, ApJ, 508, 576
Yorke H.W., 1977, A&A, 58, 423
Yun M.S. et al., 1997, ApJ, 479, L9

Sylvain Veilleux and Pepa Masegosa

Carlton Baugh (left) and Alessandro Bressan (right)

THE QSO – ULTRALUMINOUS INFRARED GALAXIES CONNECTION

Sylvain Veilleux
Department of Astronomy, University of Maryland
veilleux@astro.umd.edu

Dong-Chan Kim
Institute of Astronomy and Astrophysics, Academia Sinica
kim@asiaa.sinica.edu.tw

David B. Sanders
Institute for Astronomy, University of Hawaii
sanders@galileo.ifa.hawaii.edu

Abstract For the past several years, our group has pursued a vigorous ground-based program aimed at understanding the nature of ultraluminous infrared galaxies. We recently published the results from a optical/near-infrared spectroscopic survey of a large statistically complete sample of ultraluminous infrared galaxies (the "IRAS 1-Jy sample"). We now present the results from our recently completed optical/near-infrared imaging survey of the 1-Jy sample. These data provide detailed morphological information on both large scale (e.g., intensity and color profiles, intensity and size of tidal tails and bridges, etc) and small scale (e.g., nuclear separation, presence of bars, etc) that helps us constrain the initial conditions necessary to produce galaxies with such high level of star formation and/or AGN activity. The nature of the interdependence between some key spectroscopic and morphological parameters in our objects (e.g., dominant energy source: super-starburst versus quasar, nuclear separation, merger phase, star formation rate, and infrared luminosity and color) is used to clarify the connection between starbursts, ultraluminous infrared galaxies, and quasars.

Keywords: QSOs, Seyfert galaxies, Starburst galaxies, Origin, Evolution

Introduction

Ultraluminous infrared galaxies (ULIGs; log $[L_{IR}/L_\odot] \geq 12$ by definition) may provide the clearest observational link between galaxy mergers, starbursts and powerful AGN. However, the exact nature of ULIGs remains unclear. The most important questions, and the ones we propose to address here, are: (1) *What is the dominant energy source in ULIGs (starburst versus AGN)?*, and (2) *Is there a evolutionary connection between ULIGs and quasars?*

In recent years, several surveys to faint flux levels in the IRAS database have been carried out to search for luminous objects. Arguably the most important of these studies is the '1 Jy' survey of Kim (1995). This study provides a complete list of the brightest ULIGs with F[60 μm] > 1 Jy which is not biased toward 'warm' quasar-like objects with large F[25 μm]/F[60 μm] ratios. The '1 Jy' sample contains 118 objects with $z = 0.02 - 0.27$ and log $[L_{ir}/L_\odot] = 12.00 - 12.84$. The infrared luminosities of these objects therefore truly overlap with the bolometric luminosities of optical quasars. Other surveys have discovered objects of comparable luminosity at fainter flux levels as well as a few 'hyperluminous' objects at higher L_{ir}. However, the '1 Jy' sample contains the brightest objects at a given luminosity, hence the best candidates for follow-up studies.

The results of our analysis of the IRAS database on the '1 Jy' sample were published in Kim & Sanders (1998) and are discussed in these proceedings (Sanders contribution). The present paper summarizes the results from our ground-based follow-up surveys. In Section 1, we discuss the results from our optical and near-infrared spectroscopy of this sample, and in Section 2, the preliminary results from our optical and near-infrared imaging survey. Section 3 summarizes our conclusions.

1. Optical and Infrared Spectroscopy of the 1 Jy Sample

The results from our optical and near-infrared spectroscopic surveys were published recently in Veilleux, Kim & Sanders (1999a) and Veilleux, Sanders & Kim (1999b), respectively. The main conclusions of this analysis are the following:

1. The fraction of luminous infrared galaxies with Seyfert characteristics increases rapidly with increasing L_{ir}. About 30% of the ULIGs host a Seyfert 1 or Seyfert 2 nucleus. For $L_{ir} > 10^{12.3} L_\odot$, this fraction is nearly 50%.

2. From 50% to 70% of the Seyfert 2s in our sample show signs of an AGN in the near-infrared (e.g., broad-line region or strong [Si VI]

1.962 μm), therefore confirming the detection of genuine AGN in these objects. In contrast, none of the optically classified LINERs and H II galaxies in our near-infrared sample shows any obvious signs of an energetically important AGN.

3. Combining our optical and near-infrared results, we find that the fraction of objects with bonafide AGN is 25–30% for the 1 Jy sample of ULIGs and 35–50% for objects with $L_{ir} > 10^{12.3}\ L_\odot$.

4. Comparisons of the dereddened emission-line luminosities of the optical or obscured BLRs detected in the ULIGs of the 1-Jy sample with those of optical quasars indicate that the obscured AGN/quasar in ULIGs is the main source of energy in at least 15 – 25% of all ULIGs in the 1-Jy sample. This fraction is closer to 30 – 50% among ULIGs with $L_{ir} > 10^{12.3}\ L_\odot$.

These results are compatible with those from recent mid-infrared spectroscopic surveys carried out with *ISO* (e.g., Genzel et al. 1998). Indeed, a detailed object-by-object comparison of the optical and mid-infrared classifications shows an excellent agreement between the two classification methods (Lutz, Veilleux & Genzel 1999). These results suggest that strong nuclear activity, once triggered, quickly breaks the obscuring screen at least in certain directions, thus becoming detectable over a wide wavelength range.

2. Optical and Infrared Imaging of the 1 Jy Sample

We now have high signal-to-noise ratio, sub-arcsecond resolution R and K' images of all '1 Jy' sources. Additional spectra were obtained of several sources in the field to identify them (stars versus galaxies) and determine if they are involved in the ULIG event. Our preliminary analysis of these data suggests the following tantalizing trends:

1. As found in previous studies, the large majority (> 95%) of the optical and near-infrared images show signs of a strong tidal interaction/merger in the form of distorted or double nuclei, tidal tails, bridges, and overlapping disks.

2. The small mean nuclear separation (< 3 kpc) of the ULIGs examined so far suggests that the majority of these galaxies are in a terminal stage of a merger. These mergers generally involve *two* large (0.5 – 2 L*) galaxies. Multiple mergers are seen in only 4 of the 118 systems.

3. These galaxies span a broad range of total (= nuclear + host) luminosities. In advanced mergers, L_R(tot) $\sim$ 1.5 L* and $L_{K'}$(tot) $\sim$ 3 L* on average but with a lot of scatter.

4. Roughly 30%/10%/60% of the R-band surface brightness profiles are well fitted by a elliptical-like $R^{1/4}$-law / exponential disk / neither or both. The percentage of poor fits is likely to decrease when we examine our data at K', where the effects of dust obscuration and star formation are less important.

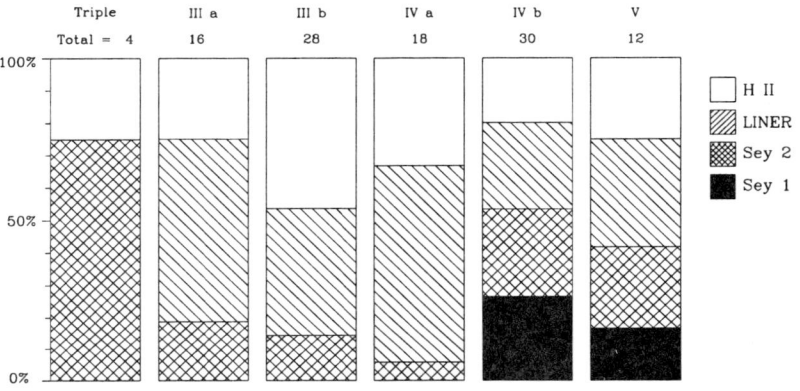

Figure 1. Optical spectral type versus morphological class. See text for a description of the morphological classification.

5. Using a classification scheme first proposed by Surace (1998) and based on the results of published numerical simulations, we classify all our objects according to morphology: I. Pre-contact – relatively unperturbed and separate disk. II. First contact – overlapping disks but no evidence for strong bars or tidal tails. III. Pre-merger – two distinct galaxies with well-developed tidal tails and brigdes. (a) apparent separation > 10 kpc and (b) apparent separation < 10 kpc. IV. Merger – single nucleus with prominent tidal tails. (a) diffuse nucleus with $L_{K'}(2\ \mathrm{kpc})/L_{K'}(\mathrm{tot}) < 1/3$. (b) compact nucleus with $L_{K'}(2\ \mathrm{kpc})/L_{K'}(\mathrm{tot}) > 1/3$. V. Old merger – no obvious signs of tidal tails but disturbed central morphology. We find that all Seyfert 1s and most of the Seyfert 2s are advanced mergers either based on their morphology (classes IVb or V; Fig. 1) or their nuclear separation (< 5 kpc, generally; Fig. 2). A similar result is found when we consider the 'warm' objects with $f_{20}/f_{60} > 0.2$ (Fig. 3).

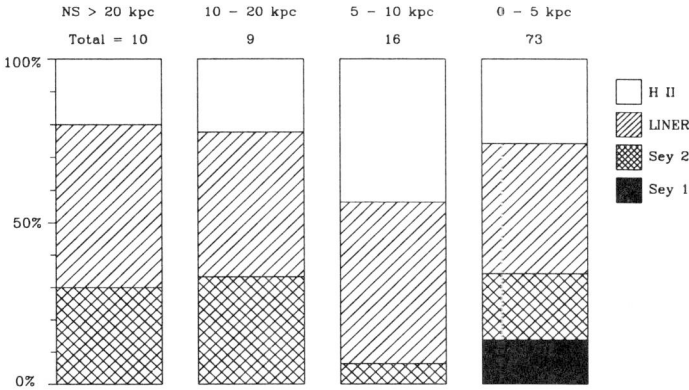

Figure 2. Optical spectral type versus nuclear separation.

3. Summary

The results from our spectroscopic survey of the 1 Jy sample of ULIGs indicate that the fraction of ULIGs powered by quasars increases with increasing infrared luminosity, reaching a value of 30 – 50 % for

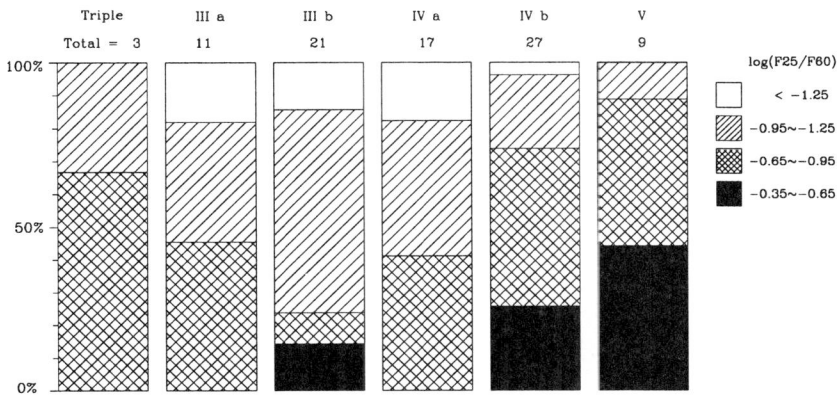

Figure 3. IRAS f_{20}/f_{60} colors versus morphological classification.

$L_{ir} > 10^{12.3}\ L_\odot$. The preliminary results from our imaging survey of the same sample suggest trends between merger phase, infrared colors, and the presence of an AGN. Objects with 'warm' quasar-like infrared colors show signs of AGN activity and are generally found in advanced mergers, based not only on the apparent nuclear separation but also on the morphology of the tidal tails. These results suggest that the evolutionary sequence 'cool' ULIGs → 'warm' ULIGs → quasars applies to at least some (though admittedly probably not all) ULIGs and quasars.

Acknowledgments

S. V. gratefully acknowledges the financial support of NASA through LTSA grant NAG 56547.

References

Genzel, R. et al., 1998, ApJ, 498, 579
Kim, D.-C., Sanders, D. B., 1998, ApJS, 119, 41
Lutz, D., Veilleux, S., Genzel, R., 1998, ApJL, 517, L13
Sanders, D. B., Mirabel, I. F., 1996, ARA&A, 34, 725
Sanders, D. B. et al., 1988, ApJ, 325, 74
Veilleux, S., Kim, D.-C., Sanders, D. B., 1999a, ApJ, 522, 113
Veilleux, S., Sanders, D. B., Kim, D.-C., 1999b, ApJ, 522, 139

RECENT STAR FORMATION IN VERY LUMINOUS INFRARED GALAXIES

Alessandro Bressan, Bianca Poggianti
Osservatorio Astronomico di Padova, Padova, IT

Alberto Franceschini
Dipartimento di Astronomia, Universita' di Padova, Padova, IT

Abstract   Among star forming galaxies, a spectral combination of a strong Hδ line in absorption and a moderate [OII] emission has been suggested to be a useful method to identify dusty starburst galaxies at any redshift, on the basis of optical data alone. On one side it has been shown that such a spectral particularity is indeed suggestive of obscured starburst galaxies but, on the other, the degeneracy of the optical spectrum has hindered any quantitative estimate of the star formation and extinction during the burst. The optical spectrum by itself, even complemented with the information on the far-IR flux, is not enough to identify univocal evolutionary patterns. We discuss in the following whether it is possible to reduce these uncertainties by extending the spectral analyses to the near infrared spectral region.

Introduction

Recent observations in the IR/mm have revealed that luminous and ultra luminous IR galaxies (LIRGs and ULIRGs) constitute a major cosmological component during the past epochs of the universe: in spite of their short duration, the transient IR-active phases are responsible for a large fraction of the energy emission and metal production at high redshifts. With luminosities spanning the range 10^{11}–10^{13} L$_\odot$ and space densities similar to those of quasars (Soifer et al. 1986) LIRGs and ULIRGs are the most luminous objects in the local Universe. Highly extinguished and strong IR emitters, ULIRGs are considered to be the local analogues of the newly discovered high-z IR luminous galaxies. Recent optical spectroscopic surveys (Wu et al. 1998, Poggianti & Wu 2000) reveal that the spectra of a large fraction of (U)LIRGs display a peculiar combination of spectral features in the optical: a strong Hδ line in absorption (EW> 4 Å) and a moderate [OII] emission (EW< 40 Å). Galaxies with this type of spectra were named "e(a)" galaxies and were found to be quite numerous in the cluster and field environments at $z = 0.4 - 0.5$ (Poggianti et al. 1999). The equivalent width of their Hδ line exceeds that of typical, quiescent spirals at low-z and their low [OII]/Hα ratios are consistent with the emission line fluxes being highly extincted by dust. Until recently, such a combination of moderate [OII] emission and strong Balmer absorption, was thought to be associated with post-starburst galaxies, with little or no star-formation. Such a claim was mostly based on dust-free models. Poggianti et al. (1999) however, proposed that this peculiar spectral signature corresponds in fact to highly obscured starburst galaxies. Clearly, one still needs to investigate the exact origin of these spectra, namely, the nature of the star formation as well as the properties of the dust. A possible explanation for this unusual combination of emission/absorption features is selective extinction (Poggianti & Wu 2000). In such a scenario, HII regions (where the [OII] emission originates) are highly embedded and thus are affected by more extinction compared to the older stellar population which is responsible for the Balmer absorption.

Surprisingly enough, the combination of moderate emission lines and strong Balmer absorption has been also detected in the spectra of high-z IR luminous ISO galaxies (e.g. Flores et al. 1999). A recent near-IR survey of the same population revealed that the ISO galaxies are in fact actively starbursting but highly obscured objects, based on the strength of the Hα emission line and high equivalent width (Rigopoulou et al. 2000). Differential extinction is again at the heart of this spectral behaviour. Rigopoulou et al. (2000) have estimated the extinction in their sample of ISO galaxies based on optical colours and found it to be $A_v \sim 3$. After correcting the Hα line flux using optical indicators of extinction, the inferred Star Formation Rates (SFR) fall significantly below the proper SFR estimated from the far-Infrared flux. This implies that optical extinction is significant, and requires additional information for a more reliable evaluation.

Poggianti, Bressan & Franceschini (2001) have recently investigated in some detail the optical spectra of LIRG and ULIRG galaxies, to constrain the recent history of star-formation and the dust extinction characteristic of various stellar populations. By examining different star formation patterns they concluded that only a starburst, selective extinction scenario could explain the e(a) spectra. In this scenario the extinction is larger in the younger populations, a fact that mimics the progressive escape of young stars from their parental molecular clouds, as they age. The observed FIR/V luminosity ratio could be explained only by models where a significant fraction of the FIR luminosity originates in regions that are practically obscured at optical wavelengths. Unfortunately this hinders any quantitative estimate of the duration and intensity of the burst and/or of the wavelength dependence of the extinction. Thus the optical-near UV spectrum by itself, even complemented with the information on the far-IR flux, are not enough to identify univocal modellistic solutions. One chance to reduce these uncertainties is provided by the extension of the spectral analyses to the near-IR regime (eg. Murphy et al. 1999).

1. Models in the near IR

In order to highlight the advantages of investigating over a wider wavelength range, we have followed the same methodology described in Poggianti, Bressan & Franceschini (2001). The integrated model spectrum, from the far UV to the near IR, has been generated as a combination of 10 stellar populations of different ages, computed with a Salpeter IMF between 0.15 and 120 $M_\odot$. The stellar SEDs have been obtained by extending the Pickles spectral library below 1000 Å and above 24000 Å with the Kurucz (1993) models. The composite (stars+gas) spectrum of each single generation has been then produced by one of us (AB) with the help of the photo-ionization code CLOUDY (Ferland 1990). The ages of the 10 populations have been chosen considering the evolutionary time scales associated with the observational constraints: the youngest generations ($10^6, 3\cdot 10^6, 8\cdot 10^6, 10^7$ yr) are responsible for the ionizing photons that produce the emission lines; the intermediate populations ($5\cdot 10^7, 10^8, 3\cdot 10^8, 5\cdot 10^8, 10^9$ yr) are those with the strongest Balmer lines in absorption, while older generations of stars have been modelled as a constant star formation rate (SFR) between 2 and 12 Gyr before the moment of the observation and can give a significant contribution to the spectral continuum, hence affecting also the equivalent widths of the lines.

Each simple population is assumed to be extincted by dust in a uniform screen according to the standard extinction law of the diffuse medium in our Galaxy ($R_V = A_V/E(B-V) = 3.1$, Cardelli et al. 1989). While a more complex picture of the extinction cannot be excluded, Poggianti et al. already have shown that the characteristics of the emerging spectrum require a significant amount of *foreground* dust (screen model). Indeed in the case of a uniform mixture of dust and stars, increasing the obscuration does not yield a corresponding increase in the *reddening* of the spectrum: the latter saturates to a value (E(B–V)$\sim$ 0.18) which is too low to be able to account for the observed emission line ratios (see also Calzetti et al. 1994). In the model used here, the extinction value E(B–V) is allowed to vary from one stellar population to another and the extincted spectral energy distributions of all the single generations are added up to give the total integrated spectrum.

Finally, the best-fit model, within a chosen star formation scenario, was obtained by minimizing the differences between selected features in the observed and model optical range: the equivalent width of four lines ([OII]λ3727, Hδ, Hβ and Hα) and the relative intensities of the continuum flux in eight almost featureless spectral windows (3770-3900Å, 4020-4070Å, 4150-4250Å, 4600-4800Å, 5060-5150Å, 5400-5850Å, 5950-6250Å and 6370-6460Å).

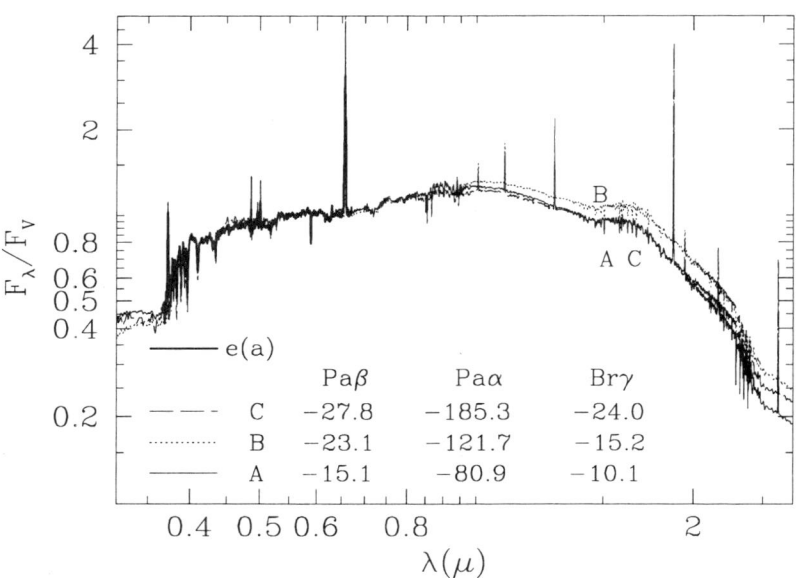

Figure 1. The observed average e(a) optical (3700Å– 6900Å) spectrum is compared with suitable models extended into the NIR region. All models reproduce well the shape of the observed continuum and equivalent widths of Hα (-62.48±5.3), Hβ (0.68±0.8), Hδ (5.64±0.5) and [OII]3727 (-12.68±2.1)(Poggianti et al 2001). The predicted EWs of selected lines in the NIR are shown. See text for more details.

2. Results

Fig. 1 depicts three particular cases that were successful in reproducing the optical spectral features of e(a) galaxies (Poggianti et al. 2001). Model A, the reference model, reproduces both the line strengths and the highly reddened continuum, but it can account for only 1/3 of the FIR emission, hence of the star formation rate. In this model the mass of young stars amounts to about 10% of the galaxy mass.

To solve the above discrepancy, some practically obscured regions were added in model B. As in model A, the starburst began $2 \cdot 10^8$ yr ago, but both the SFR during the burst and the extinction of the two youngest stellar generations are higher than in A. The recent SFR is about a factor of 10 larger than in model A and, for a standard Salpeter IMF in the range $0.1 - 100\,M_\odot$, about 60% of the total mass in stars is formed during the burst. Thus the high ratio FIR/optical in this case, is indicative of a significant fraction of the galaxy mass being formed during the burst. It is worth noticing that assuming a top-heavy IMF during the burst phase would substantially reduce the mass fraction in young stars: for example, with a Salpeter IMF lower mass limit $= 1\,M_\odot$, the starburst in model B forms about 40% of the total stellar mass. However different IMF with the same distribution of stars above say $1\,M_\odot$, would essentially produce the same integrated spectrum, and would be practically indistinguishable. Finally, in model C the observed FIR/optical ratio is reproduced by adopting an extinction law with $R_V = A_V/E(B-V) = 5$, as observed towards some dense clouds in our Galaxy (Mathis 1990). In this case the recent SFR is only a factor of two larger than that of model A and the observed FIR/optical ratio is fully reproduced because, for a given E(B–V), the optical flux is much more extinguished. In spite of the different star formation histories, the shape of the continuum in the near infrared is fairly similar in the three cases. This is striking because the inspection of the models shows that, while in case A the near infrared continuum is contributed in almost equal parts by the young burst and the old disk, in case B it is dominated by the young burst while, in case C, it is due to the old disk.

Thus it appears that the degeneracy encountered at optical wavelengths would not be removed by the NIR photometry alone.

On the other hand the simulations clearly suggest different equivalent widths of the NIR hydrogen lines. In particular, in models B and C the line intensities are about twice as strong as in model A. In the case of model B, though the SFR is much higher than in the case A, the young populations are much more heavily obscured so that line intensities which are similar in the optical, are only a factor of two larger in the NIR. A significant fraction of the SFR remains hidden even in the NIR. In the case of model C the intensity of the NIR lines simply reflects the larger SFR adopted. Finally, Table 1 shows the average E(B–V) re-

Table 1. E(B–V) from Hydrogen line ratios in the different models.

	Hβ/Hα	Paα/Brγ	Paβ/Brγ	Hβ/Brδ	Hα/Paα
A	0.96	0.20	1.15	1.08	1.19
B	1.19	0.72	1.45	1.35	1.47
C	0.90	1.37	1.98	1.38	1.67
C[a]	0.71	0.70	1.01	0.90	0.98

[a] adopting an extinction law with $R_V=5$

covered with the ratios of the Hydrogen lines, in the optical and NIR spectral region, adopting the extinction law with $R_V=3.1$. We remind that, in the case of VLIRGs where the EW Hβ is almost null, the Balmer decrement is severely affected by the assumed equivalent width of the absorption component. On the other hand the NIR ratio Paα/Brγ is affected by too small a wavelength range. As for the other three line ratios, the Paβ/Brγ ought to be preferred because it is less affected by uncertainties in the old population and unaffected by [NII] emission.

Acknowledgments

We are grateful to Pasquale Panuzzo for his help in running CLOUDY and for discussions. AB acknowledges support from the TMR grant ERBFMRXCT960086.

References

Cardelli, J. A., Clayton, G. C., Mathis, J. S., 1989, ApJ, 345, 245
Calzetti, D., Kinney, A. L., Storchi-Bergmann, T., 1994, ApJ, 429, 582
Ferland, G.J., 1996, *Hazy, a Brief Introduction to Cloudy*, University of Kentucky, Department of Physics and Astronomy Internal Report.
Flores, H., Hammer, F., Thuan, T. X., Cesarsky, C., Desert, F. X., Omont, A., Lilly, S. J., Eales, S., Crampton, D., Le Fevre, O., 1999, ApJ, 517, 148
Mathis, J. S., 1990, ARA&A, 28, 37
Murphy, T. W Jr., Soifer, B. T., Matthews, K., Kiger, J. R., Armus, L., 1999, ApJ, 525, 85
Kurucz, R., 1993, Kurucz CD-ROM No. 13. Cambridge, Mass.: Smithsonian Astrophysical Observatory.
Pickles, A. J. 1998, PASP, 110, 863
Poggianti, B. M., Wu, H., 2000, ApJ, 529, 157
Poggianti, B. M., Bressan, A., Franceschini, A., 2001, ApJ, Vol 550, astro-ph 0011160
Rigopoulou, D. et al., 2000 ApJ, 537, 85
Soifer, B.T. et al., 2000, AJ, 119, 509
Wu, H., Zou, Z. L., Xia, X. Y., Deng, Z. G., 1998b, A&AS, 132, 181

A MOLECULAR GAS SURVEY OF $Z < 0.2$ INFRARED EXCESS, OPTICAL QSOS

A. S. Evans
SUNY, Stony Brook

D. T. Frayer, J. A. Surace
SIRTF Science Center/Caltech)

and D.B. Sanders
Institute for Astronomy, University of Hawaii)

Abstract Millimeter-wave (CO) observations of optically-selected QSOs are potentially a powerful tool for studying the properties of both the QSOs and their host galaxies. We summarize here a recent molecular gas survey of $z < 0.17$ optical QSOs with infrared (IR) excess, $L_{\rm IR}(8-1000\mu{\rm m})/L_{\rm bbb}(0.1-1.0\mu{\rm m}) > 0.36$. Eight of these QSOs have been detected to date in CO($1 \to 0$), and the derived molecular gas masses are in the range $1.7 - 35 \times 10^9$ M$_\odot$. The high $L_{\rm IR}/L'_{\rm CO}$ of QSOs relative to the bulk of the local ($z < 0.2$) IR luminous galaxy merger population is indicative of significant heating of dust ($L_{\rm IR}$) by the QSO nucleus and/or by massive stars created in the host galaxy with high efficiency (i.e., per unit molecular gas, $L'_{\rm CO}$).

Introduction

Millimeter-wave (CO) observations of QSOs have not been a traditional means of studying the underlying QSO host galaxies. This is partially due to the fact that millimeter-wave telescopes and arrays with large (≥ 500 m^2) collecting areas and sensitive, low-noise receivers have only been available within approximately the last decade; prior to this time, single-dish and interferometric CO($1 \to 0$) observations of "distant" sources with faint mid- and far-IR flux densities, such as the $z < 0.16$ QSO Mrk 1014 (= PG 0157+001: Sanders et al. 1988a), were time-consuming ventures. As a result, it is not surprising that most of the advances in our understanding of QSOs have been achieved via observations obtained at other wavelengths (primarily optical and radio). The situation today is significantly different from what it was circa 1990 – millimeter arrays operated by Caltech, the Berkeley-Illinois-Maryland (BIMA) consortium, Nobeyama, and IRAM have five to six 10–15m diameter dishes with state-of-the-art receivers, and upcoming facilities such as the Smithsonian SubMillimeter Array (SMA), the Large Millimeter Telescope (LMT), the Combined Array for Research in Millimeter Astronomy (CARMA), and the Atacama Large Millimeter Array (ALMA) will provide additional telescope availability and sensitivity with which to conduct routine, large millimeter-wave surveys of distant galaxies in a variety of different molecular tracers (e.g., CO, HCN).

For the past few years, we have made use of the Owens Valley Millimeter Array (OVRO) to conduct a CO(1 → 0) survey of QSOs from the Palomar-Green (PG) Bright Quasar Survey (Schmidt & Green 1983). In the context of this meeting, the motivation for undertaking such a survey is three-fold: First, the rotational transitions of CO are tracers of star-forming molecular gas, thus strong CO emission from QSOs is an indication that their host galaxies have a significant cold molecular component to their interstellar medium (ISM). Massive galaxies that are known to contain such an ISM are spiral galaxies and ongoing IR-selected mergers, which typically have molecular gas masses in excess of 10^9 $M_\odot$. In contrast, optically-selected, massive elliptical galaxies are intrinsically poor in molecular gas and dust, with typical molecular gas masses $< 10^8$ $M_\odot$. Second, in addition to fueling star formation, molecular gas in the host galaxy is a potential source of fuel for QSO activity. Thus, correlations between the distribution, kinematics and amount of molecular gas and the level of QSO activity may eventually aid in understanding the nature of mass accretion processes of QSOs. Third, there exists the possibility that some ultraluminous IR galaxy mergers (ULIGs: defined as having IR luminosities, $L_{IR}[8-1000\mu m] \geq 10^{12}$ $L_\odot$) are the evolutionary precursors of QSOs (Sanders et al. 1988b). In such a scenario, ULIGs evolve from a molecular gas-rich, dust-enshrouded cool (i.e., $25\mu m$ to $60\mu m$ flux density ratio, $f_{25}/f_{60} < 0.2$) ultraluminous phase where vigorous star formation and accretion of significant amounts of mass unto the supermassive nuclear black hole have commenced, to a warm ($f_{25}/f_{60} \geq 0.2$) phase in which AGN signatures are visible (i.e., bright nuclei with near-IR colors consistent with a reddened QSO nucleus and Seyfert-like emission-line spectrum), then finally to an UV-excess QSO. This latter stage can only occur after significant consumption or clearing of molecular gas and dust from the nuclear region has occured, thus revealing the optical QSO nucleus. Given this,

TABLE 1
$z < 0.2$ PG QSOs Detected in CO(1 → 0) to Date

Source	$z_{opt}{}^a$	$\log L_{IR}{}^b$ ($\log L_\odot$)	$\log L_{bol}{}^b$ ($\log L_\odot$)	L_{IR}/L_{bbb}	z_{CO}	Δv_{FWHM} (km s^{-1})	$M(H_2)^c$ ($10^9 M_\odot$)
PG 0007+106	0.089	11.63	12.21	0.36	...	...	< 3.5
PG 0050+124	0.061	12.00	12.32	2.01	0.061	400	20
PG 0157+001	0.163	12.49	12.68	2.57	1.163	400	35
PG 0838+770	0.131	11.60	12.01	0.63	0.132	90	8.4
PG 1119+120	0.050	11.17	11.47	1.00	0.050	260	1.7
PG 1126-041	0.060	11.47	11.94	0.50	...	...	< 1.4
PG 1202+281	0.165	11.78	12.07	1.05	...	...	< 9.2
PG 1351+640	0.088	11.82	12.37	0.40	0.088	230	5.2
PG 1402+261	0.164	11.85	12.33	0.49	...	...	< 4.9
PG 1415+451	0.114	11.48	11.83	0.80	0.114	50	6.1
PG 1440+356	0.079	11.62	11.95	0.87	0.078	370	8.0
PG 1613+658	0.129	11.99	12.20	1.58	0.129	490	20

[a]Redshifts Based on Optical Emission Lines.
[b]Calculated assuming $H_0 = 75$ km s^{-1} Mpc^{-1} and $q_0 = 0.5$.
[c]Calculated assuming $\alpha = 4 M_\odot$ [K km s^{-1} pc^2]$^{-1}$. 3σ root mean square (rms) upper limits are calculated assuming $\Delta v = 280$ km s^{-1}.

CO(1 → 0) observations of QSOs designed to search for residual amounts of molecular gas from an earlier ULIG phase may enable a direct comparison with the molecular gas content of IR luminous galaxy mergers.

1. IR-Excess, Optical QSOs

In order to better facilitate comparisons with IR luminous galaxies, it was necessary to select a QSO sample for which accurate determinations of L_{IR} could be made (i.e., QSOs that have *IRAS*[1] detections at mid and far-IR wavelengths). The sample was thus chosen from a $z < 0.17$ IR-excess (i.e. IR to "big blue bump" luminosity ratio, $L_{IR}/L_{bbb}[0.10 - 1\mu m] > 0.36$) sample of 18 PG QSOs compiled by Surace & Sanders (2001).[2] The IR-excess criteria of the sample thus selects the most likely "transition" candidates between the dust-enshrouded ULIG phase and the optical, UV-excess QSO phase.

While selecting QSOs with IR excesses introduces a bias, it must be remembered that 20–40% of the bolometric luminosity, L_{bol}, of PG QSOs is emitted at IR wavelengths (Sanders et al. 1989), thus the present sample of 18 QSOs simply populates one extreme of a fairly narrow "IR luminosity fraction" distribution.[3] Still, the *IRAS* detection rate

---

[1]I.e., the Infrared Astronomical Satellite
[2]The sample was selected by Surace & Sanders for the purpose of optical and near-IR imaging and is summarized in the Surace & Sanders contribution to this proceedings.
[3]See the Müller contribution to this proceedings for an updated discussion of PG QSO spectral energy distributions.

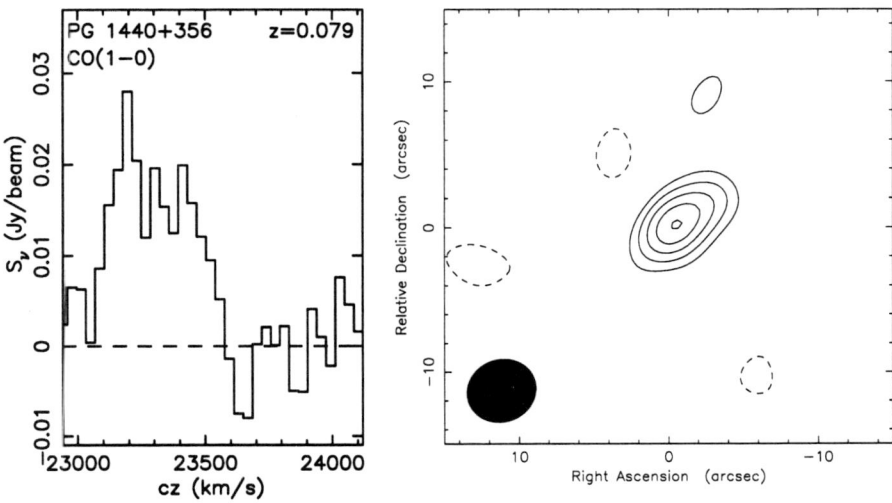

Figure 1. CO(1 → 0) spectrum and integrated intensity map of the IR-excess QSO PG 1440+356. For the map, contours are plotted as $1\sigma \times (-2.3, 2.3, 3.3, 4.3, 5.3, 6.3)$; the peak intensity is 0.016 Jy beam^{-1}, and corresponds to the position RA=14:42:07.48 dec=+35:26:22.33 (J2000.0).

of spiral galaxies and ongoing mergers versus optically-selected elliptical galaxies is very high, and if the host galaxies of QSOs are a mixture of the above galaxy types, the "IR-excess" criteria will be biased towards selecting QSOs hosts that are spiral galaxies and ongoing mergers.

The sample of 18 QSOs contains two QSOs (Mrk 1014 = PG 0157+001 and I Zw 1 = PG 0500+124) that have been observed in CO multiple times with different millimeter telescopes, and two QSOs (PG 0838+770 and 1613+658) that have been observed once with the IRAM 30m telescope (e.g. Sanders et al. 1988a, Barvainis et al. 1989, Alloin et al. 1992). Thus far in the survey, PG 0838+770 and 1613+658 have been reobserved, and 8 additional IR-excess QSOs have been searched for CO(1 → 0) emission for the first time (see Table 1). Two transits were typically done per source; in terms of sensitivity, this corresponds to a 3σ rms molecular gas mass detection limit of 1×10^9 $M_\odot$ (assuming $\alpha = 4$ $M_\odot$ [K km s^{-1} pc^2]$^{-1}$) for a $z \sim 0.1$ galaxy with a CO velocity line width of 280 km s^{-1}.

2. Molecular Gas Properties

Table 1 lists several properties of the 12 IR-excess QSOs observed in CO to date, and a CO(1 → 0) spectrum and integrated intensity map

A Molecular Gas Survey of $z < 0.2$ Infrared Excess, Optical QSOs   181

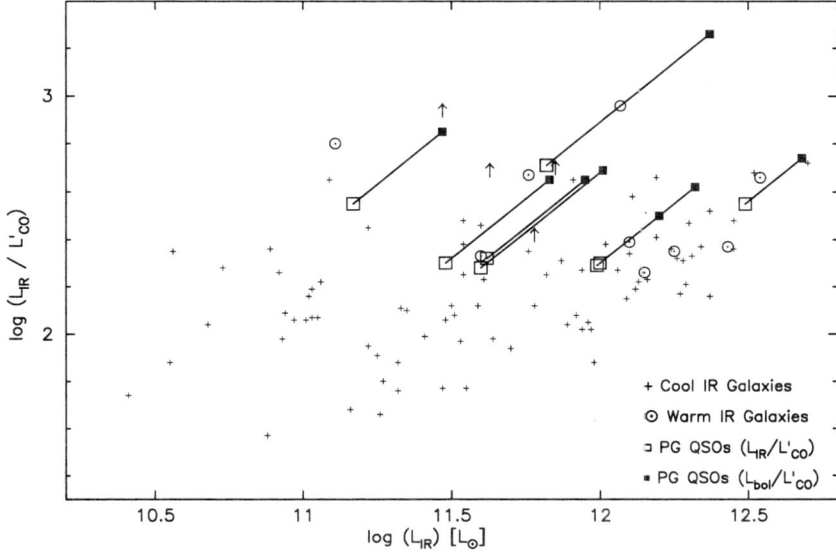

Figure 2. A plot of L_{IR}/L'_{CO} vs. L_{IR} for the low-z QSO sample, a flux-limited sample ($f_{60\mu m} > 5.24$ Jy) of IR luminous galaxies and a sample of ultraluminous IR galaxies. Arrows denote 3σ lower limits on L_{IR}/L'_{CO} of PG 1126-041, PG 1202+281, and PG 1402+261.

of one QSO (PG 1440+356) is shown in Fig. 1. Eight QSOs have been detected thus far, with molecular gas masses in the range $1.7 - 35 \times 10^9$ $M_\odot$ and velocity line widths ranging from approximately 50 to 500 km s^{-1}. The molecular gas mass range of the detected QSOs indicates the presence of host galaxies with massive cold molecular components, but the upper limits of the remaining 4 QSOs yield inconclusive results.

In order to compare the CO($1 \to 0$) and IR properties of these QSOs and IR luminous galaxies, L_{IR}/L'_{CO} versus L_{IR} for both a sample of cool and warm IR galaxies, as well as the IR-excess QSOs, are plotted in Fig. 2. The ratio L_{IR}/L'_{CO} is commonly referred to as the star formation efficiency; in starburst galaxies, it is a measure of the cumulative luminosity of massive stars responsible for heating the dust (L_{IR}) relative to the amount of fuel available for star formation (L'_{CO}). The QSOs occupy the upper portion of the L_{IR}/L'_{CO} distribution of IR luminous galaxies for a given value of L_{IR}; this is an indication that the QSOs contribute significantly to heating the dust in their host galaxies (thus increasing L_{IR}) and/or that the dust is heated by massive stars formed in the host galaxy with high efficiency (per unit molecular gas mass). The latter possibility may indicate that star formation accompanies mass accretion unto supermassive nuclear black holes, and might support recent observed correlations between black hole mass and both stellar bulge mass and velocity dispersion in nearby quiescent galaxies (e.g. Magorrian et al. 1998, Ferrarese & Merritt 2000, Gebhardt et al. 2000).

Also plotted on Fig. 2 are the positions the QSOs would occupy if the optical QSO nucleus was completely enshrouded in dust (i.e., if $L_{\rm IR} = L_{\rm bol}$, as is the case for ULIGs). If QSOs are the evolutionary products of ULIGs, and if ULIGs are powered primarily by embedded QSO nuclei, then the area of Fig. 2 covered by the lines connecting $L_{\rm bol}/L'_{\rm CO}$ to $L_{\rm IR}/L'_{\rm CO}$ for the QSOs show the possible paths that dust-enshrouded QSOs may follow as they evolve towards UV-excess QSOs.

3. The Future

This CO survey is the first attempt to detect molecular gas in a complete sample of QSOs, and it builds upon previous observations of $z < 0.17$ observations of QSOs done by other groups. There are several ways in which such a survey might be improved:

- The sample size is presently very small, and it is biased towards IR-excess QSOs. A large CO survey of a volume-limited sample of QSOs would give a more accurate assessment of the diversity in the molecular gas content of QSOs.
- These observations were done in low-resolution mode ($4''$ beam) for the simple purpose of making CO detections of these QSOs. Higher resolution ($0.5''$) CO($2 \to 1$) observations of the detected QSOs are required to determine the spatial distribution and detailed kinematics of the molecular gas. Recent CO observations by Schinnerer et al. (1998) have shown the CO in I Zw 1 (PG 0050+124) to be extended.

Acknowledgments

We are indebted to the OVRO staff and postdoctoral scholars for their assistance. ASE was supported by RF9736D and AST 0080881.

References

Alloin, D., Barvainis, R., Gordon, M. A., Antonucci, R. R. J., 1992, A&A, 265, 429
Barvainis, R., Alloin, D., Antonnuci, R., 1989, ApJ, 337, L69
Evans, A. S., Frayer, D. T., Surace, J. A., Sanders, D. B., 2001, AJ, in press
Ferrarese, L., Merritt, D., 2000, ApJ, 539, L9
Gebhardt, K. et al., 2000, ApJ, 539, L13
Magorrian, J. et al., 1998, AJ, 115, 2285
Sanders, D. B., Phinney, E. S., Neugebauer, G. et al., 1989, ApJ, 347, 29
Sanders, D. B., Scoville, N. Z., Soifer, B. T., 1988a, ApJ, 335, L1

Sanders, D.B., Soifer, B.T., Elias, J.H. et al., 1988b, ApJ, 325, 74
Schinnerer, E., Eckart, A., Tacconi, L. J., 1998, ApJ, 500, 147
Schmidt, M., Green, R. F., 1983, ApJ, 269, 352
Surace, J. A., Sanders, D. B., 2001, AJ, in preparation

Aaron Evans

Stephane Leon (left) and Jeremy Lim (right)

MOLECULAR GAS IN NEARBY POWERFUL RADIO GALAXIES

S. Leon [1], J. Lim [2], F. Combes [3] and D. Van-Trung [2]

[1] *University of Cologne, Germany*
[2] *ASIAA, Academia Sinica, Taipei, Taiwan*
[3] *Observatory of Paris, France*

Abstract We report the detection of ^{12}CO$(1 \to 0)$ and ^{12}CO$(2 \to 1)$ emission from the central region of nearby 3CR radio galaxies ($z <$ 0.03). Out of 21 galaxies, 8 have been detected in, at least, one of the two CO transitions. The total molecular gas content is below 10^9 M$_\odot$. Their individual CO emission exhibit, for 5 cases, a double-horned line profile that is characteristic of an inclined rotating disk with a central depression at the rising part of its rotation curve. The inferred disk or ring distributions of the molecular gas is consistent with the observed presence of dust disks or rings detected optically in the cores of the galaxies. We reason that if their gas originates from the mergers of two gas-rich disk galaxies, as has been invoked to explain the molecular gas in other radio galaxies, then these galaxies must have merged a long time ago (few Gyr or more) but their remnant elliptical galaxies only recently (last 10^7 years or less) become active radio galaxies. Instead, we argue that the cannibalism of gas-rich galaxies provide a simpler explanation for the origin of molecular gas in the elliptical hosts of radio galaxies (Lim et al. 2000). Given the transient nature of their observed disturbances, these galaxies probably become active in radio soon after the accretion event when sufficient molecular gas agglomerates in their nuclei.

Introduction

Bright radio sources served as the first signposts for highly energetic activity in galaxies. The nature of these galaxies, and the reason for their luminous radio activity, have since been subjects of detailed investigation. Although the vast majority resemble luminous elliptical galaxies, observations (Smith & Heckman 1989) showed that a significant fraction of the most powerful radio galaxies at low redshifts exhibit peculiar optical morphologies suggestive of close encounters or mergers between galaxies. Nevertheless, all of the powerful radio galaxies examined by Smith et al. (1990), mostly selected from the 3C catalog, lie within the fundamental plane of normal elliptical galaxies.

Some radio galaxies possess so much dust that they can be detected in the far-infrared by *IRAS*. This discovery has motivated many subsequent searches for molecular gas in radio galaxies, in all cases by selecting those with known appreciable amounts of dust. At low redshifts, eleven radio galaxies have so far been detected in ^{12}CO. The detected galaxies span nearly three orders of magnitude in radio luminosity, and comprise those exhibiting core-dominated radio sources as well as classical double-lobed FR-I (edge darkened) and FR-II (edge brightened) radio sources. All have inferred molecular-gas masses between 10^9 M$_\odot$ and 10^{11} M$_\odot$, except for the very nearby radio galaxy Centaurus A which has a molecular-gas mass of $\sim 2 \times 10^8$ M$_\odot$ (Eckart et al. 1990). In four of the five cases the CO gas is found to be concentrated in a compact (diameter of a few kpc or smaller) ring or disk around the center of the galaxy. The molecular gas in radio galaxies may therefore comprise the reservoir for fueling their central supermassive black holes.

At low redshifts, the only other type of galaxy to commonly exhibit molecular gas masses as high as $\geq 10^{10}$ M$_\odot$ are infrared-luminous galaxies. These gas and dust rich galaxies exhibit vigorous star formation thought to be triggered by galaxy-galaxy interactions, and indeed the majority of ultraluminous infrared galaxies (i.e., those with $L(60\mu m) \geq 10^{12} L_\odot$) are found to be merging systems of gas-rich disk galaxies. In the latter systems, the CO gas is found to be preferentially concentrated in a ring or disk around the nuclear regions of the merging galaxies (e.g Bryant & Scoville 1999). The observed similarities have prompted the suggestion that radio galaxies also originate from the mergers of two gas-rich disk galaxies, and in many cases comprise their still disturbed E/S0 merger products (Mirabel et al. 1989, Evans et al. 1999a,b).

Given the predisposition of the above mentioned surveys towards relatively dust-rich objects, are the radio galaxies so far detected biased towards those with unusually large amounts of molecular gas? It is not clear whether the IR-bright radio galaxies so far detected represent the most gas-rich members of a population that all possess substantial amounts of dust and gas, or extreme members of a population that possess a broad range of dust and gas masses. This issue is of importance for a proper understanding of the nature of radio galaxies, and what fuels their supermassive black holes.

To address this issue, we have initiated a deep survey of all the previously undetected radio galaxies at redshifts $z \leq 0.031$ in the revised 3C

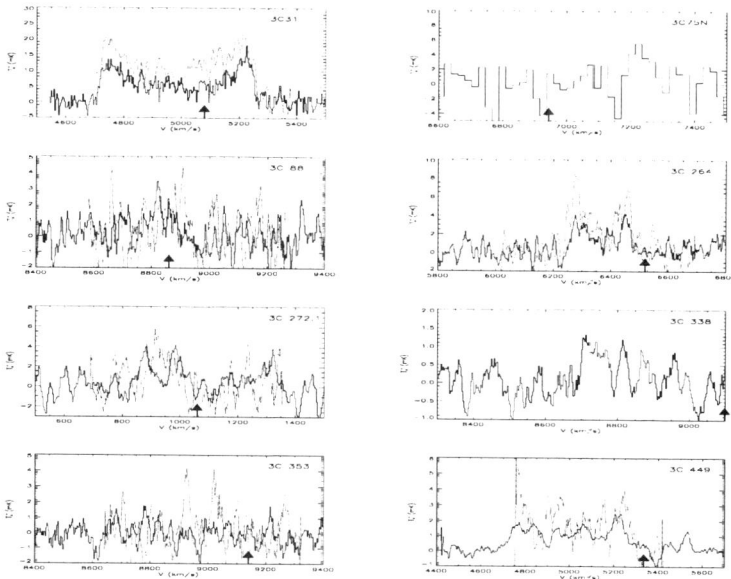

Figure 1. CO(1-0) spectra (solid line) with CO(2-1) spectra (dotted line) towards a sample of 3CR radio galaxies detected in, at least, one of the both transition lines. The vertical arrows show the optical velocity.

catalog (Spinrad et al. 1985). The objects in this catalog represent the most luminous radio galaxies at their particular redshifts in the northern hemisphere. At low redshifts the vast majority can be clearly seen to be luminous elliptical galaxies, which in most cases comprise first ranked galaxies in poor clusters of roughly ten members although a number reside in much richer environments (Zirbel 1997). Only a small minority have been detected in the infrared by *IRAS*; indeed, only three objects in this catalog have previously been detected in CO, one of which lies in the redshift range of our survey.

1. Molecular gas

Out of the 21 sources observed, 8 have been detected in ^{12}CO(1 $\to$ 0) and 3 in ^{12}CO(2 $\to$ 1) (see Table 1). The line profiles of the ^{12}CO(1 $\to$ 0) and ^{12}CO(2 $\to$ 1) emission are shown in Fig. 1. Because of the half-power beam of the IRAM-30m, we are mainly sensitive to emission from the central part of the radio-galaxies, typically the inner 10/5 kpc for the $(1 \to 0)/(2 \to 1)$ transitions. We emphasize nevertheless that the CO beam sizes are larger than the dust features seen in these radio-galaxies (Martel et al. 1999).

Table 1. Radio-galaxies detected in ^{12}CO transitions in our survey.

Name	Velocity (km.s^{-1})	Transition	I(CO) (K.km.s^{-1})	M(H$_2$) (Log(M$_\odot$))
3C31	5071	^{12}CO (1 → 0)	3.87	9.02
		^{12}CO (2 → 1)	4.80	
3C75N	6816	^{12}CO (1 → 0)	0.37	8.26
3C88	8859	^{12}CO (1 → 0)	0.18	8.19
3C264	6523	^{12}CO (1 → 0)	0.45	8.30
		^{12}CO (2 → 1)	0.93	
3C272.1	1060	^{12}CO (1 → 0)	0.35	6.61
		^{12}CO (2 → 1)	0.77	
3C338	9100	^{12}CO (1 → 0)	0.10	7.94
3C353	9150	^{12}CO (2 → 1)	0.20	7.98
3C449	5345	^{12}CO (1 → 0)	0.86	8.37

The double-horned line profile is clearly observed in 5 radio-galaxies and is characteristic of an inclined rotating disk with a central depression in CO emission at the rising part of its rotation curve (Wiklind et al. 1997). The rotational velocities observed are quite high and if they are corrected for the inclination of the dust disk they reach high values which are more typical of the nuclear molecular-gas disks or rings in IR-luminous galaxies. Using a standard CO-to-H$_2$ conversion factor we estimate the molecular content (see Table 1) within these galaxies. In the case of 3C253, where we have only a detection in ^{12}CO (2 → 1), we compute the molecular mass using the mean average line ratio.

The line ratio of ^{12}CO(2 → 1)/^{12}CO(1 → 0), with the individual line intensities measured provides information on the opacity and excitation of the gas. Line ratio less than 0.7 imply that the gas has different filling factor in the two transitions or is subthermally excited, whereas line ratio greater than unity imply that the gas is optically thin. If both transitions originitate from a region comparable in size with their individual dust features, as is the case in all the galaxies so far mapped in CO, the line ratios are between 0.6 and 0.8. These values are close to the extreme upper limits measured for many inactive elliptical galaxies (Wiklind et al. 1995), but close to the average value of 0.9 measure at the centers of both inactive and active disk galaxies (Braine & Combes 1992).

2. Discussion

The total/upper-limit molecular mass found in this 3CR sample of radio galaxies is well below the molecular mass found in a typical galaxy like the Milky Way (several 10^9 M$_\odot$) for most of the case. We detected only 4×10^6 $M_\odot$ molecular gas in the nearby radio-galaxy 3C272.1 (M84) and an upper-limit of 3.4×10^6 in 3C270 (M87). These low values contrast with the previous high-content molecular gas found mainly in IRAS-selected radio-galaxies (Mazzarrela et al. 1993, Evans et al. 1999a,b). In Fig. 2 the distribution of the molecular gas mass in the radio galaxies show the dichotomy bettween the IRAS-selected sample of radio-galaxies and our 3C sample: a clear cut appears at $10^9 M_\odot$ between both samples. Compared to a sample of radio-quiet elliptical galaxies (Knapp & Rupen 1996, Wiklind et al. 1997) the 3C sample exhibits a statistically significant lower gas mass content than the elliptical galaxy sample.

A possible link between Ultra-Luminous Infrared Galaxies (ULIRGs) and radio-galaxies has been proposed (Evans et al. 1999b). Some case exhibits properties that place them in both categories. The question is if major mergers are responsible for the AGN phase in radio galaxies. By comparison with ULIRGs, however, radio galaxies exhibit a much broader range of molecular gas masses. Moreover the host galaxies of the 3C sample appear to be very well relaxed and lie in the fundamental plane. The timescale for relaxation after a merging is about 1-2 Gy. Nevertheless the radio emission phase is *very short*, few 10^7 years. Given the accretion rate for the AGN (< 1 $M_\odot.yr^{-1}$), only few massive GMCs ($10^6 M_\odot$) would be sufficient to fuel the AGN. Indeed in the nearby radio galaxies (M84, M87) observed, the detection or upper-limit of molecular gas are lower than 10^7 $M_\odot$. In our survey only low molecular gas content has been detected. Major mergers require radio galaxies to be very old merger remnants that have only recently become active after the remnant has relaxed and much of the gas disappeared. But minor mergers present a simpler alternative to the dust and molecular gas seen in many 3CR radio galaxies. Furthermore they can explain the presence of a molecular disk and the loss of angular momentum necessary to bring the gas towards the center.

References

Braine, J., Combes, F., 1992, A&A, 264, 433
Bryant, P. M., Scoville, N. Z., 1999, AJ, 117, 2632
Eckart, A., Cameron, M., Rothermel, H., Wild, W., Zinnecker, H., Rydbeck, G., Olberg, M., Wiklind, T., 1990, ApJ, 363, 451
Evans, A. S., Sanders, D. B., Surace, J. A., Mazzarella, J. M., 1999a, ApJ, 511, 730
Evans, A. S., Kim, D. C., Mazzarella, J. M., Scoville, N. Z., Sanders, D. B., 1999b, ApJ, 521, L107
Knapp, G. R., Rupen, M. P., 1996, ApJ, 460, 271
Lim, J., Leon, S., Combes, F., Dinh-V-Trung, 2000, ApJ, 545, 93

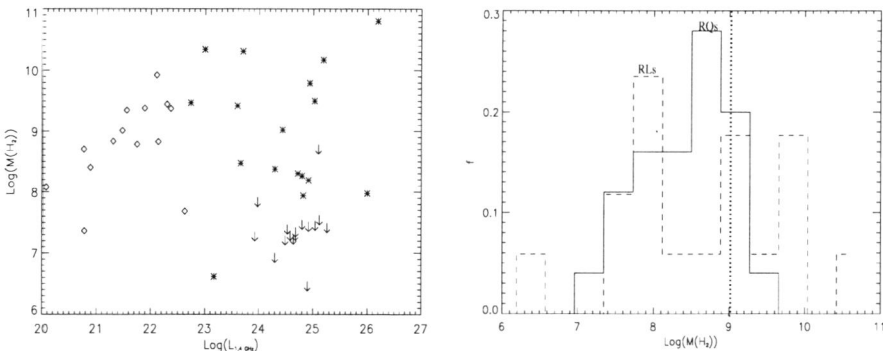

Figure 2. Left: Molecular gas content vs. radiocontinuum (1.4 GHz). Right: Histogram of the molecular content in the radio-galaxies (RGs) and elliptical radio-quiet galaxies (RQs). The vertical dash line is at the maximum molecular gas mass detected in our 3CR sample.

Martel, A. R., Baum, S. A., Sparks, W. B., Wyckoff, E., Biretta, J. A., Golombek, D., Macchetto, F. D., de Koff, S., McCarthy, P. J., & Miley, G. K. 1996, ApJs, 122, 81

Mazzarella, J. M., Graham, J. R., Sanders, D. B., Djorgovski, S., 1989, ApJ, 409, 170

Mirabel, I. F., Sanders, D. B., Kazes, I., 1987, ApJ, 340, L9

Smith, E. P., Heckman, T. M., 1989, ApJs, 69, 365

Smith, E. P., Heckman, T. M., Illingworth, G. D., 1990, ApJ, 356, 399

Spinrad, H., Marr, J., Aguilar, L., Djorgovski, S., 1985, PASP, 299, L7

Wiklind, T., Combes, F., Henkel, C., 1995, A&A, 297, 643

Wiklind, T., Combes, F., Henkel, C., Wyrowski, F., 1997, A&A, 323, 727

Zirbel, E. L., 1997, ApJ, 476, 489

HI IMAGING OF LOW-Z QSO HOST GALAXIES

Jeremy Lim
Academia Sinica Institute of Astronomy & Astrophysics
PO Box 1-87, Nankang, Taipei 11529, Taiwan
jlim@asiaa.sinica.edu.tw

Hua-ting Chuo, Shyang Wen, Wen-shuo Liao
National Taiwan University
1 Roosevelt Road, Section 4, Taipei 106, Taiwan

Paul T. P. Ho
Smithsonian Astrophysical Observatory
60 Garden Street, Cambridge, MA 02138, USA
ho@cfa.harvard.edu

Abstract   We present results from an ongoing program to image in neutral atomic-hydrogen (HI) gas the host galaxies of all QSOs at redshifts $z < 0.07$ and declinations $\delta > -35°$ compiled in the regularly updated catalog of Véron-Cetty & Véron. The vast majority of those so far observed have been detected in HI, all of which show severe disturbances in HI gas even when none can be clearly seen in optical starlight. The observed disturbances directly trace or can be naturally explained by gravitational interactions with neighboring galaxies, although in the relatively few cases where no neighboring galaxies are visible the observed disturbances may trace recent mergers. Most of the QSO host galaxies detected are richer in HI gas than our Galaxy, although not always more gas rich than their interacting companion galaxies. Where the QSO host and interacting companion galaxies are well separated and both detected in HI, the gas kinematics of the QSO host is more strongly disturbed. The severe gaseous disturbance caused by galaxy-galaxy interactions is presumably responsible for directing appreciable amounts of gas to the central supermassive black holes of these QSO host galaxies.

Keywords: Galaxies: active – Galaxies: interactions – Galaxies: Seyferts – Quasars: general – Radio lines: galaxies

Introduction

Much of our knowledge on QSO host galaxies is based on imaging of their starlight in the optical or near-infrared. More comprehensive observations of those at low redshifts occasionally reveal interacting systems of galaxies, and more frequently asymmetric morphologies or projected neighboring galaxies cited as circumstantial evidence for gravitational interactions between galaxies (e.g., Hutchings & Campbell 1983, Hutchings et al. 1984). Observations with the Hubble Space Telescope of a small subset of the same objects but at higher clarity reveal that only a small fraction actually comprise visibly interacting systems of galaxies, whereas the vast majority appear to be otherwise normal galaxies even though many have projected neighboring galaxies (Bahcall et al. 1997, Boyce et al. 1998).

To study the gas content, distribution, and kinematics in QSO host galaxies, we have begun a program to image a volume-limited sample in the 21-cm emission line of neutral atomic-hydrogen (HI) gas with the Very Large Array (VLA). To avoid any HI selection biases, we have selected all the QSOs in the optical catalog of Véron-Cetty & Véron (1998) that lie at redshifts $z < 0.07$ and which have declinations $\delta > -35°$ thus visible from the VLA. Only two radio-loud QSOs (both 3C objects with extended radio jets and lobes) lie in this redshift and declination range, reflecting the small fraction of such objects among optically-selected QSOs; the detection of any HI in emission from these objects is technically challenging, and in this contribution we will concentrate solely on the radio-quiet QSOs. Because HI is a particularly sensitive tracer of gravitational interactions between galaxies (e.g., Yun et al. 1994), we hoped that our observations also may directly reveal any such interactions in the QSO host galaxies studied (Lim & Ho 1999).

1. Observations and Results

Of the ten (radio-quiet) QSOs so far observed, we have detected the host galaxies of eight in HI. In one of the two remaining cases where the QSO host galaxy was not detected, we detected an interacting companion galaxy in HI. Optical and HI images centered on six of the QSOs in our sample are shown in Figs. 1–6, arranged in order of decreasing projected physical separation between the QSO and any visible neighboring galaxies.

Wasilewski 26 ($z = 0.063$) — The optical image from the Digitized Sky Survey (DSS) shows few if any details in this QSO host galaxy, except perhaps for a slight extension to the south-east. Bothun et al. (1989) presented an integrated one-dimensional radial light profile of the QSO host that is characteristic of a disk galaxy, but did not note any other redeeming features. In dramatic contrast, our radio image shows that the HI gas has a complex spatial-kinematic distribution extending far beyond the optical confines of the host galaxy. This HI disturbance likely traces tidal interactions with a companion disk galaxy to the south-east with approximately the same systemic velocity as the QSO host galaxy, and which also shows tidal disturbances in HI.

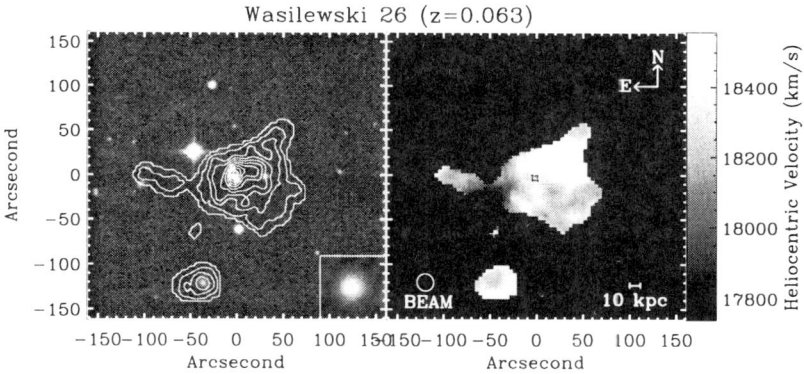

Figure 1. Left panel — Contours of the total HI intensity superposed on a grayscale optical image centered on the QSO Wasilewski 26. The immediate optical environment of the QSO is shown magnified in the lower right corner. Right panel — Grayscale-coded image of the HI intensity-weighted mean-velocity. The angular resolution is shown in the lower left corner, and the angular scale in the lower right corner for $H_o = 100$ km s^{-1} Mpc^{-1} and $q_o = 0.5$.

IRAS 17596+4221 ($z = 0.045$) — The DSS optical image shows an asymmetric host galaxy. Our HI image reveals that the QSO host is interacting with a neighboring disk galaxy: a tidal bridge can be seen connecting the two galaxies, and the companion galaxy also exhibits a tidal bulge or tail on the side opposite the QSO. The bulk of the gas in the companion galaxy shows a relatively smooth velocity gradient along its optical major axis, tracing the rotation of a HI-gas disk. By contrast, the HI kinematics of the QSO host galaxy is severely disturbed, showing no apparent systematic motion. Both the QSO host and its interacting companion galaxy have essentially the same systemic velocities, again indicating a low-velocity encounter.

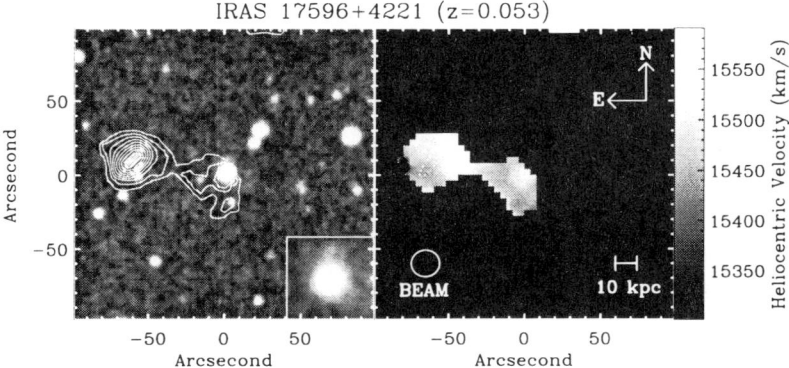

Figure 2.   Same caption as in Fig. 1, but for IRAS 17596+4221.

4U 0241+61 ($z = 0.045$) — This is the most luminous QSO in our sample, and in the DSS optical image can be seen to lie in a crowded star field at low galactic latitutes. We did not detect the QSO host in HI, but instead a neighboring inclined disk galaxy that shows a tidal bulge on the side facing the QSO.

Figure 3.   Same caption as in Fig. 1, but for 4U 0241+61.

Ton 951 ($z = 0.064$) — The DSS optical image shows an asymmetric host galaxy, which exhibits a relatively prominent extension towards a neighboring galaxy to the south-west. A clearer near-infrared image can be found in McLeod & Rieke (1994), showing also the stellar disk of the host galaxy. Our HI image reveals that the gas around the QSO host galaxy is highly disturbed, with a prominent tidal bulge or tail on the side opposite the companion galaxy. Perhaps surprisingly, the gas kinematics appear to be well ordered on a global scale, exhibiting a relatively smooth velocity gradient indicative of a large HI-gas disk aligned with the stellar disk.

Figure 4. Same caption as in Fig. 1, but for Ton 951.

I Zwicky 1 ($z = 0.061$) — The optical image from the Hubble Space Telescope shows two spiral-like arms, which also can be seen in the DSS image. One arm appears to terminate at or near a companion galaxy to the east with the same redshift as the QSO (Stockton 1981), whereas the other arm appears to terminate near a foreground star to the north. Our HI image shows that the gas is extended and has an asymmetric distribution about the QSO host galaxy, caused presumably by tidal interactions with the companion galaxy. Like in Ton 951, however, the gas kinematics appear to be well ordered on a global scale, exhibiting a relatively smooth velocity gradient tracing an extended HI-gas disk aligned with a molecular-gas disk (Schinnerer et al. 1998).

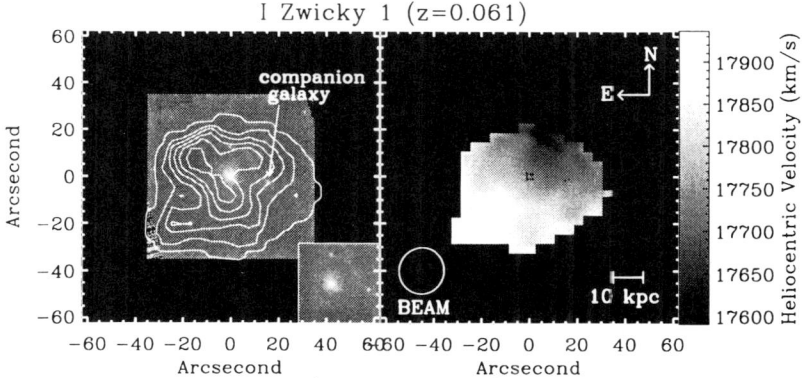

Figure 5. Same caption as in Fig. 1, but for I Zwicky 1.

Markarian 304 ($z = 0.066$) — The DSS optical image shows an apparently symmetrical host galaxy; deeper images in the optical by

Hutchings & Neff (1992) and the near-infrared by McLeod & Rieke (1994) show a smooth and circular galaxy. No projected neighboring galaxies have been reported. In dramatic contrast with its optical appearance, our HI image reveals that the gas around this object is highly disturbed. This is our best, and one of perhaps only two candidates, for a relatively recent merger.

Figure 6. Same caption as in Fig. 1, but for Markarian 304.

2. Discussion

Our HI images reveal dramatic evidence for severe gaseous disturbances in all the QSO host galaxies detected, even when no such disturbances are clearly if at all visible in optical starlight. In nearly all cases, the observed disturbances directly trace or can be naturally explained by gravitational interactions with visible neighboring galaxies. In one or two cases where no neighboring galaxies are visible, the observed disturbances probably trace recent mergers.

The QSO host galaxies detected are very rich in gas, with HI gas masses in the range $2-24 \times 10^9$ $M_\odot$ (for $H_o = 75$ km s^{-1} Mpc^{-1}); i.e., comparable or up to an order of magnitude larger than the total HI-gas mass of our Galaxy. Interestingly, however, the QSO does not always reside in the more gas rich of the two interacting galaxies (e.g., IRAS 17596+4221 and 4U 0241+61). In the two cases where the QSO host and interacting companion galaxies are well separated and both detected in HI, the gas kinematics of the QSO host is more severely disturbed (Wasilewski 26 and IRAS 17596+4221). This may provide an important clue to why a QSO, if present, usually appears in only one of the two galaxies in an interacting system (see also Liao, Lim & Ho, this volume). The severe gaseous disturbance caused by galaxy-galaxy interactions is presumably responsible for directing an appreciable amount of gas to the central supermassive black holes of the QSO host galaxies observed.

Acknowledgments

This project is supported in part by the National Science Council (NSC) of Taiwan as a grant to J. Lim, who thanks the NSC for financial support to attend this workshop.

References

Bahcall, J. N., Kirhakos, S., Saxe, D. H., Schneider, D. P., 1997, ApJ, 479, 642
Bothun, G. D., Halpern, J. P., Lonsdale, C. J., Impey, C., Scmitz, M., 1989, ApJSS, 70, 271
Boyce, P. J., Disney, M. J., Blades, J. C., Boksenberg, A., Crane, P., Deharveng, J. M., Macchetto, F. D., Mackay, C. D., Sparks, W. B., 1998, MNRAS, 298, 121
Hutchings, J. B., Campbell, B., 1983, Nature, 303, 584
Hutchings, J. B., Crampton, D., Campbell, B., 1984, ApJ, 280, 41
Hutchings, J. B., Neff, S. G., 1992, AJ, 104, 1
Lim, J., Ho, P. T. P., 1999, ApJ, 510, L7
McLeod, K. K., Rieke, G. H., 1994, ApJ, 420, 58
Schinnerer, E., Eckart, A., Tacconi, L. J., 1998, ApJ, 500, 147
Stockton, A., 1981, ApJ, 257, 33
Véron-Cetty, M. P., Véron, P. A., 1998, A Catalogue of Quasars and Active Galactic Nuclei (ESO Sci. Rept.) (8th ed.; Garching:ESO)
Yun, M. S., Ho, P. T. P., Lo, K. Y., 1996, Nature, 372, 530

Sven Müller

Jason Surace

DUST IN QUASARS AND RADIOGALAXIES AS SEEN BY ISO*

S. A. H. Müller, R. S. Chini
Astronomisches Institut der Ruhr–Universität Bochum (AIRUB), Universitätsstraße 150, 44780 Bochum, Germany
smueller@astro.ruhr-uni-bochum.de

M. Haas, K. Meisenheimer, U. Klaas, D. Lemke
MPI für Astronomie (MPIA), Königstuhl 17, 69117 Heidelberg, Germany

E. Kreysa
MPI für Radioastronomie (MPIfR), Auf dem Hügel 69, 53121 Bonn, Germany

Abstract With ISO and IRAM we have obtained sensitive infrared and millimetre photometry for a random sample of 17 PG quasars and a sample of 10 pairs of FRII radio galaxies and quasars selected from the 3CR catalogue such that they match in redshift and lobe luminosity. From the SEDs, which allow a clear distinction of the thermal bump above the synchrotron emission, we derive the following major results: 1. quasars (and radio galaxies) are dusty with thermal infrared power ranging luminous to hyperluminous values, 2. for most objects starbursts are required to produce the FIR-emission, while the AGN might be responsible for the MIR-emission 3. the MIR- and FIR-luminosities are correlated, providing evidence for a close AGN-starburst connection. 4. the dust properties of matched pairs of quasars and radio galaxies are very similar, thus providing the first successful evidence in the infrared that the aspect-angle unification plays a substantial role in the relation between quasars and radio galaxies.

Keywords: Galaxies: fundamental parameters – nuclei – photometry – Quasars: general – unification – Infrared: galaxies

---

*Based on observations with ISO, an ESA project with instruments funded by ESA Member States (especially the PI countries: France, Germany, the Netherlands and the United Kingdom) and with the participation of ISAS and NASA.

1. PG Quasars

1.1. Spectral Energy Distributions (SEDs)

The most remarkable feature of the SEDs (Fig. 1) is the steady flux increase from the NIR over the MIR peaking in the FIR regime at about 60 μm, followed by the begin of the Rayleigh-Jeans tail which steeply falls off towards the 1.3 mm data points. It demonstrates that the radiation is not due to synchrotron emission but rather originates from thermal dust emission as was previously shown for some quasars by Chini et al. (1989). The SEDs point to considerable dust emission and look like those of warm luminous or ultraluminous infrared galaxies (ULIRGs).

1.2. Quasars are dusty

The spectra shown in Fig. 1 provide an impressive confirmation for the fact that the SED of an optically selected quasar contains a strong FIR bump due to thermal emission by dust. The thermal IR luminosities span three orders of magnitude, from $10^{11} L_\odot$ to $10^{14} L_\odot$. The random sample achieves a remarkably high mid- and far-infrared detection rate of 14/17. Thus we infer a low probability that there exist dust-free naked quasars. The amount of dust lies in the order of $10^{7\pm1} M_\odot$ which is typical for dust rich spirals as well as ULIRGs. Since the dust must have been produced by stars, this supports the hypothesis that the quasars are accompanied or proceeded by a starburst. Also, optical to NIR imaging reveals partly interactions with companions, providing further arguments in favor of starbursts.

1.3. Evidence for Starbursts in Addition to AGN

The SEDs show a variety of shapes suggesting the influence of different heating mechanisms. Since current models fail to explain the high FIR emission by an AGN heating only, starbursts have to be invoked, too. We identify the following 3 typical SED features:

1) a power law like flux density increase from the NIR to FIR, making independent heating mechanisms very unlikely. These objects (e.g. PG 0050+124, PG 1613+658 and PG 1634+706) are mainly AGN-heated with a special geometry of the dust like a clumpy torus. In the cooler regions starbursts become significant.

2) a strong dominance of the MIR. The SEDs seem to have also a power law like flux increase, but only from the NIR to the MIR.

Dust in Quasars seen by ISO 201

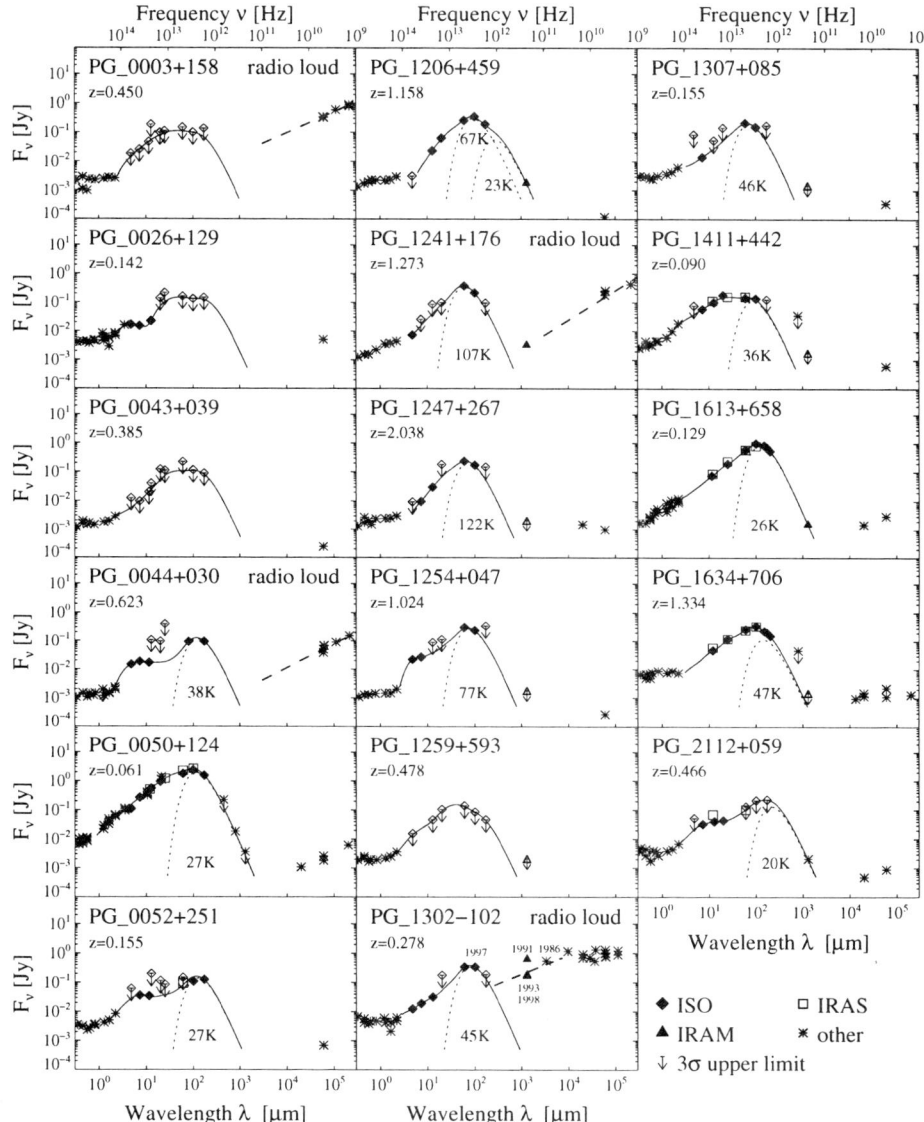

Figure 1. SEDs of the PG-quasars. The wavelength ranges are as observed and not corrected to the rest frame of the objects. Several modified blackbodies with emissivity $\propto \lambda^{-2}$ are eyeball fitted to the data (solid lines) and only the coolest components are plotted individually (dotted lines). The temperatures listed are corrected for redshift. The long-dashed lines indicate extrapolated synchrotron spectra.

These sources (e.g. PG 1411+442, PG 2112+059 and PG 1247+267) are only heated by a central AGN.

3) two maxima with a jump in between, one peak lies in the NIR/MIR and another dominant one in the FIR. The spectra of these objects (e.g. PG 0044+030 and PG 0052+251) are produced by a combination of AGN- and starburst heating.

1.4. Luminosity Correlation: Starburst - AGN Connection

The good correlations of MIR and FIR luminosities (Fig. 2) indicate a coupling between the different heating sources – AGN for the MIR and starbursts for the FIR – and can qualitatively be interpreted as proposed by Rowan-Robinson (1995): the more massive and powerful the starburst, the more likely it is for the matter to have an unstable angular momentum and drop into the black hole. The AGN then strongly heats the central dust component.

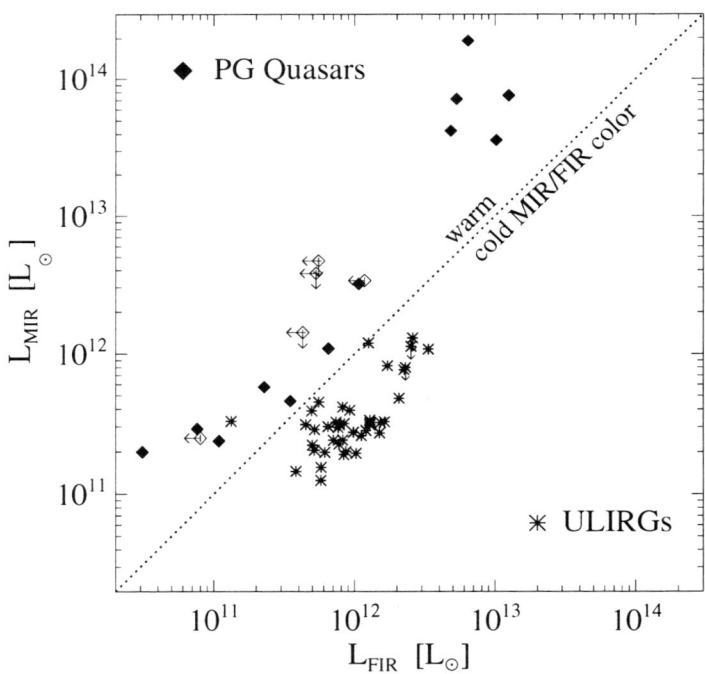

Figure 2. Restframe MIR (3-40 μm) versus FIR (40-1000 μm) luminosity of the PG-quasars (diamonds) and the ULIRGs (asterisks) from Klaas et al. (2001).

1.5. Relation to ULIRGs: Edge-on PG-Quasars

Concerning evolution, consensus is growing that the precursors of quasars have to be searched for among ULIRGs (Sanders et al. 1988). The high detection rate of PG-quasars by ISO, however, adds to the evolutionary scheme a geometric one: PG-quasars have blue optical colors and low extinction, but are simultaneously dust rich. Thus the dust might be distributed in a non-spherical disk/torus geometry which is seen nearly face-on. Then from the dust mass a high edge-on optical depth is derived even in the MIR. Hence, any candidate for an edge-on PG-quasar will appear like an ULIRG. But not every ULIRG needs to evolve into a quasar or to be a dust-shielded quasar. This subject is further discussed by Haas et al. (1998, 2000) and Klaas et al. (2001).

2. 3CR Radio Galaxies and Radio Loud Quasars

In order to test the unified scheme between radio galaxies and radio loud quasars (Barthel 1989) via their presumably isotropic dust emission, IR to millimeter SEDs have been obtained by ISO and IRAM for a sample of 10 galaxy-quasar pairs with similar redshifts and 178 MHz radio fluxes, which are independent of aspect angle (Müller et al. 2001, Meisenheimer et al. 2001). The SEDs (Fig. 3) reveal similar thermal dust emissions for galaxies and quasars showing up as clear bumps above the synchrotron radiation. The detection statistics galaxies/quasars = 6/7 is well balanced and also the luminosities, which are in the same order as for the PG-quasars, fit well. Thus the observations provide evidence in favour of the galaxy-quasar unification.

References

Barthel P., 1989, ApJ, 336, 606
Chini R., Kreysa E., Biermann P.L., 1989, A&A, 219, 87
Haas M., Chini R., Meisenheimer K. et al., 1998, ApJ, 503, L109)
Haas M., Müller S.A.H., Chini R. et al., 2000, A&A, 354, 453
Klaas U., Haas M., Müller S.A.H., et al., 2001, A&A, submitted
Meisenheimer K., Haas M., Müller S.A.H., et al., 2001, A&A, in press
Müller S.A.H., Haas M., Meisenheimer K., et al., FIRSED2000, Elsevier, in press
Rowan-Robinson M., 1995, MNRAS, 272, 737
Sanders D.B., Soifer B.T., Elias J.H. et al., 1988, ApJ, 325, 74

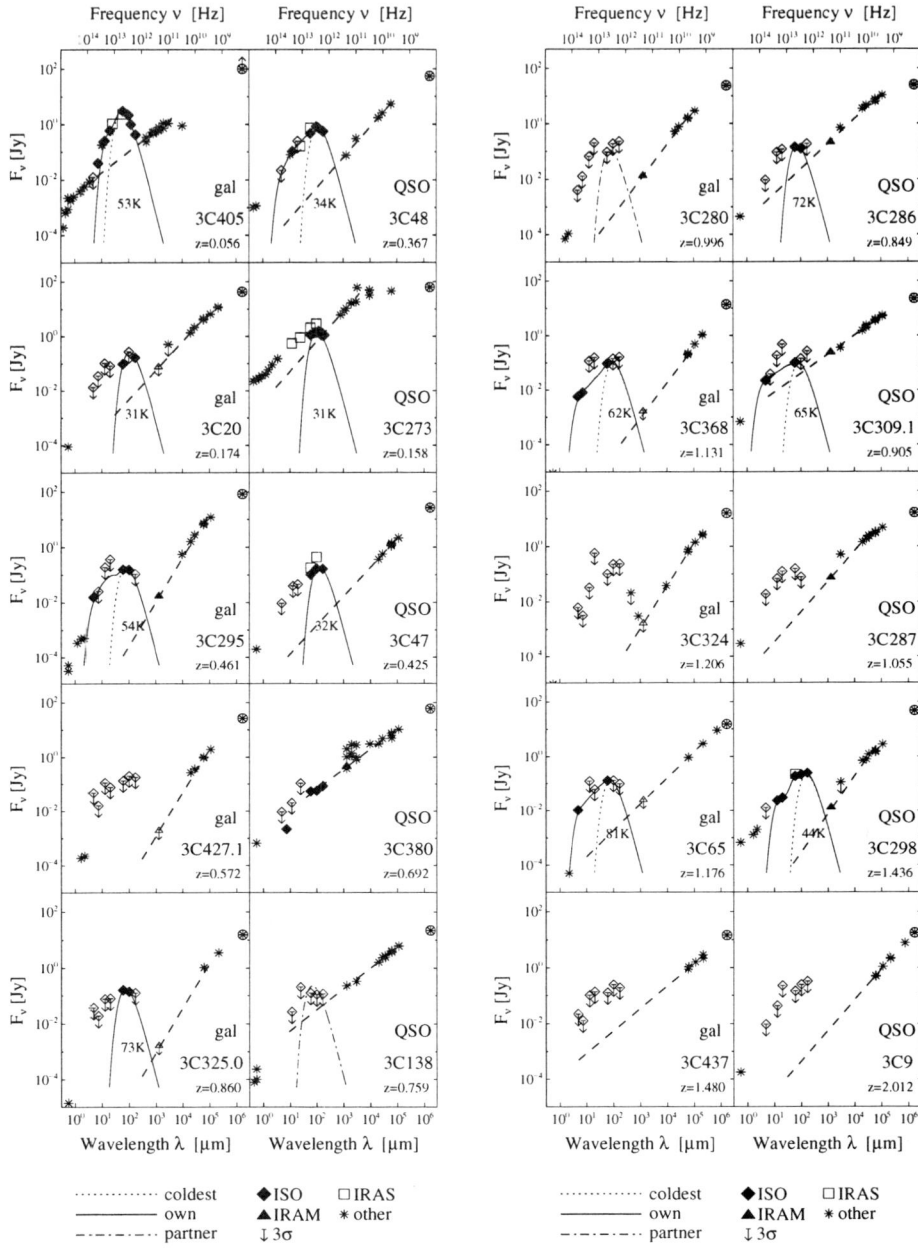

Figure 3. Observed SEDs of 3CR radio galaxies and quasars. The wavelength ranges are as observed and not corrected to the rest frame of the objects. The left column contains the radio galaxies, the right column the radio loud quasars. The long-dashed lines indicate the extrapolated synchrotron spectra contribution. Several modified blackbodies with emissivity $\propto \lambda^{-2}$ have been fitted to the data (drawn lines) and only the coolest components are plotted individually (dotted lines). The temperature values listed are in the restframe of the object.

IMAGING OF A COMPLETE SAMPLE OF IR-EXCESS PG QSOS

Jason A. Surace
SIRTF Science Center/Caltech

D.B. Sanders
Institute for Astronomy, University of Hawaii

1. Background

Sanders et al. (1988) proposed an evolutionary connection between Ultraluminous Infrared Galaxies (ULIGs) and optically-selected QSOs. In this scenario, mergers of dust and gas-rich galaxies provide the fuel to create and/or fuel an AGN and circumnuclear starburst. This dust completely enshrouds the AGN, and subsequent reradiation of the short-wavelength AGN emission produces the high far-IR luminositiy that defines ULIGs as a class. Dust clearing by superwinds eventually begins to unveil the central AGN, which is then perceived as a QSO. Surace et al. (1998, 2000) carried out a comprehensive program of multi-wavelength high spatial resolution observations using *HST* and ground-based tip/tilt in order to characterize the morphology and colors of ULIGs. Particular emphasis was given to ULIGs with "warm" mid-IR colors, whose spectral energy distributions (SEDs) and emission line features suggested that they were the most evolved ULIGs in the process of becoming QSOs. These studies showed that ULIGs were the merger of two L^* galaxies, and that they had compact central sources whose luminosity and colors are similar to reddened AGN. They also possess "knots" of star formation distributed in the circumnuclear regions and along the merger-generated tidal debris.

2. Infrared-Excess QSOs

During the evolutionary process, the SEDs of these mergers must evolve from the ULIG far-IR dominated spectra towards those with a strong relative optical/UV component. Therefore, QSOs with strong contributions to their total luminosity from far-infrared emission should be less evolved and more similar to the ULIGs than QSOs in general.

A complete sample of 18 QSOs was selected from the Palomar-Green Bright Quasar Survey (Schmidt & Green 1983), which at the time had the most complete published infrared data. They lie at the same distances ($z<0.16$) as previous samples of ULIGs examined by Surace et al., thus alleviating resolution dependencies in interpreting the data. ULIGs are defined to have the same minimum bolometric luminosity ($10^{12} L_\odot$) as QSOs, as defined by Schmidt. Far-infrared data from IRAS was used to evaluate the contribution to the bolometric luminosity of the "big blue bump" (0.1—1 μm; L_{BBB}) relative to that emitted in the far-IR at 8—1000 μm (L_{IR}). All the QSOs with far-IR excesses (L_{IR}/L_{BBB}) as great as the least far-IR active ULIG (3C 273; $L_{IR}/L_{BBB}=0.46$) were selected. A campaign of high resolution observations at B, I, H, and K' using a fast tip/tilt guider on the UH 2.2m telescope was carried out on 17 out of the complete sample of 18 objects. Typical spatial resolutions at H and K' were 0.25" while those at B and I were 0.7". The data were either photometric or were tied to that of Neugebauer et al. (1987) so that colors could be derived.

3. Results

1) All of the IR-excess QSOs have readily detectable hosts. Their morphologies are varied, but at least 50% are spiral-type systems, as evidenced by the presence of spiral arms, and just over half of these are barred. This is similar to results for Seyfert galaxies and radio quiet quasars (McLeod & Rieke 1994, 1995, Taylor et al. 1996). The variance between this result and other recent imaging studies (McLure et al. 1999), which have found a predominance of elliptical-like hosts, is probably due to selection effects. Our infrared criterion selects galaxies with significant reservoirs of gas and dust (Evans et al., this vol.) and is known from *IRAS* to strongly select spiral galaxies. Furthermore, the sample does not match the luminosity *distribution* of the Surace et al. ULIGs. Instead, the demand for low redshift systems selects the least-luminous QSOs, as opposed to the higher-luminosity samples used by others.

2) The mean H-band luminosity of the (point-source subtracted) host galaxies is 2.2 L*. The distribution of host galaxy luminosities for "warm" ULIGs and QSOs is nearly identical.

3) 22% are in major merger systems. An additional 25% have barred morphologies that are consistent with late-stage minor merger morphologies. The total number of host galaxies where mergers may be impli-

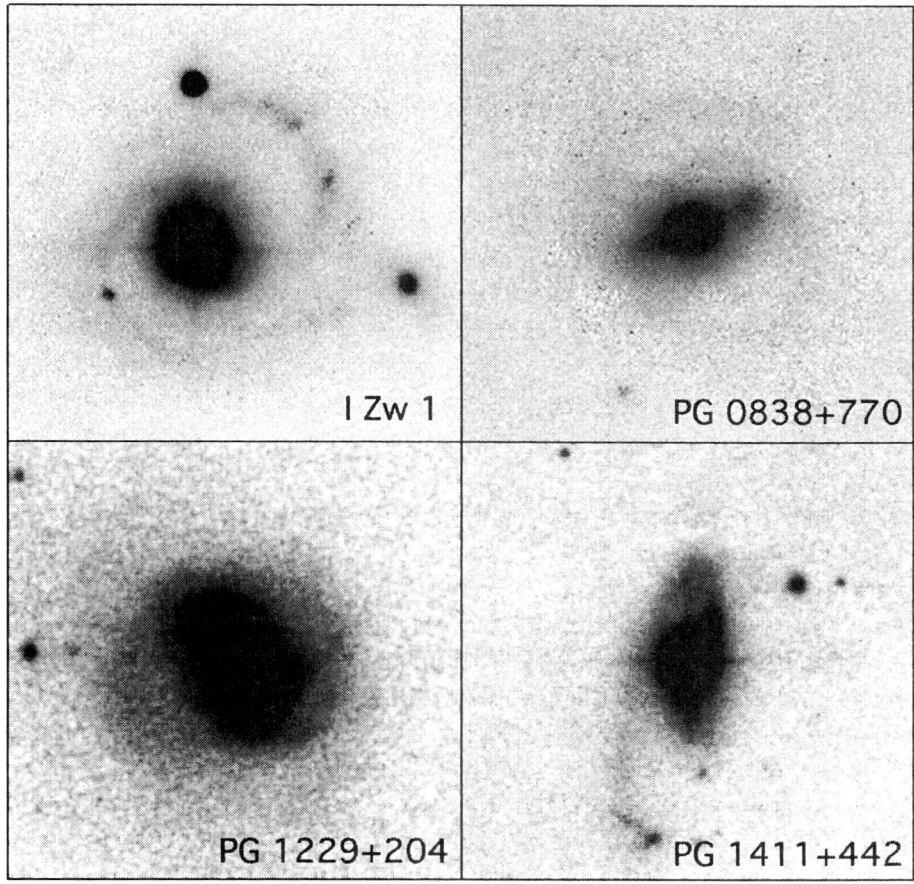

Figure 1. Optical images showing the diversity in host galaxy morphology of Infrared-Excess PG QSOs, including a spiral galaxy, two barred spirals, and a major merger with an 80 kpc tidal tail.

cated is thus between 25—50%. The remaining 25% with indeterminate or elliptical-like features may also be merger remnants.

4) Several have small-scale (<1 kpc) structure similar to that seen in ULIGs; these knots are typically very red and are similar to dust-enshrouded star formation with an age of ≤ 100 Myrs. The QSO nuclei also have near-IR excesses (beyond dust-reddening) which may be the result of small amounts of hot thermal dust emission.

5) These results imply that while some 25% of IR-excess QSOs have morphologies consistent with advanced mergers, another 25% (apparently unperturbed spiral galaxies) cannot have had their infrared activity

triggered via the major mergers implicated in the formation of ULIGs. A more thorough understanding of these statistics will necessitate similar obervations of the complementary sample of non-IR-excess QSOs. A larger complete sample of QSOs with greater counting statistics would also be useful.

References

McLeod, K., Rieke, G., 1994, ApJ, 420, 58
McLeod, K., Rieke, G., 1995, ApJ, 441, 96
McLure, R.J., Kukula, M.J., Dunlop, J.S. et al., 1999, MNRAS, 308, 377
Neugebauer, G., Green, R.F., Matthews, K. et al., 1987, ApJS, 63, 615
Sanders, D.B., Soifer, B.T., Elias, J.H. et al., 1988, ApJ, 325, 74
Schmidt, M., Green, R.F., 1983, 269, 352
Surace, J.A., Sanders, D.B., Vacca, W.D. et al., 1998, ApJ, 492, 116
Surace, J.A., Sanders, D.B., Evans, A.S., 2000, ApJ, 529, 170
Taylor, G.L., Dunlop, J.S., Hughes, D.H. et al., 1996, MNRAS, 283, 930

INTERACTION PATTERNS IN A COMPLETE SAMPLE OF COMPACT GROUPS

Valentina Zitelli[1], Paola Focardi[2], Birgit Kelm[2] and Carla Boschetti[3]

[1] *Osservatorio Astronomico di Bologna, Via Ranzani 1, 40127 Bologna - I*
zitelli@bo.astro.it
[2] *Dep. of Astronomy, Univ. of Bologna, Via Ranzani 1, 40127 Bologna - I*
[3] *Dep. of Astronomy, Univ. of Padova, Vicolo dell'Osservatorio 2, 35122 Padova -I*

Abstract Applying an automatic code to the 3D UZC galaxy catalogue, we have extracted a complete sample of about 400 physical compact groups having cz≤10000 km/s and at least 3 member galaxies within a projected distance radius of 200 h^{-1} kpc. We find interaction pattern to be more common among triplets than among higher multiplicity compact groups, which nicely matches predictions based on the lower velocity dispersion and higher overdensity displayed by triplets. Whilst morphological disturbance are rather common among CG members, we retrieve a relative low fraction of AGNs. Interestingly only Seyfert 2 are retrieved, thus confirming once more that locally dense environments discriminate against Seyfert 1.

Keywords: Compact Groups, Environment, Morphological parameters, Seyfert

Small galaxy systems, such as pairs and compact groups (CGs), are ideal sites in which to investigate the effect of interaction on galaxy formation and evolution. They show, in fact, both high compactness and low velocity dispersion, conditions which would lead to strong interaction and merging within few crossing times. Theoretical models reconcile the rather high number of detected CGs, which is in excess of predictions based on merging rate, by simply varying the fraction of dark matter distribution within the CGs (Mamon 1987, Athanassoula et al. 1997, Zabludoff & Mulchaey 1998) or by assuming a continous formation of galaxies from the surrounding loose groups (Governato et al. 1996, Diaferio et al. 1994).

Up to now existing CG samples (Hickson 1993, Prandoni et al. 1994, Coziol et al. 2000, and Barton et al. 1996) are essentially based on Hickson selection criteria and may thus reflect some of the systematic biases intrinsic to the identification criteria.

Application of an automatic neighbour search algorithm (Kelm et al. 2000) to the UZC galaxy catalogue (Falco et al. 1999), which includes about 14000 northern galaxies having $m_{Z_w} \leq 15.5$ and cz between 1000 and 10000 km/s, has provided a complete sample of 427 CGs including at least 3 galaxies within a projected distance of r=200h^{-1} kpc, and velocity distribution $\Delta V=\pm 1000$ Km/s. Surrounding large scale galaxy density has been computed within a radius of R=1h^{-1}Mpc. A detailed description of the CG sample is given in Focardi & Kelm (in preparation).

The CG sample contains 283 triplets (Ts), 77 quartets, 22 quintets and 45 multiplets with more than 6 galaxy members. Detailed information from DSS images and recent available literature has allowed to confirm the real richness of the CG, to check and search for faint members and to classify the interaction/disturb level for each component of CG. We have presently morphological information for all the compact groups with more than 3 members (Ms) and half part of Ts sample. The analysis we report is thus limited to 100 Ts and 144 Ms.

Fig. 1 shows the distribution of the overdensity of each structure ($\delta\rho / \rho$) versus ΔV_{max}. The overdensity computes the numerical excess of galaxies within the defined radius on small scale with respect to the number of galaxies within the defined radius on large scale. ΔV_{max} is the maximum difference in velocity between a member and the CG center. Following Focardi & Kelm the sample has been splitted in 4 subclasses to mimic homogeneous samples, covering different luminosity range.

Fig. 1 clearly shows that Ts (triangles) display higher density contrast and, on average, lower velocity dispersion than Ms (square). This evidence strongly supports the hypothesis of triplets being real structures well detached from the underlying background. Interaction (filled symbols) is detected in about 50 % of Ts, again supporting the physical status of these structures, and for about 25 % of Ms.

Seyfert incidence within CGs has also been determined by cross correlation of CGs galaxies and Véron-Cetty & Véron catalogue (issue 2000) and by inspection of the more recent available literature. No significant difference is retrived in the occurrence of Seyfert between Ts and higher multiplicity CGs ($\approx$ 2%). Moreover no Sy1 are found among Triplets and higher multiplicity members. The dominance of Sy2 that inhabit

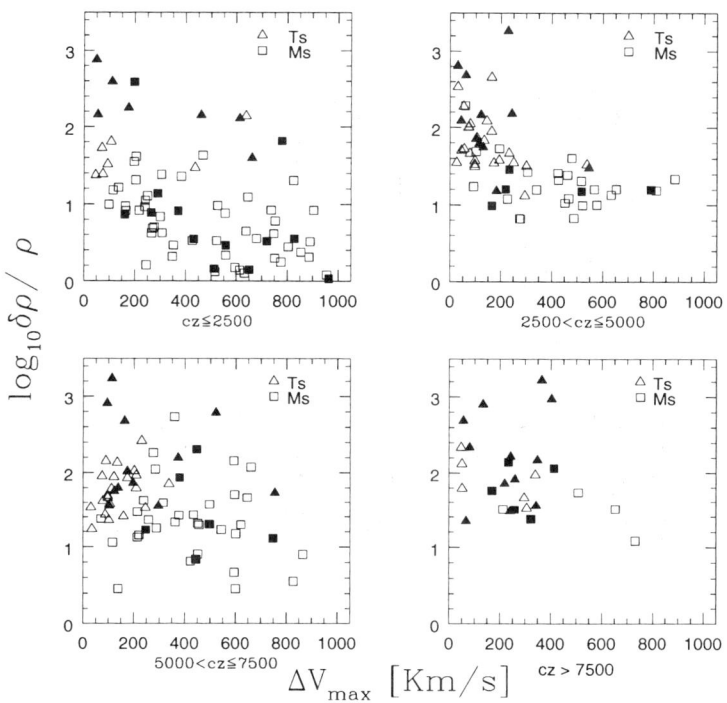

Figure 1. Overdensity distribution for triplets (triangles) and higher multiplicity class (square) with respect to the maximum CG velocity dispersion. Filled simbols display CG whose members show disturbed morphology or interaction patterns.

groups more often than Sy1 is in agreement with what previously found for Hickson CGs (Kelm et al. 1998 and Laurikainen et al. 1995).

References

Athanassoula, E. et al., 1997, MNRAS, 286,825
Barton, E. et al., 1996, AJ, 112, 871
Coziol, R. et al., 2000, AJ, 120, 47
Diaferio et al., 1994, AJ, 107,868
Falco, E.E. et al., 1999, PASP, 111, 438
Governato, F., Tozzi, P., Cavaliere, A., 1996, ApJ, 458,18
Hickson, P., 1993, ApJL&C, 29,1
Kelm, B., Focardi, P., Palumbo, G.G.C., 1998, A&A, 335,912
Kelm, B., Focardi,P., Palumbo,G.G.C., 2000,in *Small Galaxy Groups ASP Conf. Series* Valtonen and Flynn eds., 209, 90
Mamon,G.A., 1987, ApJ, 321,622
Prandoni, I.,Iovino, A., MacGillivray, H.T., 1994, AJ, 107, 1235
Zabludoff, A.I., Mulchaey, J.S., 1998, ApJ, 496, 39

Dario Trèvese (left), Valentina Zitelli (center) and Birgit Kelm (right)

José M. Vilchez and José Antonio de Diego (backwards)

IV

GALAXIES HOSTING LOWER LEVEL AGN ACTIVITY

From left to right, Birgit Kelm, David Schade, Sebastian Rabien and Wolfgang Duschl

From left to right, Antxon Alberdi, Lucas Lara and Heino Falcke

AGN HOST GALAXIES: HST * AT $Z \sim 0.1$ AND GEMINI[†] ADAPTIVE OPTICS AT $Z \sim 2$

David Schade
National Research Council Canada, Dominion Astrophysical Observatory

Scott Croom, Brian Boyle
Anglo-Australian Observatory

Michael Letawsky
Subaru Telescope

Tom Shanks
University of Durham

Lance Miller, Nicola Loaring
Oxford University

Robert Smith
Liverpool John Moores University

Abstract

A study of 76 host galaxies of Active Galactic Nuclei (AGN) in the local universe ($z \sim 0.1$) carried out with Hubble Space Telescope shows that these hosts are remarkably similar to typical field galaxies. They have normal distributions of disk and bulge size and surface brightness

---

*Based on observations with the NASA/ESA Hubble Space Telescope, obtained at the Space Telescope Science Institute, which is operated by the Association of Universities for Research in Astronomy, Inc. under NASA contract No. NAS5-26555

[†]Based on observations obtained at the Gemini Observatory, which is operated by the Association of Universities for Research in Astronomy, Inc., under a cooperative agreement with the NSF on behalf of the Gemini partnership: the National Science Foundation (United States), the Particle Physics and Astronomy Research Council (United Kingdom), the National Research Council (Canada), CONICYT (Chile), the Australian Research Council (Australia), CNPq (Brazil) and CONICET (Argentina)

and do not show an obvious preponderance of peculiar structures or morphologies which might suggest recent strong interactions. The outstanding characteristic is that early-type galaxies constitute 55% of the AGN host sample compared to 10-20% of the field population suggesting a spheroid-black hole linkage. This AGN sample was selected on the basis of X-ray flux from the Einstein Medium Sensitivity Survey and is expected to show no bias with respect to the host galaxy properties.

There are some challenges associated with comparing this local sample with one at higher redshift. The first requirement is that the two samples be observed at equivalent restframe wavelength and the second is that similar physical resolution in the galaxies frame should be achieved. These two requirements can be met in a satisfactory (but not perfect) manner using adaptive optics on a large-aperture telescope and observing in the K-band. Although the resolution is lower by a factor of a few at $z \sim 2$ relative to HST observations at $z \sim 0.1$, it is possible to derive the global properties of the host galaxies in an equivalent manner given sufficient signal-to-noise ratio.

A third and very difficult problem is selecting comparable and unbiased samples at high and low redshift. This problem requires a careful analysis of selection effects. We present Gemini Quick Start observations of two AGNs at $z \sim 1.7$ with the University of Hawaii Hokupa'a adaptive optics system which delivered resolution of 0.12 arcseconds (FWHM) in the K-band. The guide-star requirement has driven most earlier observers to guide upon the AGN itself which restricts one to highly luminous quasars. This constraint was removed by selecting candidates for observation from the 2dF QSO Redshift Survey (Boyle et al. 2000) allowing us to image typical $L \sim L^*$ galaxies at high redshifts and thus greatly strengthen the validity of the comparison between the low and high redshift AGN samples.

Keywords: Active Galactic Nuclei, Quasars, Galaxies, Host Galaxies, Cosmological Evolution

Introduction

The relationship between active galactic nuclei and the formation and evolution of galaxies may prove to be profound. The variation of star-formation with redshift (Madau et al. 1996) has a similar form to the evolution of the density of luminous quasars (Boyle & Terlevich 1998). The correlation of spheroidal properties with black hole mass (Magorrian et al. 1998, Ferrarese & Merritt 2000) suggests formation histories with substantial linkage. Evidence that the hosts galaxies of AGN at $z \sim 2$ may have star-formation rates much higher (up to 200 solar masses per year; Aretxaga, Boyle & Terlevich 1995, Hutchings 1995) than typical galaxies at $z \sim 2$ suggests that AGN hosts may evolve in a manner more closely resembling the dramatic evolution of the AGN itself than the milder evolution of the field galaxy population. We have completed a study of the host galaxies of nearby AGN and are in the process of compiling a comparable sample at $z \sim 2$ to investigate this evolution.

1. HST observations of AGN at $z \sim 0.1$

A sample of 76 low redshift ($z \leq 0.15$) AGNs selected on the basis of soft (0.3–3.5 keV) X-ray emission from the Einstein Medium Sensitivity Survey (Stocke et al. 1991) was analyzed by Schade, Boyle & Letawsky (2000). A snapshot survey with HST produced I band (F814W) images of 600 seconds each and these were supplemented by B and R band imaging from ground-based telescopes. The X-ray selection is a key property of this study because it ensures that the sample is free from bias with respect to the characteristics of the host galaxies.

The median ratio of the AGN to host galaxy luminosities is 0.2 and the host galaxy luminosities range from $M_B(AB) \sim -18$ to $M_B(AB) \sim -23$ and the AGN luminosities are low (largely at $M_B(AB) < -23$). There is no clear evidence for frequent mergers or of strong interactions in this sample although a carefully constructed control sample would be needed to do an effective study of the inner structural features (which display an interesting variety) of these galaxies.

A large fraction (55%) of the host galaxies in this sample are early-type (E/S0) galaxies based on either two-dimensional profile fitting or on visual classifications. Compared to a matching field galaxy sample from the Autofib survey (Ellis et al. 1996) early-type galaxies are over-represented and the host galaxy luminosities are skewed toward higher luminosities. The luminosity skew is consistent with being produced purely by the morphological bias of the sample. In other words, a sample of field galaxies with the same distribution of morphological types would present the same distribution of luminosities as the AGN host galaxy sample.

2.   Observations at higher redshifts

A meaningful comparison between AGN host galaxy properties at two different redshifts requires that the two samples be selected in such a way as to avoid redshift-dependent biases in the host galaxy properties. Furthermore, the observations themselves need to be comparable in terms of the measurement of galaxy properties. To achieve the latter goal it is necessary to observe the two samples at the same restframe wavelength with the same resolution in physical units at the same signal-to-noise ratio. The HST observations in the I-band at $z = 0.1$ translate roughly into K-band observations (2.2μm) at $z = 2$. The resolution of

HST is roughly 0.1 arcseconds corresponding to $0.25h_{50}^{-1}$ kpc at $z = 0.1$ ($q_o = 0.1$). A comparable physical resolution at $z = 2$ requires resolution of ~ 0.03 arcseconds which is smaller than the diffraction limit of an 8-meter telescope. In addition, the sky brightness at K-band makes it difficult to achieve the same signal-to-noise as the HST data.

Adaptive optics makes it feasible to approach the diffraction limit in the infrared with large telescopes but there are several challenges to this approach. First, bright guide stars must be found near the object to be observed. Second, it is essential, if quantitative information is to be derived from the luminosity profile, to have a good understanding of the delivered point-spread function (PSF) and the errors associated with it. The first problem was overcome by using the 2dF QSO Redshift Survey (Boyle et al. 2000) which contains over 13,000 objects so that a sample of quasars with typical luminosities and with nearby guide stars can be compiled despite the rarity of such configurations. The second challenge can be overcome by devoting time to characterizing the PSF from observations of star-star pairs at similar separations and orientation as the star-AGN pairs in the science program.

Pioneering work (Chapman, Morris & Walker 2000, Aretxaga et al. 1998, Wizinowich et al. 2000) clearly demonstrates the scientific promise of adaptive optics but the technical challenges are also brought into focus (e.g., Chapman, Walker & Morris 1999). There may be an unexpected variety of small-scale artifacts in the PSF which complicates the interpretation of the imaging.

The problem of selecting samples by equivalent criteria at low and high redshift is difficult. In the present case, the low-redshift sample of AGNs has been selected on the basis of X-ray emission which is independent of the properties of the galaxy itself (stars and gas) while the high-redshift sample (Boyle et al. 2000) has been selected on the basis of color, a quantity that is affected to some degree by the characteristics of the stellar populations in the host galaxy. Therefore, we cannot claim that the samples are selected in an identical manner. The effect of possible biases needs to be taken into account in the interpretation of the observational results.

3. Gemini Adaptive Optics observations

A pilot program (Principal Investigator: Croom) to observe AGN host galaxies was granted time by the Australian Gemini time allocation committee as part of the Quick Start program. Observations were obtained September 25 and 30 and October 1, 2000. Clouds compromised the observations on the first and last nights. Two AGNs were observed. The first was observed three times with either 12 or 16 images of 180 seconds each, and the second was observed once for 5400 seconds (30×180 seconds). In 3 of the 4 cases a point-spread-function (PSF) star was observed immediately after the AGN observation.

The most effective strategy for characterizing the PSF is to interleave the PSF observations with the AGN observations. Then the variation of the PSF during the observations can be directly estimated. During the Quick Start observations the PSF stars were observed for 5 or 6 × 30 seconds following the AGN observations. During the 5 minutes that the PSFs were observed the variation in the gaussian-equivalent fullwidth at half-maximum was as large as 20% so that this can be taken as a minimum variation in the PSF during the 45 minutes to 90 minutes that elapsed during the AGN observations. The variation could be much larger and this is a major uncertainty in the results discussed here.

The resolution of the observations was estimated by fitting a gaussian to the core of the PSF and was found to vary from 0.24 to 0.12 arcseconds. The separate images were stacked using the offsets read from the header of the fits images. These could be verified to be precise using the PSF observations where the star positions could be measured on the individual frames. The luminosity profiles were analyzed using the same bulge-disk-point source fitting procedure applied to the HST images (Schade, Boyle & Letawsky 2000). This includes the convolution with the PSF before computing goodness-of-fit.

Objects TQS247_284 and TQS247_340 can both be distinguished statistically from pure point sources using our best estimates of the PSF. In other words, host galaxies are detected in both cases although this result could be modified if our adopted PSF—constructed from the observations following the AGN observations—is in serious error. There is no formal way at present to estimate the likelihood of such an occurrence. It was not possible to distinguish formally between bulge-dominated and disk-dominated host galaxies because of insufficient signal-to-noise ratio.

Table 1. Gemini Adaptive Optics Observations

Date	Object	Exposures	Resolution
Sept. 25, 2000	TQS247_284	16 × 180s (2880s)	0.20"
Sept. 30, 2000	TQS247_284	16 × 180s (2880s)	0.12"
Sept. 30, 2000	TQS247_340	30 × 180s (5400s)	0.24"
Oct. 01, 2000	TQS247_284	12 × 180s (2160s)	0.12"

A primary goal of the Gemini observational program is to test for strong evolution in the luminosity of the host galaxies. To investigate the sensitivity and resolution of the Gemini AO observations a number of galaxies from the low-redshift AGN sample were used as input for simulating AO observations. The goal was to estimate the data quality that we should expect if the high-redshift host galaxies were similar in morphology to those at low redshift but had evolved only in luminosity. The sensitivity was derived from standard star observations from the Quick Start program and the sky brightness and delivered resolution were also estimated from the Gemini observations. The angular sizes at $z = 1.7$ were estimated by scaling from the nearby galaxies and adopting $q_o = 0.1$. K-corrections were derived from the spectral energy distributions of Coleman, Wu & Weedman (1980).

In the presence of strong evolution (2-3 magnitudes) we will be able to clearly distinguish a variety of morphological structure in these host galaxies at $z \sim 2$. Both visual classification and quantitative morphological analysis will be feasible and profitable. In the absence of evolution it will be difficult to derive detailed measurements of the galaxy properties by any means although we will be able to distinguish disks and bulges in some cases and will be able to estimate host galaxy luminosities fairly well.

The ultimate goal of this observational program is to produce high-quality adaptive optics imaging for a set of ~ 70 AGN at $z \sim 2$ to compare with the local sample of similar size. The observations will be challenging in a number of ways but the final product will be a view of the galaxy population at $z \sim 2$ based on a selection criteria (the presence of active nuclei) that differs from other approaches to population sampling and thus has the potential for providing us with a very different viewpoint on the galaxy evolution process.

Acknowledgments

The authors wish to thank the Gemini staff and the Hokupa'a team for the efforts that produced these observations.

References

Aretxaga, I., Le Mignant, D., Melnick, J., Terlevich, R., Boyle, B.J., 1998, astro-ph/9804322
Aretxaga, I., Boyle, B.J., Terlevich, R., 1995, MNRAS, 275, L27
Boyle, B., Shanks, T., Croom, S., Smith, R., Miller, L., Loaring, N., Heymans, C., 2000, MNRAS, 317, 1014
Boyle, B.J., Terlevich, R.J., 1998, MNRAS, 293, 49
Chapman,S., Morris, S., Walker, G., 2000 MNRAS, 319, 666
Chapman, S., Walker, G., Morris, 1999, in *Astronomy with adaptive optics : present results and future programs*, Proceedings of an ESO/OSA topical meeting, held September 7-11, 1998, Sonthofen, Germany, Publisher: Garching, Germany: Eu-

ropean Southern Observatory, 1999, ESO Conference and Workshop Proceedings, vol. 56, Edited by Domenico Bonaccini, p.73

Coleman, G., Wu, C., Weedman, D., 1980 ApJS, 43, 393

Ellis, R., Colless, M., Broadhurst, T., Heyl, J., Glazebrook, K., 1996 MNRAS, 280, 235

Ferrarese, L., Merrit, D., 2000 ApJ, 539, L9

Hutchings, J., 1995, AJ, 100, 994

Madau, P., Ferguson, H.C., Dickinson, M.E., Giavalisco, M., Steidel, C.C., Fruchter, A., 1996, MNRAS, 283, 1388

Magorrian,J. Tremaine, S., Richstone, D. Bender, R., Bower, G., Dressler, A., Faber, S. M., Gebhardt, K., Green, R., Grillmair, C., Kormendy, J., Lauer, T., 1998, AJ, 115, 2285

Schade, D. Boyle, B.J., Letawsky, M., 2000, MNRAS, 315, 498

Stocke, J. et al., 1991, ApJS, 76,813

Wizinowich, P., Acton, D. S., Shelton, C., Stomski, P., Gathright, J., Ho, K., Lupton, W., Tsubota, K., Lai, O., Max, C., Brase, J., An, J., Avicola, K., Olivier, S., Gavel, D., Macintosh, B., Ghez, A., Larkin, J., 2000, PASP, 112, 315

Susan Ridgway (left), Valentina Zitelli (center) and Birgit Kelm (right)

From left to right, José M. Rodríguez Espinosa, Jeremy Lim, Rosa González Delgado, Johan Knapen and Clive Tadhunter

HOST GALAXIES AND NUCLEAR STRUCTURE OF AGN WITH H$_2$O MEGAMASERS AS SEEN WITH *HST*

Heino Falcke, Christian Henkel
Max-Planck-Institut für Radioastronomie, Auf dem Hügel 69, 53121 Bonn, Germany
hfalcke@mpifr-bonn.mpg.de, chenkel@mpifr-bonn.mpg.de

Andrew S. Wilson[*]
Astronomy Department, University of Maryland, College Park, MD 20742-2421, USA
wilson@astro.umd.edu

James A. Braatz
National Radio Astronomy Observatory, P.O. Box 2, Green Bank, WV 24944, USA
jbraatz@nrao.edu

Abstract We present results of an *HST* survey in Hα and continuum filters of a sample of H$_2$O megamaser galaxies compiled by Braatz et al., all of which contain AGN. These observations allow us to study the AGN/host-galaxy connection, e.g. study the relation between the parsec scale masing disk/torus, bipolar outflows, and large scale properties of the galaxies such as dust lanes, signs for interaction, and galaxy types. A number of galaxies indeed show large-scale bi-polar Hα structures which, however, are more reminiscent of outflows than excitation cones. Most megamaser galaxies are found in spiral galaxies. Only one galaxy in the original sample is known to be an elliptical and one galaxy imaged by us shows clear signs of interactions. In all cases we see evidence for obscuration of the nucleus (e.g. dust-lanes) in our color maps and the disk galaxies are preferentially edge on. This suggests that the nuclear masing disk has some relation to the large scale properties of the galaxy and the dust distribution.

---

[*]Adjunct Astronomer, Space Telescope Science Institute

Introduction

After the discovery of extragalactic water vapor masers in star-forming regions (Churchwell et al. 1977) a search started for other possible extragalactic H_2O masers. In the late seventies and early eighties, this work lead to the discovery of five so-called megamasers in the nuclei of NGC 4945, the Circinus galaxy, NGC 1068, NGC 4258, and NGC 3079 which are 10^6 times more powerful than the commonly observed galactic masers. Only much later, after an intense search of a distance limited galaxy sample, Braatz, Wilson & Henkel (1994, 1996) were able to increase the number of known megamasers by a factor of three, thus making for the first time a statistical analysis of this rare phenomenon possible.

It is obvious that the H_2O megamaser phenomenon must have something to do with nuclear activity as all such megamaser sources are either in Seyfert or Liner galaxies and the emission is always concentrated at the nucleus. Interestingly, Seyferts of type 1 are absent from this group. The standard interpretation for the difference between Sy 1 and Sy 2 is that a torus surrounds the nucleus which obscures the central engine (black hole and accretion disk) for large inclination angles.

It therefore appears reasonable to infer that the masers trace molecular material associated with this torus or an accretion disk that feeds the nucleus. This, and the small number of H_2O megamasers known, suggests that the maser emission occurs for certain, very restricted, viewing angles, perhaps where the line of sight is along the plane of the molecular disk or torus.

This notion was recently confirmed in great detail by VLBI observations of the megamaser in NGC 4258 (Miyoshi et al. 1995; Greenhill et al. 1995). The positions and velocities of the central, the red- and the blue-shifted H_2O maser lines show that the masing region is in a thin disk structure, rotating on Keplerian orbits around a central mass of $3.6 \cdot 10^7 M_\odot$ at a distance of 0.13 pc from the nucleus. This observations proved for the first time the existence of a small scale molecular disk surrounding an active nucleus. Recently, VLBI maps have been obtained for even more maser sources: while NGC 4945 (Greenhill et al. 1997) and NGC 3079 (Trotter et al. 1998) also show a structure interpreted as a disk, the masers in NGC1068 show weak emission from a region presumably shocked by the nuclear jet at a projected galactocentric distance of ~ 30 pc from the nucleus (Gallimore et al. 2001) in addition to the masers in the disk (Greenhill et al. 1996).

Although plausible scenarios for the megamaser phenomenon exist (e.g., Neufeld & Maloney 1995) it is by no means clear how the obscuring torus and the masing disk are related. The masing disk may be the innermost part of a molecular accretion disk adjacent to the torus or just be the thin, central plane of the torus in which the column density is high enough for strong amplification. Alternatively, it might also be dynamically independent from "the torus" with a strong misalignment between their axes. In the most simple-minded picture, however, one would hope to find that masing disk, obscuring torus and the extended molecular cloud distribution form one coherent accretion structure feeding the central engine.

We have therefore investigated a sample of megamaser host galaxies with the Hubble-Space-Telescope (HST) to look for large scale evidence of the molecular material, some of which we describe in the following.

1. Observations

The sample was selected from the list of 16 known megamasers known in 1996 and published by Braatz, Wilson & Henkel (1996). We observed all galaxies that had not been previously observed with HST, which included: NGC449 (Mrk 1), NGC1386, NGC2639, NGC3079, IC2560, NGC4945, IRAS18333-6528, TXS 2226-184, and IC1481. Observations were made with WFPC2 to cover a red and a green continuum filter, as well as the Hα+[NII]$\lambda\lambda$6548,6583 emission lines. The filters used were F814W (red continuum), F547M (green continuum), F673N and the Linear-Ramp-Filters (LRF) for galaxies at higher redshifts. The pixel scales were either $0\rlap.{''}1$/pixel or $0\rlap.{''}0455$/pixel depending on whether LRF usage required imaging on the Wide Field or Planetary Camera. All observations were made within one orbit.

The images were processed through the standard Wide-Field and Planetary Camera 2 (WFPC2) pipeline data reduction at the Space Telescope Science Institute. Further data reduction was done in IRAF and included: cosmic ray rejection, flux calibration, and rotation to the cardinal orientation. The galaxy continuum near the Hα+[NII] line was determined by combining the red and green continuum images, scaled to the filter width of the narrow-band filter and weighted by the relative offset of their mean wavelengths from the redshifted Hα+[NII] emission. From the two broad-band images and the emission-line image, we constructed three-color (RGB) maps, where, for better contrast, the Hα+[NII] image was assigned the green channel, and the F814W and F547M filter images were assigned the red and blue channels respectively. Galaxy profile fits were made to the red continuum images. More details and first results for one galaxy (TXS 2226-184) are given in Falcke et al. (2000b).

Figure 1. Gray-scale representation of three-color *HST* images of the megamaser galaxies NGC1386 (left) and NGC2639 (right). North is up and East to the left. The images are 18."25 and 32."35 on each side respectively.

2. Results

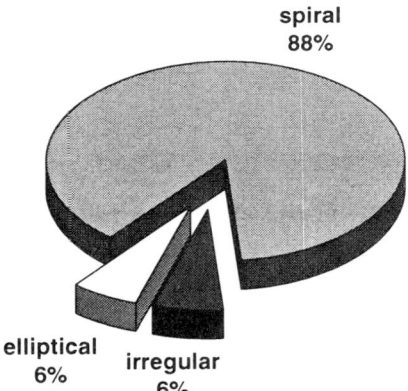 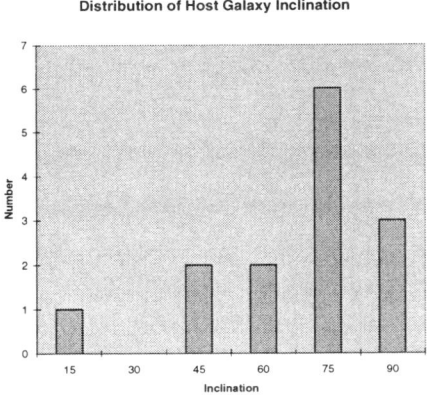

Figure 2. Left: The distribution of host galaxy types for all megamasers in the Braatz et al. (1996) sample determined from our *HST* observations and the literature. Right: Distribution of host galaxy inclination. The majority of megamaser galaxies are edge-on spirals.

The *HST* images contain a wealth of information. In many cases dust lanes and spiral arms are clearly visible and in basically all galaxies we resolve the narrow-line region (NLR). The most spectacular structures are the well-known Hα-bubble in NGC3079 (see also Cecil et al. 2001), a fan-shape region in NGC4945 (Marconi et al. 2000), and linear jet-like features in TXS 2226-184 (Falcke et al. 2000b) and NGC 1386 (Ferruit et al. 2000). In these cases the structure might be mainly related to an outflow rather than an excitation cone. The water vapor maser emission in these galaxies has been imaged with VLBI (either published or private communication: L. Greenhill and J. Braatz). In those four cases the presumed axis of the nuclear disk (even though the disk interpretation is not secure in all cases) seems to be perpendicular within less than 30° to the closest nuclear dust lane, as would be the case if the maser disks and the dust lanes tend to align. On the other hand the large scale radio or Hα axis is not always perpendicular to the disk axis (see, e.g., NGC3079, Trotter et al. 1998).

In all galaxies of our sample, we see clear evidence for the presence of large scale dust obscuration. In most, i.e. seven, cases we directly see relatively well-defined narrow dust-lanes passing through the nucleus; in the remaining two cases we see an elongated nucleus and reddening on one side of the inner galaxy.

The profile fitting of the continuum images yielded a significant preference for disk models over bulge-only/elliptical models. We also see clear signs of spiral arms in a number of the disk galaxies. Only for one galaxy, IC1481, do we obtain inconclusive results. The optical appearance is very irregular suggesting that this might be the site of an ongoing galaxy merger. One galaxy, TXS 2226-184, was re-classified as a spiral (see Falcke et al. 2000b – it was originally thought to be an elliptical). If we supplement our sample with literature data we find that only one out of 16 galaxies is an elliptical and one is irregular (see Fig. 2a) – the vast majority of megamaser host galaxies are indeed spirals.

From the observed ellipticity we can derive inclinations for the spiral galaxies in our sample. This shows that the majority of the galaxies are highly inclined with only a few face-on odd-balls (Fig. 2b).

3. Conclusions

Braatz et al. (1997) suggested that the nuclear properties of AGN with megamasers might be connected with their large scale structure. Our investigation seems to confirm this trend. Megamaser galaxies seem to be preferentially edge-on and in spiral galaxies. After TXS 2228-184 was reclassified there is only one elliptical galaxy left (NGC 1052), which, however, has masers along the jet (Claussen et al. 1998) and hence may be different from the majority of galaxies which have been interpreted in terms of masers from molecular disks[1].

The fact that we always see some obscuring material in the *HST* images directly on scales of several tens to hundreds of parsecs could indicate that the masing structure (the 'torus') on the milli-arcsecond (i.e., sub-pc) scale is related to large scale dust lanes and is thus not an isolated nuclear feature. Confirmation of this hypothesis may come from further VLBA-observations which could reveal the orientation and geometry of the nuclear disk in relation to the dust lanes seen with *HST*.

Acknowledgments

This work was supported by DFG grant 358/1-1 & 2, NASA grants NAG8-1027 and *HST* GO 7278, and NATO grant SA.5-2-05 (GRG 960086)318/96.

References

Braatz J. A., Wilson A. S., Henkel C., 1994, ApJL, 437, L99
Braatz J. A., Wilson A. S., Henkel C., 1996, ApJSS, 106, 51
Braatz J. A., Wilson A. S., Henkel C., 1997, ApJSS, 110, 321
Cecil G., Bland-Hawthorn J., Veilleux S., Filippenko A. V., 2001, ApJ, in press
Churchwell E., Witzel A., Pauliny-Toth I. et al., 1977, A&A, 54, 969
Claussen M. J. et al., 1998, ApJL, 500, L129
Falcke H., Henkel C., Peck A. B. et al., 2000a, A&A, 358, L17
Falcke H., Wilson A. S., Henkel C., Brunthaler A., Braatz J. A., 2000b, ApJL, 530, L13
Ferruit P., Wilson A. S., Mulchaey J., 2000, ApJSS, 128, 139
Gallimore J., Baum S., Henkel C., et al., 2001, ApJ, submitted
Greenhill L. J., Gwinn C. R., Antonucci R., Barvainis R., 1996, ApJL, 472, L21
Greenhill L. J., Jiang D. R., Moran J. M. et al., 1995, ApJ, 440, 619
Greenhill L. J., Moran J. M., Herrnstein J. R., 1997, ApJL, 481, L23
Marconi A., Oliva E., van der Werf P. P. et al., 2000, A&A, 357, 24
Miyoshi M., Moran J., Herrnstein J. et al., 1995, Nature, 373, 127
Neufeld D. A., Maloney P. R., 1995, ApJL, 447, L17
Trotter A. S. et al., 1998, ApJ, 495, 740

---

[1] The same may be true for the recently discovered megamaser in Mrk 348 (Falcke et al. 2000a), the galactic disk of which is very much face-on. VLBA observations of this galaxy show the masers to be at the tip of an expanding jet (Peck et al., in preparation).

STATISTICS OF SEYFERT GALAXIES IN CLUSTERS

Birgit Kelm and Paola Focardi
Univ. of Bologna, Dep. of Astronomy, Via Ranzani 1, 40127 Bologna - I
kelm@bo.astro.it, pfocardi@bo.astro.it

Abstract Cross correlation between the ACO cluster sample and the AGN catalogue by Véron-Cetty & Véron (2000) indicates that as much as 9% of all nearby Seyfert galaxies might be physical cluster members. The fraction reduces to $\approx 3\%$, when membership in the central Abell radius is demanded. No differences between Sy1 and Sy2 can be retrieved concerning occurence in clusters. The morphology of Seyfert galaxies in clusters is consistent with the morphology of Seyfert galaxies outside, with the vast majority ($\approx 70\%$) of sources being associated to Spiral hosts. Sy1 and Sy2 in clusters display similar luminosity. However, whilst the luminosity of Sy2 is the same in cluster and outside, bright Sy1 in cluster appear to be deficient. Available data indicate that roughly 35% ACO clusters include at least one Seyfert, the fraction of those which host more than one being nearly 10%.

Keywords: Galaxy Clusters, Environment, Morphological parameters, Seyfert

Introduction

Galaxies of different morphological type inhabit different evironments, following a well established morphology-density relation (Dressler et al. 1980, Postman & Geller 1984) or a morphology-radius relation (Whitmore et al. 1993). As a consequence they display significant different clustering properties, with Ellipticals and S0's dominating the cores of rich clusters, and Spirals constituing the typical field population.

Seyfert galaxies, whose host galaxies are mostly Spirals (Malkan et al. 1998) should then be expected to avoid galaxy clusters. Additionally, typical cluster galaxy velocities are on the order of ≈ 1000 km/s, much too high for effective tidal interactions to occur.

Actually previous papers pointed towards a general tendency of emission line galaxies and/or AGNs to avoid rich clusters (Gisler 1978, van den Bergh 1975, Osterbrock 1960) eventhough opposit claims can be retrieved (Petrossian 1982). More recently Dressler et al. (1985) derived the fraction of AGNs in a sample of 1095 galaxies in 14 nearby clusters and 173 field galaxies and stated that AGNs occur at a frequency of 5% in the field sample but only $\approx 1\%$ in the cluster sample. Biviano et al. (1997), investigating the occurence of emission line galaxies in clusters, assess that the fraction of AGNs in clusters is $\approx 0.7\%$, compared to a value of $\approx 2\%$ retrieved in their field sample. Way et al. (1998) warn that blue spectra-based studies of AGNs using equivalent width selection criteria (or the OIII/Hβ flux ratio) are largely contaminated by HII galaxies and that therefore the true relative fraction of cluster-to-field AGN is presently not known.

The cluster environment, however, is not necessarily an hostile one for gas rich galaxies such as Seyferts, as demonstrated by the excess of post-starburst galaxies (E+A or k+A galaxies) retrieved in the optically selected MORPHS cluster sample (Poggianti et al. 1999). Actually, the several mechanisms (interaction with the intracluster medium or with the tidal field of the cluster, high velocity interactions with many neighbours) which may be responsible for enhanced star formation in gas rich galaxies falling into a cluster (Moore et al. 1996, 1998, Dressler & Gunn 1983, Fujita 1998, Ghigna et al. 1998) could also, occasionaly, trigger an active nucleus (Lake et al. 1998). If true this scenario predicts high fractions of AGNs to be retrieved in intermediate redshift clusters, which contradicts observations showing that the environment of (at least radioquiet) QSOs has not undergone significant evolution with redshift (Smith et al. 1995, 2000, Croom & Shanks 1999, Ellingson et al. 1991). However, whether Seyfert and high luminosity QSOs undergo similar evolution with redshift (Lake et al. 1998) has still to be investigated.

1. Seyfert galaxies in clusters

To check whether statistical analysis confirms a deficiency of faint AGNs such as Seyferts ($M_B > -23$) in clusters we have cross correlated the Véron-Cetty & Véron AGN catalogue (2000) with the the ACO redshift catalogue (Struble & Rood 1999). We define a Seyfert to be a physical cluster member if the redshift distance to the cluster center is less than 1000 km/s and the projected distance less than 3 R_{Abell} ($4.5 h_{100}^{-1} Mpc$). Only the redshift range between $z = 0.01$ and $z = 0.1$ is analysed, to avoid contamination by peculiar motions and severe incompleteness in the Seyfert sample. Seyfert galaxies of type 1 and 2 have been analysed separately to investigate whether the excess of type 2 objects, which are more likely to be retrieved in locally dense environments (Kelm et al.

1998, Laurikainen & Salo 1995, Zitelli et al. these proceedings), can be confirmed also in large and dense regions such as clusters.

In **Fig. 1** the redshift distribution of Sy1 and Sy2 galaxies (panel a) and of ACO clusters (panel b) is shown. Clusters whose redshift is computed on 5 or more redshifts are denoted by a solid line, those whose redshift is based on less than 5 redshifts by a dashed line. As many as 54 Sy1 and 48 Sy2 appear to be physically associated to a cluster or at least to a cluster subclump. However, Fig. 1 shows that most Sy2 lie at cz<20 000 km/s suggesting that fair comparison between Sy1 and Sy2 should be performed within this limit. Moreover, figures could be restricted to only those Seyfert galaxies which are members in "bona fide" clusters, thereby excluding membership in subclumps and in cluster whose redshift is based on less than 5 data. Numbers reduce slightly, and we get that **37 Sy1** out of 401, and **39 Sy2** out of 459 are cluster members. Hence statistics indicate that nearly **9%** of the Seyfert population is a cluster population.

Actually more than half of the clusters in the sample are poor (richness 0), so that the 3 Abell radii out to which we compute cluster membership might be too large a value when computing physical association. If only those Seyferts which inhabit the central part of a cluster ($1R_A=1.5$ $h_{100}^{-1} Mpc$) enter computation a fraction of $\approx$ **3%** (12 Sy1 and 15 Sy2 respectively) is retrieved. For comparison the fraction of (all) galaxies which are ($R\geq 0$) cluster members (within $1R_A$) is $\approx 5\%$ (Bahcall 1999).

Distribution of Seyfert galaxies in clusters. Not only do we find that Sy1 and Sy2 display similar occurence in clusters, but also that their distribution relative to the distance to the cluster center **R** is the same (**Fig. 1c**). At distance $\approx 3R_A$ there is an excess of Sy galaxies compared to the expected projected density distribution ($N \propto R^{-1}$) of galaxies in clusters. Distributions for Sy1 and Sy2 both indicate a possible hole (wich is significant for Sy1 only) in the occurence of Seyferts at $R \approx 2R_A$. This might suggest the existence of two populations of low luminosity AGNs, triggered by two distinct mechanism.

In Fig. 1d the velocity difference (Δv) of Seyfert galaxies with respect to the cluster mean redshift is shown. Seyferts are mainly found to be at rest with the cluster mean redshift, thus indicating that they are not simple projections. A significant difference between Sy1 and Sy2 only concerns an excess of Sy1 displaying Δv higher than 800km/s.

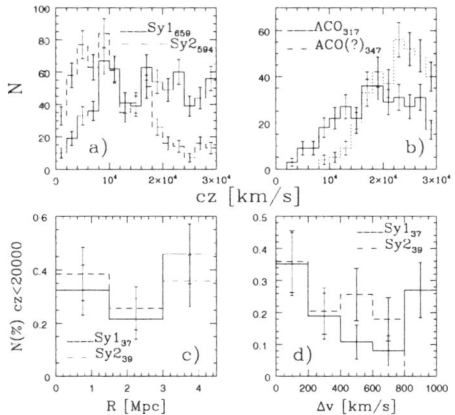

Fig. 1: Redshift distribution of Sy1 and Sy2 galaxies in V&V's compilation (panel a) and of ACO clusters (panel b). The solid line shows clusters (ACO) whose redshift is computed based on at least 5 redshifts, the hatched line shows clusters (ACO?) whose redshift is computed with less than 5 redshifts. Panel c) and d) display projected distance R and velocity difference Δv of Sy1 and Sy2 with respect to the cluster center.

Luminosity of Seyfert galaxies in clusters. As is evident in Fig. 2, Sy1 and Sy2 in clusters (panel a) display similar luminosity, whilst comparison of all Seyfert galaxies in V&V's catalogue shows that Sy1 are typically more luminous than Sy2 (panel b). The KS test indicates that the luminosity of Sy2 is unaffected by the cluster environment. Conversely bright Sy1 appear to be deficient in cluster (1σ).

Morphology of Seyfert galaxies in clusters. We have searched for Hubble type classification in NED. Presently, morphological classification is available for 18 Sy1 and 28 Sy2 only (Figs. 2c and 2d). Filled symbols trace morphological distribution of the Seyfert subsample retrieved within 1 R_A. Among sources with clear classification 12 out of 18 Sy1 and 20 out of 28 Sy2 are spirals. Eventhough incomplete, morphological classification of Sy1 and Sy2 in clusters yields for both classes a fraction of Spirals close to 70%. Late spirals are rare compared to early spirals, in accordance with the morphological distribution of all V&V Seyferts.

Relative frequency of clusters hosting AGNs. Available data also allow to adress a complementary question related to the cluster AGN connection concerning the frequency of $z = 0$ clusters including AGN members. Among the 182 ACO clusters liyng between redshift 3 000 and 20 000 we get that 63 ($\approx 35\%$) include one or more Sy members, that 15 ($\approx 8\%$) host at least 2 and that only 2 ($\approx 1\%$) host more than 2. If statistics is limited to the cluster subsets at $cz \leq 10\,000$ km/s, 25 (64%) out of 39 clusters do include a Sy member, while 12 (31%) host at least 2 active members. Whether the clusters hosting AGNs display characteristics different compared to those lacking an AGN member has to be investigated. If the same mechanism triggers both AGNs and starburst episodes we expect most of the clusters hosting a Seyfert not to be X-ray luminous (Balogh et al. 1999, Poggianti et al. 1999).

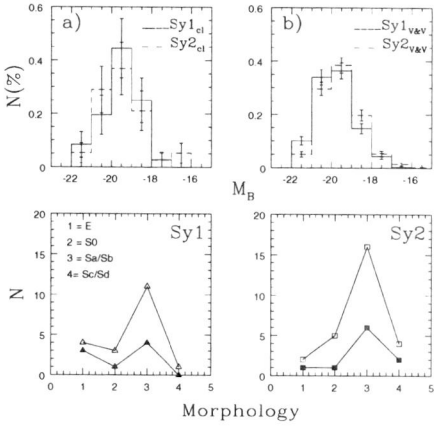

Fig. 2: Absolute magnitude of Sy1 and Sy2 galaxies in clusters (panel a) and of all Sy1 and Sy2 in V&V's compilation (panel b). The lower panels display the morphological type (NED data) of Sy1 and Sy2 in clusters. The filled symbols indicate distributions of the Sy sub-samples located within $1R_{Abell}$.

2. Discussion

Statistics based on large samples of Seyfert galaxies and clusters of galaxies show that there is no clear deficiency of Seyfert galaxies in clusters. No major differences between luminosities and morphological properties of Seyfert inside and outside clusters are retrieved, except for a possible lack of luminous Sy1. Hence, (in accordance with earlier finding by Biviano et al. concerning emission line galaxies) data could support the claim that Seyfert galaxies are infalling objects that have not yet traversed the virialized cluster core.

However, the hole in the distribution of Seyfert galaxies at distance $\approx 2R_A$ from the cluster center might suggest an alternative scenario. Whilst the upturn in the distribution at $\approx 3R_A$ is consistent with an enhanced rate of low velocity galaxy-galaxy encounters, the upturn within the central R_A requires a different triggering mechanism, which acts in loci. In fact, given the short predicted lifetimes of AGNs ($\approx 10^7$-10^8 yrs, Haehnelt & Rees 1993) central cluster Sy galaxies can not simply be infalling from the cluster outskirts. Their presence is compatible with models claiming either gas rich galaxies jumping into clusters to undergo short periods of enhanced activity (prior being depleted of their gas reservoir), or alternatively with models claiming that groups, rather than single galaxies are accreated by clusters, thus allowing low velocity interactions to occur even in the central parts of clusters.

References

Bahcall, N.A., 1999, in *Formation of structure in the universe* ed. Dekel A., Ostriker J.P. Cambridge Un. Press
Balogh, M.L., Morris, S.L., Yee, H.K.C., Carlberg, R.G., Ellinson, E., 1999, ApJ, 527, 54
van den Bergh, S., 1975, ApJ, 198, L1
Biviano, A. et al., 1997, A&A, 321, 84
Croom, S.M., Shanks, T, 1999, MNRAS, 303, 411
Dressler, A., 1980, ApJ, 236, 351
Dressler, A., Gunn, J.E., 1983, ApJ, 270, 7
Dressler, A., Thompson, I.B., Shectman, S.A., 1985, ApJ, 288, 481
Ellingson, E., Yee, H.K.C., Green, R.F., 1991, ApJ, 371, 49
Fujita, Y., 1998, ApJ, 509, 587
Ghigna, S. et al., 1998, MNRAS, 300, 146
Gisler, G.R., 1978, MNRAS, 183, 633
Haehnelt, M.G., Rees, M.J., 1993, MNRAS, 263, 168
Kelm, B., Focardi, P., Palumbo, G.G.C., 1998, A&A, 335, 912
Lake, G., Katz, N., Moore, B., 1998, ApJ, 495, 152
Laurikainen, E., Salo, H., 1995, A&A, 293, 683
Malkan, M.A., Gorjian, V., Tam, R., 1998, ApJS, 117, 25
Moore, B., Katz, N., Lake, G., Dressler, A., Oemler, A., 1996, Nature, 379, 613
Moore, B., Lake, G., Katz, N., 1998, ApJ, 495, 139
Osterbrock, D.E., 1960, ApJ, 132, 325
Petrossian, A.R., 1982, Astrofizika, 18, 548
Poggianti, B.M. et al., 1999, ApJ, 518, 576
Postman, M., Geller, M.J., 1984, ApJ, 281, 95
Struble, M.F., Rood, H.J., 1999, ApJS, 125, 35
Smith, R.J., Boyle, B.J., Maddox, S.J., 1995, MNRAS, 277, 270
Smith, R.J., Boyle, B.J., Maddox, S.J., 2000, MNRAS, 313, 252
Véron-Cetty, M.P., Véron P., 2000, *A Catalogue of Quasars and Active Galaxies* ESO Sc. Rep.
Way, M.J., Flores, R.A., Quintana, H., 1998, ApJ, 502, 134
Whitmore, B.C., Gilmore, D.M., Jones, C., 1993, ApJ, 407, 489

MORE BARS IN SEYFERT THAN IN NON-SEYFERT GALAXIES

Johan H. Knapen
*Isaac Newton Group of Telescopes, E-38700 Santa Cruz de la Palma, Spain, and
University of Hertfordshire, Department of Physical Sciences, Hatfield AL10 9AB, UK*
knapen@ing.iac.es

Isaac Shlosman
Department of Physics and Astronomy, University of Kentucky, Lexington, KY 40506-0055, USA

Reinier F. Peletier
School of Physics and Astronomy, University of Nottingham, Nottingham, NG7 2RD, UK

Seppo Laine
Space Telescope Science Institute, 3700 San Martin Drive, Baltimore, MD 21218, USA

Abstract

We use isophote analysis performed on near-infrared images of matched samples of Seyfert and non-active galaxies to study the overall bar fractions, and properties of the bars. We conclude that there is a small but significant excess of bars among the host galaxies of Seyferts as compared to non-Seyfert control galaxies. Whereas 59% ± 9% of control galaxies are barred, 79% ± 8% of Seyfert hosts are barred. We also find that the bars in Seyfert galaxies are statistically less elliptical, thus "thicker" or "weaker", than those in non-active galaxies. Implications for models of fuelling of AGN, as well as possible caveats in the interpretation of our results in that context, are briefly discussed.

Keywords: Galaxies: evolution – Galaxies: nuclei – Galaxies: Seyfert – Galaxies: spiral – Galaxies: statistics – Infrared: galaxies

Introduction

Gravitational torques from a non-axisymmetric potential of a bar on the interstellar gas in the galaxy disk have been proposed as a powerful mechanism to remove angular momentum from the inflowing disk gas, capable, in principle, of delivering material to the nuclear regions (e.g., Shlosman, Frank & Begelman 1989). It has long been debated whether the fraction of bars in active galaxies is larger than in non-active galaxies, as might be assumed if bars indeed are the main driver of gas inflow. Early studies (e.g. Adams 1977, Heckman 1978, Simkin, Su & Schwarz 1980, Dahari 1984a) cannot be taken as conclusive because the control samples used were not properly matched to the samples of active galaxies. For instance, Seyferts occur preferentially in early-type spiral galaxies, so a control sample consisting of a selection of spirals of all types cannot be used for statistical analysis.

More recent studies (e.g., Moles, Márquez & Pérez 1995, Ho, Filippenko & Sargent 1997, Mulchaey & Regan 1997) consider properly matched samples of active and non-active galaxies and conclude that there is no statistically significant difference in bar fractions. Knapen, Shlosman & Peletier (2000, hereafter KSP) also studied well-matched samples of Seyfert and non-Seyfert control galaxies, but improved on previous work by the use of near-infrared (NIR) imaging at high spatial resolution, and of a set of well-defined and objective criteria for bar classification. From our ground-based NIR imaging study of 29 Seyfert and 29 non-active control galaxies we conclude that there is a difference in bar fractions, namely about 80% of the Seyferts versus 60% of the control galaxies are barred (KSP).

Whereas the results presented here indicate the prevalence of barred morphologies in disk galaxies in general and in active galaxies in particular, there are additional factors which are required to trigger non-stellar nuclear activity (see also Shlosman et al. 1989). These additional factors can include the availability of fuel in the central regions, the efficiency of star formation there, and self-gravitating effects in the gas, but a discussion of these is outside the scope of this paper.

1. Galaxy samples and analysis

We summarize our sample selection method here, but refer to KSP for further details. We start by selecting the complete CfA sample of Seyfert galaxies (Huchra & Burg 1992). A control sample of non-active galaxies was selected from the RC3, matched to the Seyfert sample in absolute B-magnitude, distance, axis ratio, and morphological type. All galaxies from both samples were observed by us in the K' band using 4m class telescopes such as the Canada-France-Hawaii and William Herschel Telescopes. The image quality is good in all cases, with fields of view of around 2 arcmin squared, and spatial resolution around 0.7 arcsec. The data were published by Peletier et al. (1999) in the form of images and radial profiles.

In order to be able to recognize bars, we only selected galaxies which were not strongly inclined ($\epsilon_{disk} < 0.5$), not strongly disturbed morphologically due to, e.g., an ongoing merger, and not too small angularly ($\log r_{H,19} > 0.8$). Both Seyfert and control samples as used in our bar analysis thus consist of 29 galaxies.

Radial profiles of major axis position angle and ellipticity were constructed from our images and from other NIR and optical images, so that the profiles reflect a range in radius from less than 1 arcsec to the edge of the optical disk. Bars were classified using the criterion that the ellipticity as a function of radius should exhibit a significant ($\Delta\epsilon > 0.1$) rise and subsequent fall, at roughly constant position angle.

2. Results

In this study, we are interested in all galaxies which host a bar, irrespective of the size of the bar, and irrespective of whether the galaxy hosts more than one bar. Our results on bar fractions are that of the 29 Seyfert galaxies, 23 are barred (79% ± 7.5%), whereas of the 29 non-active control galaxies, 17 are barred (59% ± 9%).

The errors were calculated assuming purely Poissonian number distributions, and critically depend on the size of the parent population. A 60% bar fraction (as in the control sample) will thus have an uncertainty of 9% for a control sample size of 29 (KSP), of 5% for a sample of 100, and of 2% for a sample of 500 galaxies (in the latter case, a 25% difference between populations would be a 9 σ result). Insisting on the use of a well-defined and reproducible criterion based upon profiles fitted to ideally NIR images covering the surface brightness out to large radii, it is clear that beating the errors down in comparative studies such as the present one is not an easy task.

We are, however, at present finishing a very similar study but based upon *HST* NICMOS NIR imaging of well-matched Seyfert and non-Seyfert control samples. Preliminary results from this study, which roughly doubles the sample size as compared to KSP, indicate that the bar fractions found in KSP are reproduced very closely, but of course with significantly smaller uncertainties, in the *HST* study (Laine et al. 2001).

3. Are all galaxies barred?

One can wonder whether all disk galaxies are barred at some level, and it is just a matter of obtaining the right images to find the bars. This question has arisen mainly as a result of the discovery of bar structure in galaxies that were previously thought to be non-barred, especially the use of NIR imaging (e.g., Thronson et al. 1989), and/or through the use of high-resolution imaging (e.g. Regan & Mulchaey 1999, Martini & Pogge 1999). But while these examples are easily remembered, they are not representative. The overall bar fraction as estimated from, e.g., the RC3 catalogue can be estimated at roughly 60% of all disk galaxies (Sellwood & Wilkinson 1993, Knapen 1999, KSP). Using careful isophote analysis of high-quality NIR imaging of samples of galaxies, this fraction indeed goes up, but only slightly, to maybe 70% (Mulchaey & Regan 1997, KSP).

The question which remains is whether all Seyfert host galaxies have been barred in the (recent) past, as a prerequisite for setting up the radial gas flows needed to start the nuclear non-stellar activity. Shlosman, Peletier & Knapen (2000) found a hint that bars in Seyfert hosts are statistically thicker, or weaker, than those in non-active galaxies, and suggested that this is due to orbit rounding under the influence of a central mass concentration, statistically higher in active than in non-active galaxies. This process can also destroy bars completely, or make them undetectable, which implies that as soon as bars have formed a central mass concentration, and possibly an AGN, the bars can be destroyed by the monster they have just created. It is thus possible that all currently unbarred Seyferts have had a bar in the recent past, but we would not be able to confirm this by imaging.

4. Role of interactions

Apart from the bar-driven inflow (e.g. Shlosman et al. 1989) that the current paper is mostly concerned with, tidal interactions (e.g. Adams 1977, Dahari 1984b, Byrd et al. 1986, Hernquist & Mihos 1996) have been proposed as external mechanisms that may be related to the occurrence of Seyfert activity in galaxies.

Márquez et al. (2000) criticized the results from KSP by implying that the excess of bars among Seyfert hosts is due to the inclusion in our samples of galaxies that may be undergoing gravitational interaction. In KSP, we exclude obviously merging systems because any bar structure would not be recognizable, but we do not reject galaxies that may have companions. In fact, we are interested in the detailed relationship between the occurrence and properties of bars in Seyfert galaxies as compared to a control population, and much less in what physical processes actually brought that bar about, or maintain it.

Márquez et al. reduce the KSP sample to 13 (Sy) and 11 (control) galaxies by rejecting all galaxies that have companions within a cylindrical volume of 0.4 Mpc in radius and 2×500 km/s in cz (length, corresponding to 2×6.7 Mpc when assuming a Hubble flow with $H_0 = 75$km/s/Mpc). Such numbers are too small to reach meaningful conclusions on barred fractions in different samples, and the numbers derived by Márquez et al. (2000) cannot be interpreted as evidence against higher bar fractions in the Seyferts, or even in isolated Seyferts. It must also be pointed out that although Seyfert activity occurs in interacting and merging galaxies, there is no significant evidence for an excess of companions to Seyfert galaxies as compared to non-active control galaxies (Fuentes-Williams & Stocke 1988; later confirmed by, e.g., de Robertis, Yee & Hayhoe 1998; earlier work by e.g. Adams 1977 or Dahari 1984b was plagued by poor control sample selection). Interestingly, a rather marginal effect was found in that Seyferts of class 2 have an excess of companions less than 100 kpc away as compared to Seyfert 1's. All of this implies that there is no a priori reason to reduce samples of Seyfert and non-active galaxies artificially by excluding all galaxies with companions - the only effect would be to reduce both samples in equal measure and thus effectively only increase the error bars on the final results, exactly as shown by Márquez et al. (2000).

5. Conclusions

We analyse the fractions of barred galaxies as determined from ground-based NIR imaging of matched samples of Seyfert and non-active control galaxies. We find that Seyfert hosts are barred more often than non-Seyfert galaxies: 79% ± 8% for the Seyferts vs. 59% ± 9% for the non-Seyferts (KSP). This result is confirmed by a larger study based upon *HST* NIR imaging, which we are finalising at present (Laine et al. 2001). We argue that the presence of companions to our sample galaxies cannot influence our results to a significant degree.

Acknowledgments

IS acknowledges support from NASA grants NAG 5-3841, WKU-522762-98-6 and *HST* GO-08123.01-97A.

References

Adams, T.F., 1977, ApJS, 33, 19
Byrd, G. G., Valtonen, M. J., Sundelius, B., Valtaoja, L., 1986, A&A, 166, 75

Dahari, O., 1984a, Ph.D. Thesis, University of California at Santa Cruz
Dahari, O., 1984b, 1984, AJ, 89, 966
De Robertis, M. M., Yee, H. K. C., Hayhoe, K., 1998, ApJ, 496, 93
Fuentes-Williams, T., Stocke, J.T., 1988, AJ, 96, 1235
Heckman, T., 1978, PASP, 90, 241
Hernquist, L.E., Mihos, C. J., 1996, AJ, 471, 115
Ho L.C., Filippenko, A.V., Sargent, W.L.W., 1997, ApJ, 487, 591
Huchra, J.P., Burg, R., 1992, ApJ, 393, 90
Knapen J. H., 1999, in Beckman J. E., Mahoney T. J., eds, The Evolution of Galaxies on Cosmological Timescales. Astron. Soc. Pac., San Francisco, Vol. 187, 72
Knapen, J.H., Shlosman, I., Peletier, R.F., 2000, ApJ, 529 (**KSP**)
Laine, S., Shlosman, I., Knapen, J.H., Peletier, R.F., 2001, in preparation
Márquez, I. Durret, F., Masegosa, J., Moles, M., González Delgado, R. M., Marrero, I., Maza, J., Pérez, E., Roth, M., 2000, A&A, 360, 431
Martini, P., Pogge, R. W., 1999, AJ, 118, 2646
Moles, M., Márquez, I., Pérez, E. ,1995, ApJ, 438, 604
Mulchaey, J., Regan, M., 1997, ApJ, 482, L135
Peletier, R.F., Knapen, J.H., Shlosman, I., Pérez-Ramirez, D., Nadeau, D., Doyon, R., Rodriguez-Espinosa, J.M., Pérez-García, A.M., 1999, ApJS, 125, 363
Regan, M. W., Mulchaey, J. S., 1999, AJ, 117, 2676
Sellwood, J.A., Wilkinson, A., 1993, Rep. Prog. Phys., 56, 173
Shlosman, I., Frank, J., Begelman, M.C., 1989, Nature, 338, 45
Shlosman, I., Peletier, R.F., Knapen, J.H., 2000, ApJ, 535, L83
Simkin, S.M., Su, H.J., Schwarz, M.P., 1980, ApJ, 237, 404
Thronson, H. Jr., Hereld, M., Majewski, S., Greenhouse, M., Johnson, P., Spillar, E., Woodward, C.E., Harper, D.A., Rauscher, B.J., 1989, ApJ, 343, 158

STELLAR POPULATIONS IN THE HOST GALAXY OF AGNS

Monique Joly, Catherine Boisson and Didier Pelat
Observatoire de Paris-Meudon
F - 92195 Meudon
monique.joly@obspm.fr

Abstract

The relationship of an AGN to its host galaxy is one crucial question in the study of galaxy evolution. We perform stellar population synthesis in the central regions of galaxies of different levels of activity. Quantification of the stellar contribution is carried out in the visible range and in the near IR using the equivalent widths of the absorption features throughout the spectrum. The synthesis is performed using a method which, contrary to previous ones, gives a unique solution.

We find quite different stellar populations in the nucleus for the different types of activity, which seems to be indicative of an age sequence. In particular, Seyfert 2 nuclei present a younger population than LINER nuclei. Circum-nuclear regions in active galaxies show the contribution of a young population, less intense than in starburst galaxies but metal rich.

Keywords: Active galactic nuclei – Stellar content

Introduction

The relationship between the properties of the host galaxy and the nuclear activity is an important clue to the formation and evolution of the Active Galactic Nuclei (AGN) as well as of their host galaxy. Activity, interactions and star formation are often found concomitant in the central region of galaxies.

Is there a consequence of this relationship on the age and metallicity of the stellar population of the host galaxy ? Is there a link between the activity level of the AGN and the starburst activity ? Is this starburst activity nuclear or circum-nuclear ?

In order to answer to some of these questions, we perform stellar population synthesis of the integrated starlight emitted in the central

10 arcsec of nearby Active Galaxies in the optical and near-IR range. The good spatial resolution now achievable allows to detect gradients of population and starburst activity in the central region of the host galaxy.

1. OBSERVATIONS of the host galaxies

We got long slit spectroscopy of a sample of AGN, including 3 Seyfert 1, 3 Seyfert 2 and 3 LINERs.

In the optical, observations were obtained with the Herzberg at CFHT (Serote Roos et al. 1998) and with EMMI at NTT, in the range 5000-9000Å with a resolution of FWHM = 10Å. Spectra are extracted from the central 3-4 arcsec, as well as from surrounding regions located at 5 to 7 arcsec from the center, i.e. within the bulge at the distance of the galaxies. These spectra are compared to a library of 31 star spectra (see Section 2).

In the near IR, observations in the H band are favoured as this domain has the highest contribution of stellar light. It is remarkably rich in strong metallic features and exhibits powerful luminosity indicators. ISAAC at VLT, in Medium Resolution mode, provides a resolution of FWHM = 5Å. In the range $1.56-1.64\mu m$, 30 metallic stellar features are identified, among which some line ratios are strongly dependent on the luminosity class of the stars. A flux calibrated stellar library at a resolution R=2000 is available (Dallier et al. 1996) and is being extended for a better coverage of the H-R diagram.

Fig. 1 displays the spectrum of a circumnuclear region, located at 360 pc from the center, of the Seyfert 1 galaxy MCG-6-30-15. The spectral features are similar to those in the nucleus (within a radius R=180 pc) although quite different in strength, indicating that the stellar population may be the same in both regions but that in the nucleus stellar features are diluted by a continuous emission. This diluting continuum may be due to dust emission, in which case if one assumes no gradient of population, the comparison of the overall shape of the spectra may indicate a dust temperature of 1000 K to 1200 K.

Spectra of AGN display a great diversity of line strength in this wavelength range. The Seyfert 1 MCG-6-30-15 and the Seyfert 2 NGC 3185 have similar stellar features but different dilution factor while the Seyfert 1 NGC 3783 exhibits quite different line ratios indicating a different stellar population (in addition to a large dilution of the absorption lines).

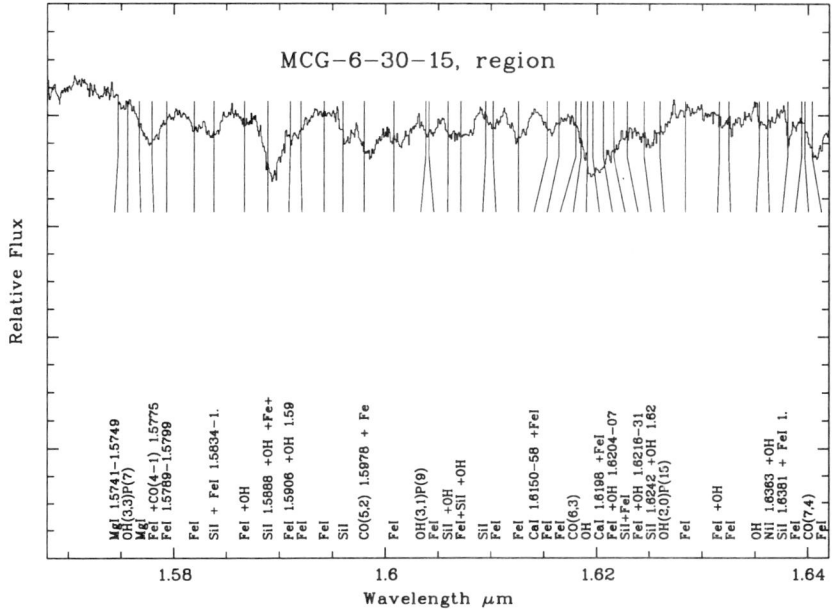

Figure 1. H band spectrum at a resolution R=3200 of a circumnuclear region of MGC-6-30-15. Numerous metallic stellar features can be identified.

2. STELLAR POPULATION SYNTHESIS

To compute stellar population synthesis, we compare the equivalent width (EW) of about 40 stellar features observed in the galaxy to the EW of the same lines measured in a library of star spectra. Recall that these quantities are independent of reddening.

We get a system of non-linear equations solved using a method developped by Pelat (1997, 1998) and Moultaka & Pelat (2000) which determines the best solution. The advantage of this method is that no hypothesis on the IMF, or on the history of star formation is made, and no evolutionary track models are preferred. It gives the contribution of each stellar type to the total radiation at a reference wavelength and the standard deviation to these contributions.

Synthetic spectra are computed using the stellar library. Intrinsic reddening is deduced from the comparison of the overall shape of the observed and synthetic spectra.

An example of the results obtained in the optical range is shown Fig. 2 where the synthetic spectrum of each region extracted from one of the

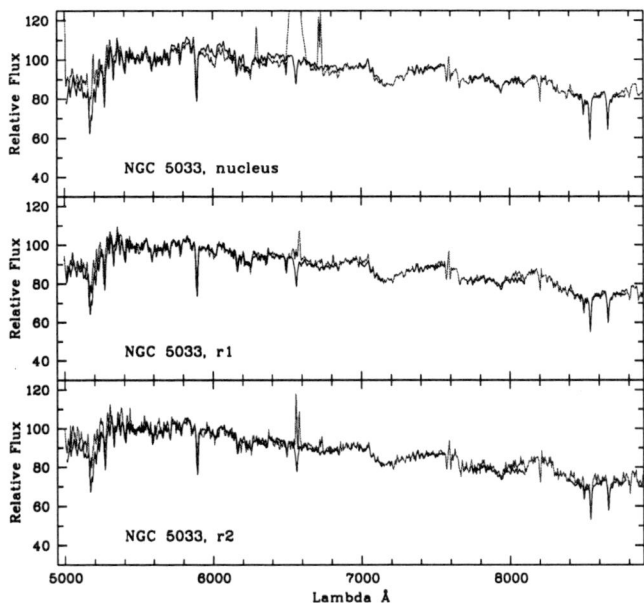

Figure 2. Synthetic spectra (black line) superimposed on the observed one (grey line)

LINER galaxies are superposed to the observed spectrum after correction of the internal reddening.

Two noteworthy results from our sample of objects are i) the homogeneity of the stellar populations within a class of AGN activity whatever the morphological type of the host galaxy and ii) that these populations are different in the very central regions of Seyfert 2 and LINERs (Boisson et al. 2000). LINERs have a very old metal rich population while in Seyfert 2 a contribution of weak or old burst of star formation is observed in addition to the old high metallicity component (in comparison, starbursts have a stronger and younger stellar population).

In the circum-nuclear regions (200 pc $\leq D \leq$ 1 kpc), i.e. in the bulge of the host galaxies, the trend is different. Table 1 shows the contribution to the total radiation (at the reference wavelength 5450Å) of the main components of the stellar populations as a function of the distance. These contributions are compared to those observed at the same distance in the bulge of a normal galaxy, NGC 3379. Within a region of ~500 pc around the nucleus stellar populations are similar in all active objects (except M 81 and Mrk 3) and reveal a non negligible

Table 1. Contribution of the main components in the circumnuclear regions

D pc	type	name	intermediate A type %	dwarfs %	giants %
200	LINER	NGC 4278	16	70	13
	LINER	NGC 5033	17	66	17
	LINER	M 81	0	74	26
	normal	NGC 3379	0	67	33
500	LINER	NGC 5033	19	63	19
	LINER	NGC 4278	13	74	12
	Sy 2	NGC 2110	3	77	20
	Sy 2	Mrk 620	14	88	5
	Sy 1	NGC 3516	10	68	22
	Sy 1	MCG-6-30-15	10	67	15
	normal	NGC 3379	0	62	38
1000	Sy 2	Mrk 3	12	45	43
	Sy 1	NGC 1275	47	37	14
	Sy 1	NGC 1275	65	25	6

component of A-type stars, i.e. an intermediate age component, ≤ 1 Gyr old. Such a component is not visible in the normal galaxy suggesting that the presence of the AGN influences the star formation or that the occurence of starbursts is linked to the AGN event.

3. CONCLUSIONS

The stellar populations in the inner regions of Active Galaxies (inside a radius less than 200 pc) are related to the level of activity of the AGN.

Farther away (within ~ 500 pc) differences vanish. The stellar populations are very homogeneous and younger than in the bulge of a normal galaxy. A relationship between the nucleus activity and circum-nuclear starburst is advocated.

References

Boisson C., Joly M., Moultaka J., Pelat D., Serote Roos M., 2000, A&A, 357, 850
Dallier R., Boisson C., Joly M., 1996, A&AS, 116, 239
Moultaka J., Pelat D., 2000, MNRAS, 314, 409
Pelat D., 1997, MNRAS, 284, 365
Pelat D., 1998, MNRAS, 299, 877

Monique Joly

Rosa González Delgado

THE STELLAR POPULATION OF POWERFUL SEYFERT 2 GALAXIES: IMPLICATIONS FOR QSOS

Rosa M. González Delgado
Instituto de Astrofísica de Andalucía (CSIC), Granada, Spain
rosa@iaa.es

Abstract

Violent events of star formation and QSO activity coexisted together very often in the past, suggesting a Starburst-AGN connection. This contribution investigates this issue searching for the direct spectroscopic signature of massive stars in the nucleus of the brightest type 2 Seyfert galaxies. The goal is to probe the role of nuclear starbursts in the Seyfert phenomenon. Starburst signatures are found in over half of the sample. In contrast, in 40% of the sample the dominant stellar population is an old one. The Seyfert 2 galaxies with young stellar populations have larger mid and far-IR luminosities, cooler mid/far-IR colors, and smaller excitation than the 'old' ones. These differences are consistent with a starburst playing a significant energetic role in the former class.

Introduction

One of the most important questions concerning the AGN phenomenon is the connection between Starburst and QSO activity. Recent results in the fields of galaxy formation, black hole and evolution of QSOs suggest that violent events of star formation and QSO activity coexisted together in the past even more often that we observe today. The ubiquity of the super-massive black hole in the nuclei of normal galaxies (Kormendy & Ho 2000), the proporcionality between the black hole and the spheroidal masses (Ferrarese & Merrit 2000, Gerbhard et al. 2000), and the correlation between the ratio mass of the black hole to spheroidal component with the age of the galaxy's stellar population (Merrifield et al. 2000), suggest that the creation of a black hole was an integral part of the formation of ellipticals and the bulge of spirals. Thus, to understand the processes that drove the cosmic evolution of QSOs seems strongly related to understand the Starburst-AGN connection.

On the other hand, Seyfert galaxies are considered the low luminosity version of QSOs and are the most plentiful class of powerful AGN in the local universe. Therefore, they are perfect laboratories where to investigate the Starburst-AGN connection with the maximal spatial resolution. In particular, Seyfert 2 galaxies offer the oportunity of studying easily the Starburst-AGN connection because their favorable geometry (the nucleus is blocked away by the dusty torus) facilitates the detection of stellar features from the circumnuclear starbursts.

This contribution presents strong evidence of the presence of nuclear starbursts in powerful Seyfert 2 galaxies, and the implications of this study for the circumnuclear stellar population of QSOs. A more extended discussion is in González Delgado et al. (2001).

1. The sample and Observations

The sample comprises the brightest Seyfert 2 nuclei selected from the compilation of Whittle (1992). They have been selected by their nuclear [OIII] $\lambda4959,5007$ emission line flux and by their non-thermal radio continuum emission at 1.4 GHz. They all satisfy at least one of the two following criteria: $\log F_{[OIII]} \geq -12.0$ (erg cm^{-2} s^{-1}) and $\log F_{1.4} \geq -15.0$ (erg cm^{-2} s^{-1}). These criteria guarantee that the sample is unbiased with respect to the presence or absence of a nuclear starburst. These objects are also very bright at the mid and far-infrared, with $L_{12\mu} \geq 10^{10} L\odot$ and with a mean $L_{FIR} \simeq 10^{11} L\odot$. Thus, these galaxies are also amongst the most powerful and brighest Seyfert 2 at mid and far-infrared. The observations were carried out using the Richey-Chrétien spectrograph attached to the 4m Mayall telescope at Kitt Peak National Observatory in the spectral range $\sim$ 3400–5500 Å.

2. Method to detect nuclear Starbursts in Seyfert galaxies

We have observed four of the sample targets (those with the brightest UV emission) at UV wavelengths with GHRS and the FOC aboard of the *HST*. The morphology of the UV light suggests that the emission is dominated by a nuclear starburst that is spatially resolved, showing sub-arcsec structures many of them similar to super-star clusters detected in typical starburst galaxies. The effective radius of the UV emission (50-200 pc) is similar to the size of the NLR. However, the strongest evidence of the presence of nuclear starbursts in these Seyfert 2 comes from the direct detection of strong stellar wind lines like NV $\lambda1240$, SiIV λ1400 and CIV λ1550 in the UV spectra. Using evolutionary synthesis models, we estimate that the bolometric luminosity of the starburst is $\sim 10^{10}$–10^{11} L⊙ which is similar to the estimated luminosity of the hidden Seyfert 1 nucleus. A detailed analysis of these data is in Heckman et al. (1997) and González Delgado et al. (1998).

Other diagnostics to look for starburst characteristics in Seyfert galaxies are at near-ultraviolet and optical wavelengths. Massive stars also show photospheric absorption features (most notably the H Balmer series and HeI lines). In the Wolf-Rayet phase, they show broad wind emission features at ~4660 Å, mainly due to NIII λ4634-4642 and HeII λ4686. Even though many of the photospheric lines in starbursts could be masked by the nebular lines, the high-order Balmer series (HOBS) and some of the HeI lines could be detected in absorption. This is because the strength of the Balmer series in emission decreases rapidly with decreasing wavelength, whereas the equivalent width of the stellar absorption lines is constant or increases with wavelength (González Delgado, Leitherer & Heckman 1999). In fact, three (NGC 5135, NGC 7130 and IC 3639) of our four S2 that show stellar wind resonance lines in the UV, also show the HOBS and HeI (λ3819, 4387 and 4922) in absorption. In the other object, Mrk 477, that also shows stellar wind resonance lines in the UV, the Balmer and HeI lines are overwhelmed by the corresponding nebular lines in emission. But it shows very broad emission at 4680 Å due to Wolf-Rayet stars. Starbursts go through the Wolf-Rayet phase when stars more massive than 40 M⊙ evolve from the main sequence a few Myrs after the onset of the burst. On the other hand, the optical metallic lines, as CaII K and G band, of these four galaxies are very diluted toward the nucleus. This strong dilution can not be produced by scattered light from the hidden Seyfert 1 because the required contribution of a power law is larger than 50% of the total optical flux, and contributions larger than 20% are ruled out by the Unification Model of AGN. Thus, our method to infer the existence of starbursts in S2 galaxies is based mainly on the detection of: 1) HOBS and HeI lines in absorption; 2) Wolf-Rayet features at optical wavelengths; 3) strong dilution of the metallic lines toward the nucleus.

3. Results

HeI and/or HOBS in absorption have been directly detected in 6 nuclei of the sample. They are: Mrk 273, Mrk 1066, Mrk 1073, NGC 5135, NGC 7130 and IC 3639 (see Fig. 1). They have relatively weak nebular emission, EW(Hβ) ≤ 25 Å, allowing the direct detection of some photospheric Balmer and HeI lines. They show the G band and CaII K lines weaker than 4 Å and 7 Å, respectively. The spatial variation of these lines indicates that there is a significant change of the stellar population with distance, being younger at nucleus than in the surroundings. The dilution factor at the nucleus ranges between 45% and $\sim 10\%$. Three other galaxies (Mrk 1, Mrk 477 and Mrk 533) show the higher order terms of the Balmer series in absorption in their circumnuclear regions, and in their nuclear spectra when the nebular emission contributions are subtracted. A broad emission feature centered at around 4680 Å is present in three nuclei in the sample: Mrk 1, Mrk 463E, and Mrk 477. It is most plausibly ascribed to a population of Wolf-Rayet stars. The analysis of the stellar lines using evolutionary synthesis models suggests that the dominant stellar population in these nuclei consists of young and intermediate age stars. Therefore, it is expected that young massive stars will dominate the ultraviolet spectra in half of the sample nuclei. These nuclear starbursts seem to be responsible for the blue continuum detected in these Seyfert 2 galaxies.

What is the dominant nuclear stellar populations in the other half of the sample? In 8 (40%) nuclei (Mrk 3, Mrk 34, Mrk 348, Mrk 573, NGC 1386, NGC 2110, NGC 5929 and NGC 7212) the stellar lines are similar to those of an old population. The presence of scattered light produced by the hidden nucleus is required to fit the spectra in only four of them, Mrk 3, Mrk 34, Mrk 573 and NGC 7212. However, only Mrk 3 shows a significant dilution of the stellar lines in its nucleus compared to its circumnuclear region. The lack of extra dilution in the nuclei of Mrk 34, Mrk 573, and NGC 7212 suggests the presence of a spatially-extended "featureless" continuum. In the other two galaxies (Mrk 78 and NGC 1068), a significant contribution of an intermediate age population in addition to the old bulge component is required to fit the spectral energy distribution and the strength of the stellar lines in the nuclei and circumnuclear regions.

4. Discussion and Implications for QSOs

In general, the most powerful Seyfert 2 nuclei can be classified in two groups: a) nuclei that have yound and intermediate age populations; then, they harbor a starburst (S2+SB); b) nuclei dominanted by old stars. Even though the two groups show similar [OIII] $\lambda 5007$ and 1.4 GHz emission, the excitation, the mid and far-infrared luminosities, and the far-infrared colors are quite different (Table 1). The S2+SB nuclei show lower excitation, larger far-infrared luminosities, and cooler $f_{25\mu}/f_{60\mu}$ IRAS colors than nuclei dominated by an old stellar population. These differences are as expected if a starburst makes a substantial

The Stellar Population of powerful Sy 2 galaxies: Implications for QSOs

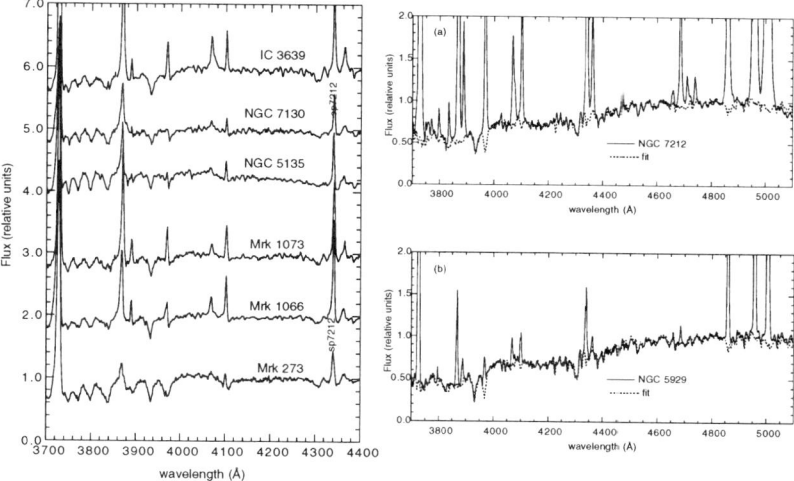

Figure 1. a) Nuclei than shows directely the HOBS and HeI in absorption. b) Two of the nuclei with old stellar populations

contribution to the heating of the dust and the ionization of the gas in the former set of nuclei, but not the latter. In fact, the Hβ and [OIII] nebular lines in the S2+SB nuclei are resolved in two components. The broad component has excitation much larger than the narrow one and it is similar to the excitation measured in the Seyfert 2 nuclei with old stellar population. If the narrow component is produced only by the starburst, the Hβ fluxes indicate that the starburst contributes between 30% and 80% of the total ionization. Thus, these starbursts have a significant impact on the emission line ratio of these Seyfert 2 nuclei. Another conclusion obtained is that the Seyfert 2 with most powerful nuclei are associated with most powerful starbursts, because the 12 μm luminosities (which arguably measures the hot gas heated by the AGN, Spinoglio & Malkan 1989) are also larger in the S2+SB. This luminosity link between the AGN and the starburst activity may be related to a common causal effect, that could be the presence of companions. In fact, 11 galaxies of the total sample show close companions or are in a group. Of these, 8 belong to the group of S2+SB, and 3 galaxies to the Seyfert 2 with old stellar population. Thus, 80% of the S2+SB have close companions against 30% of the S2 with old stellar population. The largest incidence of companions in S2+SB suggests that interactions are the cause of the triggering of the star formation in the nuclei of these Seyfert 2 galaxies, and the mechanism responsible to bring the gas to the center feeding the AGN.

Table 1. Mean values and dispersion of the luminosities (erg/s), far-IR colors, and excitation of Seyfert 2 with young (S2+SB) and old nuclear stellar populations

group	L([OIII])	L(1.4 GHz)	L(12 μm)	L(FIR)	f_{25}/f_{60}	[OIII]/Hβ
S2+SB	41.55±0.21	39.09±0.16	10.37±0.11	10.83±0.17	0.2	6
S2 old	41.51±0.32	38.63±0.34	9.84±0.19	10.05±0.20	0.4	10

The seggregation of the most powerful Seyfert 2 nuclei in two classes poses important questions: are there two different classes of Seyfert 2?, do all Seyfert 2 contain a hidden Seyfert 1 nucleus?, do all the Seyfert 2 nuclei contain nuclear starbursts, but undetectably faint in half the cases due to low relative luminosity? Most of the nuclei in the sample have been observed spectropolarimetrically at optical wavelengths, 9 of them have revealed broad recombination lines (HBLRs) suggesting the presence of hidden Seyfert 1 nuclei (Tran 1995). At least in four of the ten S2+SB (Mrk 463E, Mrk 477, Mrk 533 and IC 3639) HBLRs have been also detected, these compare with other four nuclei (Mrk 3, Mrk 348, NGC 7212 and possibly Mrk 573) with old stellar populations. This comparison suggests that there is no strong anti-correlation between the presence of a hidden Seyfert 1 nucleus and a nuclear starburst. However, these results suggest an evolutionary scenario in which the starburst and the AGN are relatively long-lived, and S2+SB can evolve to Seyfert 2. Nuclei like Mrk 78, that has a pronunced post-starburst population, may represent the intermediate state.

Could the Starburst-AGN connection found in Seyferts be extended to more powerful AGN, i.e. QSOs? Young and intermediate massive stars have been found in radio galaxies (Aretxaga et al. 2001) and in high-redshift radio galaxies selected to be very strong at the far-infrared (ULIRGs, Tran et al. 1999). They have been found also in the host of nearby QSOs (Canalizo & Stockton 2000). However, the most spectacular case is found in the QSO UNJ1025-0040 (Brotherton et al. 1999). Its ultraviolet spectrum shows broad MgII λ2800, while the HOBS are detected in absorption. This QSO could be interacting and its companion galaxy shows also strong Balmer absorption lines (Canalizo et al. 2000). I have re-analyzed the spectrum and I found that after removing the contribution of the QSO, the UV-optical spectral energy distribution and the strength of the higher-order Balmer series are compatible with a 400-600 Myr old starburst. These results also suggest a strong connection between the starburst and QSO phenomenon, and an evolutionary sequence from ULIRGs to QSOs, UNJ1025-0040 representing an intermediate state.

Acknowledgments

I thank my collaborators Tim Heckman and Claus Leitherer for their contribution to this work. I am very grateful to M. Brotherton for sending me the spectrum of UNJ1025-0040.

References

Aretxaga, I. et al., 2001, MNRAS, in press
Brotherton, M.S. et al., 1999, ApJ, 520, 87
Canalizo, G., Stockton, A., Brotherton, M.S., van Breugel, W., 2000, AJ, 119, 59
Canalizo, G., Stockton, A., 2000, AJ, 120, 1750
Ferrarese, L., Merrit, D., 2000, ApJ, 539, L9
Gebhardt, K. et al. 2000, ApJ, 539, L13
González Delgado, R.M., Heckman, T. et al., 1998, ApJ, 505, 174
González Delgado, R.M., Leitherer, C., Heckman, T., 1999, ApJS, 125, 489
González Delgado, R.M., Heckman, T., Leitherer, C., 2001, ApJ, 546, 845
Heckman, T.M., González-Delgado, R. et al., 1997, ApJ, 482, 114
Heisler, C.A., Lumsden, S.L., Bailey, J.A., 1997, Nature, 385, 700
Ho, L.C., Kormendy, J., 2000, Encyclopedia of A & A, in press
Merrifield, M.R., Forbes, D.A., Terlevich, A., 2000, MNRAS, 313, L29
Spinoglio, L., Malkan, M.A., 1989, ApJ, 342, 83
Tran, H.D., 1995, ApJ, 440, 565
Tran, H.D. et al., 1999, ApJ, 516, 85
Whittle, M., 1992, ApJS, 79, 49

José M. Rodríguez Espinosa and Isabel Márquez

From left to right, Aaron Evans, Susan Ridgway, Jason Surace and Margrethe Wold

ANISOTROPY IN THE MID-IR EMISSION OF SEYFERT GALAXIES
Seyfert Galaxies in the mid-IR

Ana M. Pérez García and José M. Rodríguez Espinosa
Instituto de Astrof'ısica de Canarias, Spain

Abstract Based on Mid and far-IR data of a complete set of Seyfert galaxies we analyse the Spectral Energy Distributions (SED) of these objects. These SEDs can be separated into three thermal emission components. This separation can be used to introduce some sort of spatial resolution in the inherently low resolution ISO data. It can also be applied as a method for detecting galactic disc components in high-z objects. The separation into emission components allows the study of these independently. In particular differences in the emission of the type 1 and 2 Seyferts can be seen in the warm emission component supposedly originated in dust heated by the nuclear source. The colder components tell us about the extended host galaxies.

Keywords: Seyfert Galaxies, IR emission, ISO, Dust emission

Introduction

The IRAS satellite provided an extensive set of IR data for a large number of galaxies, from which it was shown that Seyfert galaxies are indeed strong far IR emitters (Rodríguez Espinosa, Rudy & Jones 1987, Edelson, Malkan & Rieke 1987). However the IRAS data alone are not sufficient to clarify the nature of the IR emission, as there are measurements only at a limited number of wavebands preventing a good definition of the shape of the Spectral Energy Distribution (SED). A proper characterisation of the mid and far SED is essential to understand the emission mechanisms that produce the high output of Seyfert galaxies in the IR. Another important issue is the understanding of the differences between the two Seyfert types. According to the unified models, Seyfert 2 nuclei are intrinsically similar to Seyfert 1 nuclei the differences observed being due solely to geometrical effects. In this work, we present a study of the spectral energy distributions (SED) of Seyfert galaxies.

We present ISO data of the entire CfA Seyfert sample consisting of 25 Seyfert 1 and 22 Seyfert 2, and a LINER.

1. The observations

Observations of the CfA Seyfert sample have been carried out with the Infrared Space Observatory (ISO; Kessler et al. 1996) through filters at 16, 25, 60, 90 120, 135, 180 and 200μm. The filter set was chosen to achieve good coverage in wavelength while producing a good sampling of the SEDs. A few bright objects were mapped at 90μm with the C100 array. Fig. 1 shows the Spectral Energy Distributions (SEDs) of some of the observed objects. We have also used IRAS data (Edelson, Malkan & Rieke 1987) at 12, 25, 60 and 100μm together with the ISO data. The agreement between the ISO and IRAS data is in almost all cases very good.

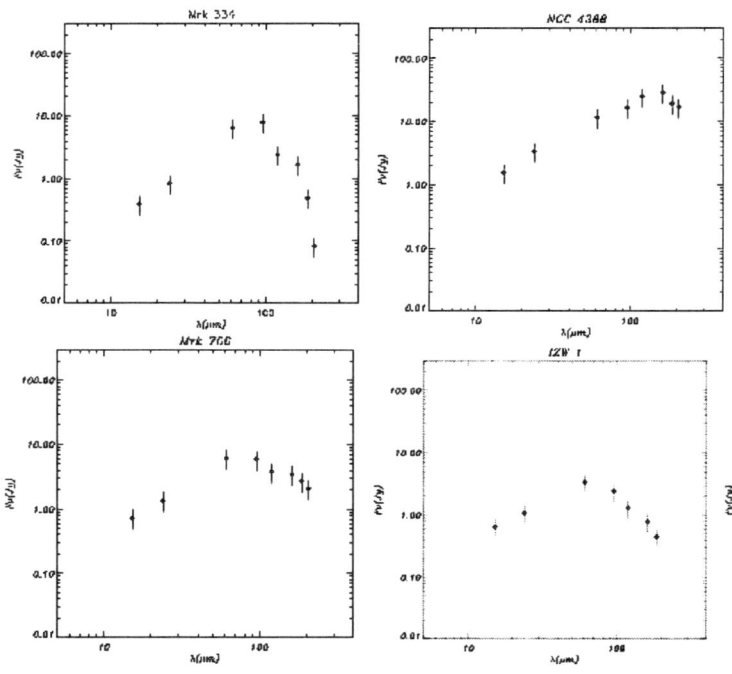

Figure 1.   Typical SED of Seyfert galaxies observed with ISO

2. The Spectral energy distribution

We have used an inversion method to analyse the SEDs of the galaxies in the sample. This has the advantage that no assumptions have to be made as to the number or location of the sources responsible for the observed spectrum. In particular, we have used an Inverse Planckian Transform algorithm, that employs an emissivity weighted Planck function kernel to switch from frequency space to the temperature domain, hence revealing the temperature distribution of the sources that originate the observed SEDs.

Fig. 2 shows an example of the results obtained after application of the Inverse Planckian Transform. For the object shown, the upper panel plots the ISO data (triangles) and the IRAS data (filled squares), as well as the best fit to the mid and far IR data (heavy solid line). The contribution of the different spectral components to the fit is shown with dashed lines. The bottom panel shows the temperature spectrum that produces these components. Similar results are obtained for the entire Seyfert sample, i.e., we recover three emission components that we have called warm, cold and very cold components. In a few cases, only two components are found.

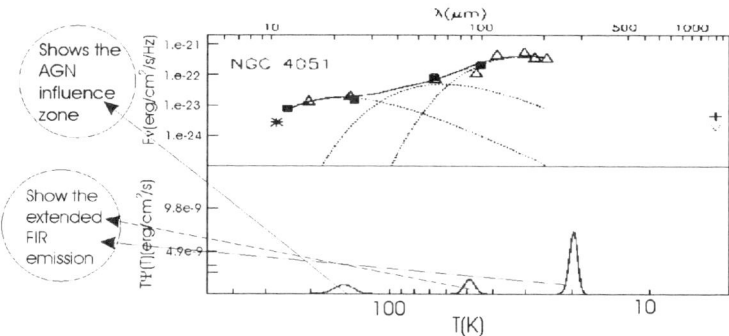

Figure 2. The spectral inversion showing the three components responsible for the Far IR SED of Seyfert galaxies. Top panel is in flux per unit frequency. The bottom panel is in flux units.

The key point now is to understand the meaning of these emission components. Previously we have shown (Rodríguez Espinosa & Pérez García 1997) that the warm emission component is related to the central regions of the galaxies while the cold and the very cold emission originate in the disk of these galaxies. We have also found a correlation between the flux produced by the warm component and the flux in high ionisation coronal emission lines such as [OIV]$\lambda 25.9\mu$m and [NeV]$\lambda 14.3\mu$m (Prieto and Viegas 2001), indicating that the warm component must be heated by the nucleus of these galaxies. Finally, Pérez García, Rodríguez Espinosa & Fuensalida (2000) have shown based on 90μm ISO maps of four nearby Seyferts, that the 90μm emission is physically extended up to radii similar or larger than those seen in optical images of the same galaxies. Furthermore, the extension of this 90μm emission has been characterised and its physical size, scale length and surface brightness profiles are typical of normal galaxy disks.

Based on the above, a scenario arises in which the warm component is associated with dust heated by radiation coming from the nuclear or circumnuclear regions of these galaxies, while the cold and very cold dust must be heated by processes occurring in the galaxy disks.

3. Anisotropy of the warm component

Recent studies of the mid IR emission from Seyfert galaxies claim that Seyfert 2 galaxies are weaker than Seyfert 1s (Heckman 1995). This result is interpreted within the framework of the unified models as an anisotropy, resulting from the presence of a molecular torus with a given optical thickness in the IR. To test this claim we have compared the fluxes of the warm component of the Seyfert SEDs to see if we find significant differences between Seyfert 1 and Seyfert 2s. We start by normalising the infrared fluxes with an isotropic property, i.e., with fluxes emitted at long enough wavelengths that they are not suspect of depending on the geometry of the sources. We have used 20 cm radio emission fluxes to normalise the IR flux, since the radio emission is not affected by selective extinction. We have used integrated radio data from Edelson, Malkan & Rieke (1987). The FWHM beamwidth used is 1.5 arcmin, directly comparable with our ISO data. The distributions of the radio normalised warm IR fluxes show the following characteristics:

$$log(F_{warm}/F_{20cm})_{Sy1} = 6.5 \sigma = 0.3$$

$$log(F_{warm}/F_{20cm})_{Sy2} = 6.1 \sigma = 0.3$$

KS tests indicate that both distributions are different at a significance level of 99.9%. Therefore, the warm flux is indeed higher in Seyfert 1s than it is in Seyfert 2s. This result suggests that at shorter wavelengths (mid IR) the emission is still anisotropic, in agreement with the molecular torus models of Pier & Krolik (1992, 1993) and Granato & Danese (1994). These and other authors have proposed different models for the absorbing material. The models proposed by Granato and Danese (1994) consist of thin and extended tori with optical depths ranging from $\tau =$ 10 to 300 in the UV band and maximum radii ranging from tens to hundreds of parsecs. On the other side, the models proposed by Pier and Krolik (1993) show thin and compact accretion disks with very large optical depths, values of $\tau \approx 1000$ in the UV band, and compact radii with dimension of a few pc. If we consider the Seyfert 1 as canonical unobscured objects, and ascribe the differences found between the two Seyfert types to absorption by the obscuring torus we obtain a mid IR optical depth of $\tau \approx 0.4$ or a $\tau_{UV} \approx 80$ for the Seyfert 2 objects. This value is indeed very mild and within the range predicted for the thin and extended tori of Granato, Danese & Franceschini (1997).

4. Summary

The analysis of the SEDs of Seyfert galaxies show a number of emission components that are independent of each other and which have direct physical interpretation, namely, the warm emission component is mostly heated by the nuclear or circumnuclear region, while the cold and very cold component are heated by processes occurring in the galaxies disks. We show that the warm emission component is different in Seyfert 1s and 2s, in agreement with dusty tori models. Finally, we note that the separation in emission components amounts to resolving the regions responsible for heating the dust in each of the components. Such an inversion could be used for instance to detect faint extended emission in high-z objects if the cold component is obtained after the inversion.

References

Edelson, R.A., Malkan, M.A., Rieke, G.H., 1987, ApJ, 321, 233
Granato, G.L., Danese, L., Franceschini, A., 1997, ApJ, 487, 147
Granato, G.L., Danese, L., 1994, MNRAS, 268, 233
Heckman, T.M., 1995, ApJ, 446, 101
Kessler, M.F. et.al., 1996, A&A, 315, L27
Pérez García, A.M., Rodríguez Espinosa, J.M., Fuensalida, J.J., 2000, ApJ, 529, 875
Pier, E.A., Krolik, J.H., 1993, ApJ, 418, 673
Pier, E.A., Krolik, J.H., 1992, ApJ, 401, 99
Prieto, M.A., Viegas, S.M., 2001, ApJ in press
Rodríguez Espinosa, J.M., Pérez García, A.M., 1997, ApJ, 487, L33
Rodríguez Espinosa, J.M., Rudy, R.J., Jones, B., 1987, ApJ, 312, 555

FIR AND RADIO EMISSION IN STAR FORMING GALAXIES

Alessandro Bressan[1], Gian Luigi Granato[1] and Laura Silva[2]
[1] *Osservatorio Astronomico di Padova, Padova, Italy*
[2] *Osservatorio Astronomico di Trieste, Trieste, Italy*

Abstract We examine the tight correlation observed between the far infrared (FIR) and radio emission in normal and starburst galaxies. We show that significant deviations from the average relation are to be expected in young starburst or post starburst galaxies, due to the different fading times of FIR and Radio power.

1. The Q-slope diagram

Observations indicate that there is a tight correlation between the FIR and Radio emission among star forming galaxies (Sanders & Mirabel 1996). At 1.4GHz, $Q = \log \frac{FIR/(3.75 \times 10^{12} erg\ s^{-1})}{L\nu(1.4GHz)/(erg\ s^{-1}Hz^{-1})} \simeq 2.3 \pm 0.2$. However, significant deviations from this relation, though with a larger scatter, are observed either in ultra luminous infrared galaxies (ULIRGS), where Q may reach values as high as 3 (Condon et al. 1991), or in the central regions of rich clusters where there seem to be a significant excess of star forming galaxies with Q below 2 (Miller and Owen 2001). To cast light on these problems, we have analysed the evolution of FIR and Radio emission of a starburst model, consisting of a superposition of an exponentially declining starburst episode to a quiescent disk galaxy. The burst is described by its strength, its age T_b and the e-folding time τ_b. Thermal radio emission is related to the ionizing photon flux from young stars, while non thermal emission, – $L_{NT}(\nu) = L_{1.4}^{SNR}(\frac{\nu}{1.4})^{-0.5} + K \times \nu_{SN}(\frac{\nu}{1.4})^{-\alpha}$– consisting of the contribution from supernova remnants and an unidentified mechanism, is normalized to the ratio between NT radio emission and supernova rate, observed in the Galaxy (Condon 1992). By adopting $\nu_{SN} \simeq 0.018$ (Cappellaro, private communication) and $L_{0.4GHz} \simeq 6.1 \times 10^{28} erg/s/Hz$ (Berkhuijsen 1984), we obtain $K \simeq 1.04\ 10^{30} erg/s/Hz/Yr^{-1}$. In order

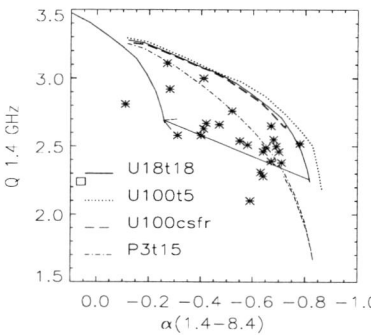 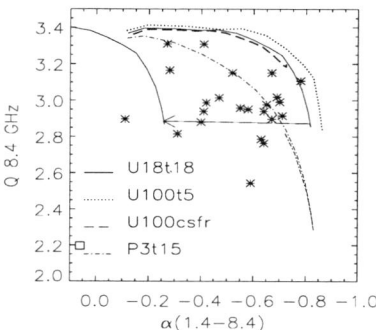

Figure 1. Predicted evolution of Q vs radio slope ($\alpha = -\frac{d\log L_\nu}{d\log \nu}$) between 1.4 and 8.4 GHz, compared with observed values (stars) in compact ULIRGS. The burst age increases from left, log t(yr)=6.5, to right, log t(yr)≤8. U18 and U100 refer to models with t_{FIR}=18 and 100 Myr while P3 to post-starburst galaxies with t_{FIR}=3 Myr. The last number is the e-folding time (Myr) of the SFR and "csfr" indicates constant SFR. The arrow shows how free-free absorption, with $\tau_{1.4GHz} = 1$, affects model U18t18.

to reproduce the observed average slope of the NT radio emission in normal spiral galaxies, $\alpha \simeq 0.9$. The infrared emission depends on the time (t_{FIR}) by which young populations get rid of their parental molecular clouds (eg. Silva et al. 1998). We have found that with an escape time of $t_{FIR} \simeq 3$ Myr and a visual extinction of A_V=1mag, we reproduce well the FIR/Radio correlation observed in normal star forming galaxies – constant SFR and Salpeter IMF between 0.15 and 120$M_\odot$, over the last few 100 Myr – namely $Q_{1.4GHz}$=2.3. The adopted values are consistent with independent results of recent infrared modelling of spiral galaxies. With this model we also obtain $SFR \simeq 0.57 \; 10^{-28} F_{1.4GHz}(erg/s/Hz)$, in good agreement with the one obtained by Carilli (2001). In the analysis of a sample of compact ULIRGS (Condon et al. 1991), we adopt larger obscuration times and a visual extinction of A_V=10 mag, as implied by optical and UV properties of observed SEDs of a few ULIRGS. With the help of a new diagnostic diagram, which contrasts the Q ratio with the radio slope (Fig. 1), we may draw the following conclusions. Our starburst models may account for the infrared and radio emission of the observed ULIRGS. The mass involved in the burst, for the sample analysed is typically between $10^9 M_\odot$ and $10^{10} M_\odot$. The range of $Q_{1.4GHz}$ is due to a combined effect of free-free absorption (see also Condon et al. 1991) and real starburst evolution. Free-free absorption induces an evident trend in the Q vs radio-slope diagram, with the higher Q being accompanied by a shallower slope. Estimated optical depths for free-

free absorption at 1.4GHz are between 0.2 and 1. The derived emission measures are consistent with the typical small sizes deduced from high resolution radio and mid infrared images (Condon et al. 1991, Soifer et al. 2000) and the typical ionized gas masses involved in the model starburst episode. At 8.4GHz free-free absorption becomes negligible ($\tau_\nu \propto \nu^{-2.1}$) and the previous trend disappears. The value of $Q_{8.4GHz}$ is a measure of the age of the starburst. Very young starbursts display an excess of FIR emission relative to the radio emission because the latter is initially contributed mainly by the free-free process. As the starburst ages, the non-thermal contribution increases and becomes the dominant source. The relative contribution of radio emission increases and the ratio $Q_{8.4GHz}$ decreases. We suggest that a similar diagram between 8.4GHz and a higher frequency would better highlight the evolutionary status of compact ULIRGS because the slope, unaffected by free-free absorption, would provide an independent estimate of the age. Models with constant star formation during the burst, always have a value of $Q_{8.4GHz}$ which remains too high at later times. As the only viable alternative, the SFR *must* decline strongly with time. Models with t_{FIR}=18 Myr and τ_{burst}=18 Myr or t_{FIR}=100 Myr and τ_{burst}=5 Myr reproduce the observed data. Apparently, the star formation in ULIRGS is strongly peaked and the burst is younger than about 60 Myr. The determination of the SFR from common *average* relations is obviously misleading in this evolutionary scenario. The open square in Fig. 1 refers to the Seyfert 1 galaxy MRK 231. By looking to the offset with respect to the other objects (or the obscured models with the lowest allowed Q) we estimate that the AGN is contributing to the 75% at least, of its total radio emission. In Fig. 1 we also show a post starburst model with the same obscuration of normal spirals, but a peaked SFR. We thus guess that a low Q in star forming cluster galaxies is a signature of a dying SF, recently ($\leq$150 Myr ago) interrupted by environmental effects.

We acknowledge discussions with C. Gruppioni, B. Poggianti, A. Franceschini and support from the TMR grant ERBFMRX-CT96-0086.

References

Berkhuijsen, E. M., 1984, A&A, 140, 431
Carilli, C.L., 2001, astro-ph/0011199v2
Condon, J.J., 1992, ARA&A, 30, 575
Condon, J.J., Huang,Z.P., Yin, Q.F., Thuan, T.X., 1991, ApJ, 378, 65
Miller, N.A., Owen, F.N., 2001, astro-ph0101158
Silva, L., Granato, G.L., Bressan, A., Danese, L., 1998, ApJ, 509, 103
Sanders, D.B., Mirabel, I.F., 1996, ARA&A, 34, 749
Soifer, B.T. et al., 2000, AJ, 119, 509

Andreas Eckart

Pepa Masegosa (left) and Elena Jiménez Bailón (right)

NO DOUBLE NUCLEUS AT THE CENTER OF ARP 220 ?

A. Eckart[1] and D. Downes[2]
[1] I.Physikalisches Institut, University of Cologne, Zülpicher Straße 77, 50937 Köln, Germany
[2] Institut de Radio Astronomie Millimétrique, 38406 Saint Martin d'Hères, France

Introduction

Millimeter interferometer observations of Arp 220 resulted in $\sim 0.5''$ resolution images that show two strong concentrations of mm-radio continuum and CO-line emission embedded in a larger molecular gas disk with a radius of about 1 kpc. The high resolution interferometer observations also show a complex velocity field with steep velocity gradients across each of the flux concentrations of $\Delta v \sim 500\,\mathrm{km\,s^{-1}}$ within $r = 0.3''$. The directions of these gradients are not aligned with each other or with that of the outer gas disk. Supported by the tidal tails visible on optical images this fact led to the conclusion that the two emission peaks represent either double nuclei with their own gas disks ($r \sim 100\,\mathrm{pc}$), which are counter-rotating with respect to each other and rotate around the dynamical center of the system or that they are two hot spots within a nuclear molecular gas disk. The overall structure of the molecular gas distribution and the corresponding complex velocity field, however, are highly symmetric, except for the unequal brightnesses of the two flux concentrations. This fact and similarities to the distribution and kinematics of the molecular gas in the central $2''$ of NGC 1068 (Schinnerer et al. 2000) motivated us to try describing the Arp 220 nucleus as a warped molecular gas disk in which the peaks in the flux and line width distributions arise due to crowding of inclined circular orbits. Detailed results are presented in Eckart & Downes (2001). *HST* images of the dust lanes flaring towards larger radii provide further support for a warped gas disk.

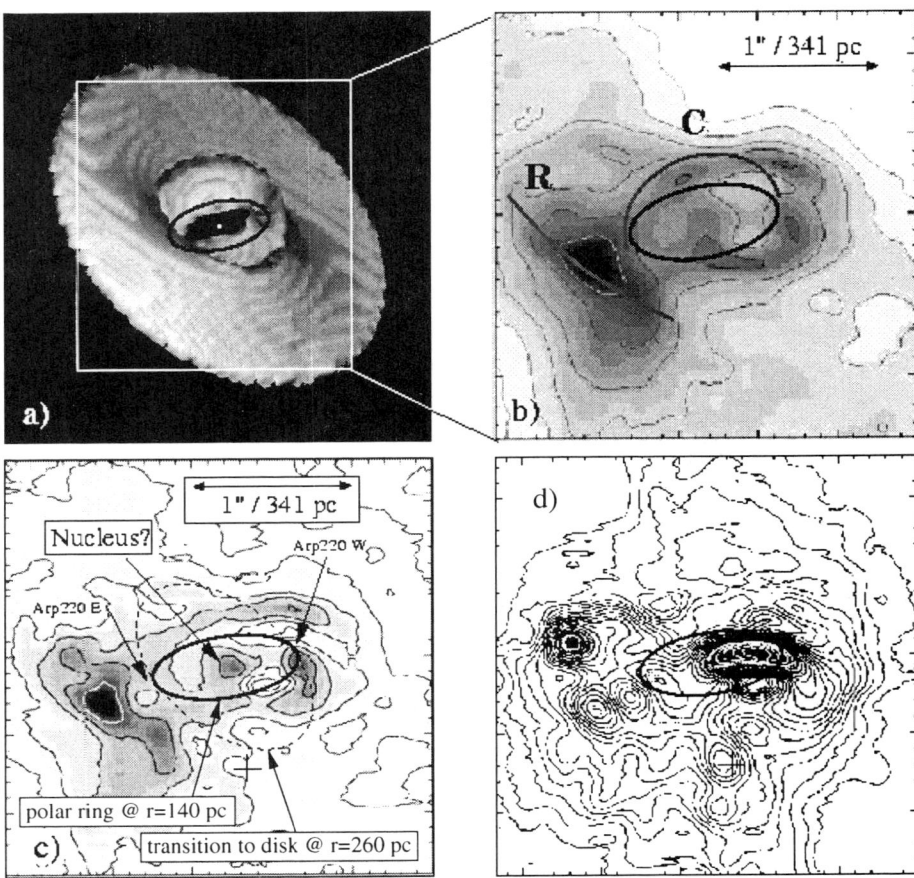

Figure 1. Comparison of the tilted ring model with near-IR *HST* data (Scoville et al. 1998). The ellipses represent the inner portion of our molecular polar ring model. **a)** A 3-dimensional view of the preferred tilted ring model. We show the surface of the warp model (not weighted by the emission). The model was calculated for ring radii between 0.41″ and 1.8″ (∼140pc and ∼600pc). Brighter sections are closer to the observer. The position of the center is indicated by a white dot. The location of the polar ring structure that is connected via the warp with the outer molecular disk is indicated by an ellipse. **b)** *HST* near-infrared 2.2μm/1.1μm color map at a resolution of 0.22″(Scoville et al. 1998). Dark areas correspond to red colors. Here we indicate with thick solid lines the approximate location of the ridge (*R*) of the northeastern bow-like structure and the northern circumference (*C*) of the nuclear warp that both should represent regions of high extinction. **c)** *HST* 2.2 μm/1.1 μm color map at a resolution of 0.14″ with source components labeled. Dark areas correspond to red colors. **d)** *HST* map of 2.2 μm continuum emission at a resolution of 0.14″.

1. Results of the Modeling

The final model is surprisingly successful in explaining the CO-line flux distribution, the velocity field, and position-velocity diagrams across the central flux concentrations. The model represents a strong 90° warp of the gas orbits out of the principal plane forming a circum-nuclear polar ring (Fig. 1a,b), and reproduces well the structures in *HST* near-IR color (Fig. 1c) and continuum maps (Fig. 1d). The large number of young stars recently formed out of the warped, polar-ring gas implies that those stars will have kinematic properties similar to that of the molecular gas. We suggest that the red near-IR source at the center of the polar ring (Fig. 1c) is the true nucleus of Arp 220. The success of the model implies that the two flux concentrations are not necessarily counter rotating nuclei, but rather the result of a combination of hot spots and orbit crowding due to the warp. The warp could be a direct consequence of the recent merger event indicated by tidal tails in optical images.

2. Implications

The new modeling has also implications for the interpretation of Arp 220 as a merger system: The two flux concentrations close to the center are not necessarily counter-rotating nuclei observed in the process of merging. We may be seeing the consequences of the merger event on the structure of the molecular gas disk, instead. There are two possible scenarii:

1) Either the several 10^9 $M_\odot$ of molecular gas have been brought in and settled in the circum-nuclear polar ring, or

2) the encounter/merger with a small companion induced the current structure and dynamics of a pre-existing nuclear gas disk.

The interaction of the polar ring with the 1 kpc molecular disk is likely to be the main driver of the currently observed intense nuclear star formation.

References

Eckart, A., Downes, D., 2001, ApJ, in press (April)
Downes, D., Solomon, P. M., 1998, ApJ, 507, 615
Sakamoto, K., Scoville, N. Z., Yun, M. S., Crosas, M., Genzel, R., Tacconi, L. J., 1999, ApJ, 514, 68
Schinnerer, E., Eckart, A., Tacconi, L. J., Genzel, R., Downes, D., 2000, ApJ, 533, 850

Yair Krongold

Paolo Marziani (left) and José Antonio de Diego (right)

STARBURST ACTIVITY IN SEYFERT 2 GALAXIES: UV–X-RAY EMISSION IN NGC 1068

E. Jiménez Bailón[1], J.M. Mas Hesse[1] and M. Santos Lleó[2]
[1] *LAEFF-INTA, VILSPA, Madrid, Spain*
[2] *XMM Science Operations Center, Astrophysics Division, ESA*

Abstract In certain galaxies with Seyfert 2 nuclei, the UV continuum has been found to be essentially dominated by starburst episodes distributed around their central regions. Our goal is to compare the overall measured X-ray emission with the expected contribution associated to the starbursts, in order to disentangle the relative intensity of both the starburst and active nucleus high energy components. The poster presents the method and discusses the preliminary results obtained on the prototypical Seyfert 2 galaxy NGC1068.

1. Description of the method

Our main goal is to quantify the relative contribution of the circumnuclear starbursts to the overall X-ray emission of the central region in Seyfert 2 nuclei. Assuming that the UV emission of the central region is dominated by a single starburst, we will use the evolutionary synthesis models developed by Cerviño et al. (2000) to compute an upper limit for the X-ray emission associated to the starburst. Comparing this prediction with the observed X-ray emission, it becomes possible to estimate the fraction of the emission originated in the central region associated to both the active nucleus and star forming processes themselves. The working method for the study of a Seyfert 2 nucleus consists of four steps:

1 **Spatial and spectroscopic analysis of the UV emission in the central region of the Seyfert 2 galaxy.** The analysis of the UV continuum will yield a first estimation of the extinction and strength of the starburst episodes.

2 **Measurement of the X-ray luminosity.** Usually, the overall emission of the galaxy will be integrated. We separate the spectral fitting in different components: a thermal one associated to the starbursts and a non-thermal due to the active nucleus.

3 **Comparison with the predictions of the evolutionary synthesis models.** Allows to constrain the properties of the starburst and provides its X-ray emission.

4 **Comparison with the measured X-ray luminosity.** Provides an upper limit to the fraction of emission attributed to the starburst.

2. NGC1068: Observational data

As NGC1068 is one of the nearest Seyfert 2 galaxies and because the galactic extinction in its line of sight is very weak, numerous and very accurate observations at different wavelength ranges have been performed. Since the discovery of its hidden Seyfert 1 nucleus by polarized light studies (Antonucci & Miller 1985), NGC1068 became the strongest observational evidence for unified models and the prototype of Seyfert 2 nuclei. In addition, NGC1068 presents very active star forming episodes at typical distances of kpc (Neff et al. 1994), and more compact circumnuclear starbursts (Thatte et al. 1997). Therefore, NGC1068 is the most appropriate target for our study, since there are numerous observational data and the central region harbors intense starbursts. NGC1068 is placed at 14.4 Mpc, ($z = 0.0038$), and has a bolometric luminosity of $2 \times 10^{11} L_\odot$.

NGC1068 was imaged in the UV by the *UIT* (Ultraviolet Imaging Telescope). This image has been used to identify the several star forming knots and the two regions considered in the analysis.

For the spectral analysis of the different regions we have combined calibrated data from the *IUE* (International Ultraviolet Explorer) and *HUT* (Hopkins Ultraviolet Telescope) archives. Taking into account the positions of the starbursts identified on the *UIT* image and the aperture of the *IUE* slit, ($10'' \times 20''$; $\sim 700 \times 1000$ pc), it can be concluded that this spectrum (1150 Å - 3350 Å) includes the overall emission of the central region, with both the active nucleus and the closest compact starburst. One *HUT* observation with a circular aperture $30''$ in diameter and a spectral range from 830 Å to 1850 Å has also been analyzed. In this case, the spectrum includes the emission originated in the innermost regions of NGC1068 and all most intense knots.

We have considered an *ASCA* observation of NGC1068. The data have been fitted to a multicomponent model, including a power law, a

Raymond-Smith plasma and the iron emission line, in order to compute the X-ray luminosity in two bands: 0.07–2.4 keV and 2–10 keV. The *ASCA* data include the whole X-ray emission of the galaxy.

To compare the observational data with the predictions of evolutionary synthesis models we have considered the UV continuum measured on two areas: the central region, as integrated within the *HUT* slit, and the external region, obtained from the difference between the *HUT* and *IUE* continua, dominated essentially by starburst activity. We have assumed that the UV continuum in both regions is dominated by single strong starbursts, and have tried to derive their properties by fitting the observed and the predicted UV continua. In this way we have obtained the starburst mass and age, and the extinction law and corresponding color excess of the star formation process that would better reproduce the observed UV properties. Based on this fit, the model gives directly the expected soft and hard X-ray emission associated to the starburst.

3. Results and conclusions

- The assumed starburst located in the central kpc could explain between 10% and 20% of the observed soft X-ray luminosity, but less than 7% of the hard X-ray emission.

- The star formation episodes located in the region between 0.7-1 kpc contributes to less than 4% to the total soft X-ray emission.

- Despite the evidence of stellar formation in the innermost region of NGC1068, (< 700 pc), its UV continuum could not be modeled by a starburst alone, indicating that the contribution of the scattered component from the nucleus has to be significant.

- The measured soft X-ray radiation in NGC1068 seems to be therefore dominated by emission reflected from the hidden active nucleus. Nevertheless, the star forming episodes in its central region could be (at least partially) responsible for the thermal soft X-ray component identified on ASCA data by Bianchi et al. (2000). On the other hand, the hard X-ray emission of NGC1068 has to be fully originated by reflected emission.

References

Antonucci, R. R. J., Miller, J. S., 1985, ApJ, 297, 621
Bianchi, S., Matt, G., Iwasawa, K., 2000, MNRAS, in press
Cerviño, M., Mas-Hesse, J. M., Kunth, D., 2000, A&A, in press
Neff, S. G., Fanelli, M., Roberts, L. J., O'Connell, R. W., Bohlin, R., Roberts, M. S., Smith, A. M., Stecher, T. P., 1994, ApJ, 430, 545

Thatte, N., Quirrenbach, A., Genzel, R., Maiolino, R., Tecza, M., 1997, ApJ, 490, 238

HOST GALAXIES AND ENVIRONMENT OF NARROW LINE SEYFERT 1 NUCLEI

Y. Krongold, D. Dultzin-Hacyan and P. Marziani[*]

Instituto de Astronomía, UNAM, Apdo. Postal 70-264, 04510 México D.F., México
yair@astroscu.unam.mx; deborah@astroscu.unam.mx; marziani@pd.astro.it

Abstract

We compared the environment and host galaxies of "Narrow Line" Seyfert 1 with those of Sy1, Sy2, and non-active galaxies. We found that NLSy1s are hosted in smaller galaxies than Sy1s. This result supports the idea of NLSy1s as Sy1s with higher Eddington ratio. NLSy1 hosts may be more isolated and farther away from bright companions than Sy2s and normal galaxies.

Introduction

Narrow-Line Seyfert 1s (NLSy1s) have been considered to be a peculiar type of Active Galactic Nuclei (AGN) because of their puzzling optical and extreme X-ray properties. They show a clear continuity with Seyfert 1s (Sy1s) and PG quasars in all the so-called Eigenvector1 parameters (Sulentic et al. 2000). The steep X-ray spectrum has led to the suggestion that, in NLSy1 nuclei, the accretion rate is systematically larger and that the black hole mass is systematically smaller than in other Sy1 nuclei. We attempt to define the relationship between NLSy1s, Sy1s, Seyfert 2s (Sy2s), and non-active galaxies by studying their morphology and their circum-galactic environment.

1. Sample Selection

The sample of NLSy1 galaxies was compiled from the catalog by Veron-Cetty and Veron (1998) considering all AGN listed as "S1n." It consists of 27 objects. We took into account only galaxies with high galactic latitude and redshift ≤ 0.061. One of the main issues when

---

[*]Osservatorio Astronomico di Padova, vicolo dell'Osservatorio 5, I-35122 Padova, Italy

comparing the environment of AGN galaxies with non-active galaxies is the proper definition of control samples that match the AGN group in all respects except the nuclear activity (as discussed by Dultzin-Hacyan et al. 1999). We generated a control sample from the CfA redshift catalog, which matches morphology, redshift, and diameter distributions.

2. Analysis

Morphology and Diameter. The morphology distribution of the samples was built using the NASA/IPAC Extragalactic Database. Diameters were (1) the standard diameters D_{25} listed in the Third Reference Catalog of Galaxies; (2) measured as isophotal diameters down to the 22 isophotal magnitude on the Digitized Sky Survey (DSS).

Environment. *(a) Identification of galaxies.* It was performed automatically on the DSS with the latest version (1998) of FOCAS (Jarvis & Tyson 1981). We searched for companion galaxies of diameters $D \geq 8$ kpc. In order to avoid the effects of differing background/noise level, objects below a surface brightness ≈ 22 mag/pix were not taken into account. The absolute magnitude for the faintest companions considered is ≈ 18.5. *(b) Estimation of the foreground/background galaxy contamination.* This is one of the most important issues in environmental studies. The fraction of galaxies with optical companions was derived from the Poisson distribution as it has been done in previous works (e.g. Dahari 1984). The fraction of NLSy1s with physical companions is defined as the fraction with observed companions, diminished by the fraction of galaxies with an optical companion: $f_{phys} = f_{obs} - f_{opt}$.

3. Results

Environment. We identified all galaxies with one or more companions within a circle of search radius equal to three times the object diameter $3D_S$, and within 100 kpc. We also computed the distribution of the nearest companion up to 700 kpc of projected linear distance. Inter-sample comparisons suggest: (a) NLSy1s vs. Non Active Galaxies: there is a statistical deficit of companions close to NLSy1 hosts. (b) NLSy1s vs. Sy1s and Sy2s: to study the environment of these objects, we used the results on Seyferts obtained by Dultzin-Hacyan et al. (1999). There is no difference between the environment of NLSys and Sy1s. The contrary holds true for Sy2 galaxies: the search shows a statistically significant excess of companions in the Sy2 sample with respect to NLSy1s.

Diameters. The diameter distribution of NLSy1s is skewed toward small values ($\leq$ 10 kpc) with respect to Sy1s (see Fig. 1). A Kolmogorov-Smirnov test suggests that NLSy1s and Sy1s are drawn from different populations to a confidence level of 95%. Consistently, there is an excess of NLSy1s classified as compact objects with respect to Sy1s.

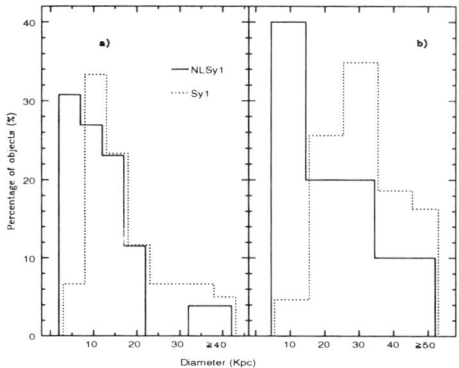

Fig. 1: Diameter distribution of NLSy1s and Sy1s. The solid line corresponds to the NLSy1 sample, while the dotted line refers to Sy1. (a) Diameters measured on the DSS down to the 22 isophotal magnitudes are binned over 5 kpc. (b) Standard diameters obtained from the Third Reference Catalog of Galaxies.

4. Conclusion

The results on the environment of NLSy1, Sy1, and Sy2 galaxies support the idea that NLSy1s are not related to Sy2s, and, in any case, they are objects similar to the Sy1 class as far as their environment is concerned. NLSy1s are believed to be the AGN with the largest Eddington ratio i.e., with the largest L/M_{BH} ratio. If there is a correlation between black hole mass in the center of galaxies and the bulge mass of the galaxy itself (Magorrian et al. 1998), smaller galaxies may host less massive black holes. The luminosity distributions of the Sy1s and NLSy1s in our samples are indistinguishable according to a K-S test. Since NLSy1s have significantly smaller diameters, but luminosity comparable to that of the Sy1s, they may indeed be radiating at higher L/M_{BH} ratios.

References

Dahari, O. 1984, AJ, 89, 966

Dultzin-Hacyan, D., Krongold, Y., Fuentes-Guridi, I., Marziani, P. 1999, ApJL, 513, L111

Jarvis, J. F. and Tyson, J. A. 1981, AJ, 86, 476

Magorrian, J. et al., 1998, AJ, 115, 2285

Sulentic, J. W., Marziani, P., Dultzin-Hacyan, D. 2000a, ARA&A, 38, 521

Véron-Cetty, M. and Véron, P. 1998, A Catalogue of quasars and active nuclei, Edition: 8th ed., Publisher: Garching: European Southern Observatory (ESO), 1998, Series: ESO Scientific Report Series vol no: 18.

THE CIRCUM-GALACTIC ENVIRONMENT OF LINERS AND BRIGHT IRAS GALAXIES

Y. Krongold, D. Dultzin-Hacyan and P. Marziani *
Instituto de Astronomía, UNAM, Apdo. Postal 70-264, 04510 México D.F., México
yair@astroscu.unam.mx; deborah@astroscu.unam.mx; marziani@pd.astro.it

Abstract We compared the environment of LINERs and Bright IRAS galaxies with those of Seyfert 1, Seyfert 2, and non-active galaxies. Interacting and merging systems are more frequently found among LINERs and Bright IRAS galaxies than among Seyfert 1 and normal galaxies. The environment of LINERs and Bright IRAS galaxies resembles that of Seyfert 2 galaxies. A notable exception are LINERs showing a broad component in Hα, whose environment is apparently similar to that of Seyfert 1s.

Introduction

Low Ionization Nuclear Emitting Regions (LINERs) have been considered to be an heterogeneous class at the low-luminosity end of Active Galactic Nuclei (AGN). While LINERs have been explained in the context of shock ionization, some of them appear as transition LINERs with properties more similar to H_{II} regions (Ho et al. 1997). Several other LINERs are definitely "dormant" Seyfert 1 galaxies. Bright IRAS (BIRAS) galaxies are objects with $60\mu m$ flux density ≥ 5.4 Jy. We attempt to define the relationship between LINERs, Sy1s, Sy2s, BIRAS, and non-active galaxies by studying their circum-galactic environment.

1. Sample Selection

We generated three LINER samples from the catalogue by Carrillo et al. (1999). The first sample consists of 103 objects and contains only pure LINERs. The second sample includes 63 transition LINERs (Ho et al. 1997). The third group includes 27 LINERs with broad $H\alpha$

---

*Osservatorio Astronomico di Padova, vicolo dell'Osservatorio 5, I-35122 Padova, Italy emission. We built the LINER control sample with objects that did not show emission lines when observed with Hubble Space Telescope. This sample consists of 61 objects. We considered only objects with redshift ≤ 0.017 and high galactic latitude. The Bright IRAS sample was compiled from the IRAS Bright Galaxy Survey by Soifer et al. (1987) and consists of 87 objects. We generated a BIRAS control sample of 90 objects from the CfA catalog. Morphology, redshift, and diameter distributions were all matched between target and control samples.

2. Analysis

The identification of companion galaxies was performed automatically on the DSS for the LINERs, and on the DSS-II for the BIRAS, with the latest version (1998) of FOCAS (Jarvis & Tyson 1981). We searched for companion galaxies of diameters $D \geq 4$ kpc. The estimation of the foreground/background galaxy contamination is one of the most important issues in environmental studies. The fraction of galaxies with optical companions was derived from the Poisson distribution as has been done in previous works (e.g. Dahari 1984). The fraction of objects with physical companions is the fraction with observed companions, diminished by the fraction of galaxies with an optical companion: $f_{phys} = f_{obs} - f_{opt}$.

3. Results

We identified all galaxies with one or more physical companions within a circle of search radius equal to three times the object diameter ($3D_S$). In addition, we computed the distribution of the nearest companion up to 140 kpc of projected linear distance. Considering also the previous work by Dultzin-Hacyan et al. (1999), we obtain the following main results. *Bright IRAS sources* – There is a significant excess of companions among BIRAS galaxies with respect to non-active objects and Seyfert 1 galaxies. Consistently, the search shows no difference in the environments of BIRAS galaxies and Seyfert 2s (see Fig. 1). *LINERs* – Again, we find that there is a significant excess of companions among pure LINERs with respect to broad emission objects (i. e., Seyfert 1 galaxies) and non active galaxies. Pure LINERs and transition LINERs do not differ in this respect. The search shows even a marginal excess of companions in the environments of LINERs with respect to Sy2s (see Fig. 1). Our analysis also shows that mergers are statistically more frequent in LINERs than in non active galaxies. On the contrary, LINERs with broad emission lines seem to be more isolated objects, like Seyfert 1s.

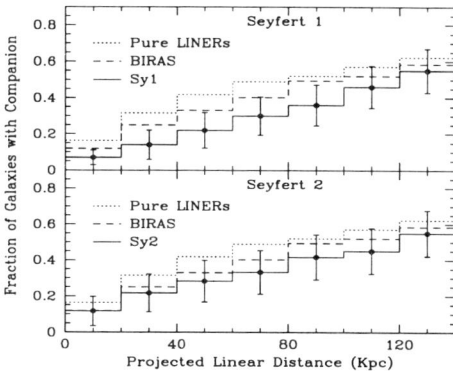

Fig. 1: The cumulative distributions of the nearest companion with diameter $D_C \geq 10$ kpc, binned over 20 kpc, up to a projected linear distance of 140 kpc, for Sy1 (upper panel, solid line), Sy2 (lower panel, solid line), "Pure" LINERs (dotted line) and Bright IRAS (dashed line). The error bars were set with a bootstrap technique and by taking an uncertainty of twice the standard deviation.

4. Conclusions

Our results indicate a clear correlation between interaction and the enhancement of FIR emission. A new result is that there is a connection between strong interactions and LINER-type activity. Both pure and transition LINERs are surrounded by a rich circumgalactic environment as found for Seyfert 2 galaxies. These are the LINER types that may be linked more directly to shock heating and/or photoionization by a thermal continuum (and hence to star formation). On the contrary, results on Seyfert 1 and LINERs with a broad Hα component suggest that strong interactions with bright companions may be poorly related to low luminosity non-thermal activity, at least within the limitations of the present study, as discussed by Dultzin-Hacyan et al. (1999).

References

Carrillo, R., Masegosa, J., Dultzin-Hacyan, D., Ordoñez, R. 1999, Revista Mexicana de Astronomía y Astrofísica, 35, 187

Dahari, O. 1984, AJ, 89, 966

Dultzin-Hacyan, D., Krongold, Y., Fuentes-Guridi, I. Marziani, P. 1999, ApJL, 513, L111

Ho, L. C., Filippenko, A. V., Sargent, W. L. W. 1997, ApJS, 112, 315

Jarvis, J. F. and Tyson, J. A. 1981, AJ, 86, 476

Soifer, B. T., Sanders, D. B., Madore, B. F., Neugebauer, G., Danielson, G. E., Elias, J. H., Lonsdale, C. J., Rice, W. L. 1987, ApJ, 320, 238

Véron-Cetty, M. -. and Véron, P. 1998, A Catalogue of quasars and active nuclei, Edition: 8th ed., Publisher: Garching: European Southern Observatory (ESO), 1998, Series: ESO Scientific Report Series vol no: 18.

Wen-shuo Liao

Ken Ohsuga

HI IMAGING OF SEYFERT GALAXIES

Wen-shuo Liao
National Taiwan University, Taiwan
wsliau@asiaa.sinica.edu.tw

Jeremy Lim
Academia Sinica Institute of Astronomy & Astrophysics, Taiwan
jlim@asiaa.sinica.edu.tw

Paul T. P. Ho
Smithsonian Astrophysical Observatory, USA
ho@cfa.harvard.edu

Abstract   We present results from an ongoing program to image Seyfert galaxies in neutral atomic-hydrogen (HI) gas. Our objectives are to study the distribution and motion of gas in a relatively large sample of Seyfert galaxies, and to search for gaseous disturbances caused by any galaxy-galaxy interaction. Many of our images reveal dramatic evidence for tidal interactions between the Seyfert and neighboring galaxies, even when no such interactions are readily visible in the optical. In a number of other cases, our images reveal the presence of gas-rich but optically-dim companion galaxies (that would otherwise have been easily overlooked in the optical) that may be interacting with the Seyfert galaxy. The gaseous disturbances caused by galaxy-galaxy interactions may be responsible for directing gas to the central supermassive black holes of these Seyfert galaxies.

Keywords: Galaxies: active – Galaxies: interactions – Galaxies: Seyferts – Radio lines: galaxies

Introduction

What triggers AGNs in Seyfert galaxies? Nuclear bars are frequently invoked to channel gas from the ISM of these galaxies to their central supermassive black holes, but the presence of such bars in most Seyfert galaxies remains uncertain (see review by Mulchaey et al. 1997). The other mechanism frequently invoked to trigger AGNs in disk galaxies is galaxy-galaxy interactions, which can produce radial gas inflows. Optical observations, however, do not reveal direct evidence for such interactions in most Seyfert galaxies, and the case for an excess of projected neighboring galaxies around Seyfert galaxies remains controversial (see review by de Robertis et al. 1998).

To study the distribution and kinematics of gas in Seyfert galaxies, we have selected 30 Seyfert galaxies at redshifts $0.015 \leq z \leq 0.017$ and declinations $\delta > 0°$ compiled in the catalog of Véron-Cetty & Véron (2000) for mapping in HI with the Very Large Array (VLA). This is the closest redshift range in which AGNs have luminosities approaching those of QSOs ($M_B < -23$ for $H_0 = 50 \text{km s}^{-1} \text{Mpc}^{-1}$); the host galaxies of QSOs are being mapped in a separate project with the VLA (Lim et al., this volume). The selected Seyfert galaxies allow us to study whether tidal interactions continue to play an important role in triggering AGNs at lower nuclear luminosities, and if not at what nuclear luminosity other mechanisms take over.

1. Results and Discussion

We have so far reduced the data for eight fields (containing nine Seyfert galaxies) in our survey. All the Seyfert galaxies were detected in HI, with gas masses ranging from about $2\text{--}40 \times 10^9$ $M_\odot$ (for $H_0 = 70 \text{km s}^{-1} \text{Mpc}^{-1}$). In Fig. 1, we show our HI images along with optical images of these galaxies

In four cases, we find that the Seyfert galaxies show disturbances in HI even where no disturbances can be seen in optical starlight (i.e., NGC 7679, NGC 7682, Akn 539, and Mark 1). In two cases where the Seyfert galaxies can be seen to be optically interacting with a nearby companion galaxy, our HI images do not have sufficient resolution to properly separate the two interacting galaxies (i.e., UGC 3995A and Mark 1040). In two other cases, we detected HI-gas-rich but optically-dim companion galaxies next to apparently optically undisturbed Seyfert galaxies; these companion galaxies would otherwise have been easily overlooked in the optical (i.e., Mark 993 and UGC 3157). One of these Seyfert galaxies is probably interacting with its companion galaxy (Mark 993), whereas in the other case observations at higher angular resolutions are required to reveal any such interactions (UGC 3157). Interestingly, the Seyfert galaxy classified as a LINER does not show any neighboring galaxies, nor does its HI appears to be disturbed (NGC 1167).

In summary, at least seven of the eight Seyfert galaxies show interactions with companion galaxies. The gaseous disturbances caused by these interactions may be responsible for directing a large amount of gas to their central supermassive black holes.

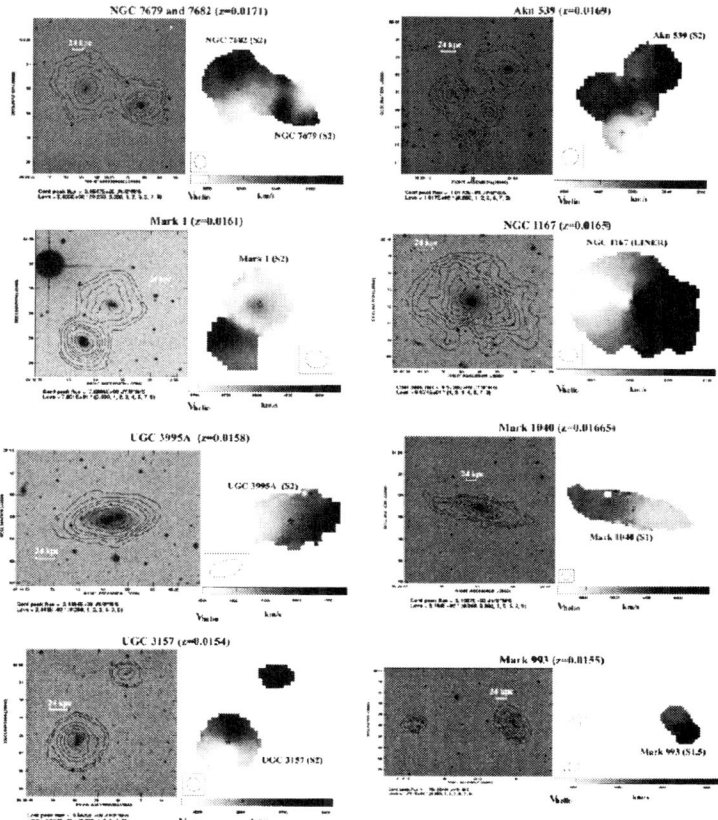

Figure 1. For each Seyfert galaxy, we show: Left panels — Contours of the integrated HI intensity overlaid on grayscale optical images from the Digitized Sky Survey; Right panels — grayscale-coded images of the HI intensity-weighted mean velocity with a velocity resolution of ~ 20 km s^{-1}. The angular resolution of the individual observations is shown by the ellipse at the lower corner side of each panel. We assume a cosmology of $H_0 = 70$km s^{-1} Mpc^{-1} and $\Omega = 1$

Acknowledgments

This project is supported in part by the National Science Council (NSC) of Taiwan as a grant to J. Lim and a Masters student stipend to Wen-shuo Liao. The latter also thanks the Foundation for the Advancement of Outstanding Scholarship of Taiwan for financial support to attend this workshop.

References

de Robertis, M. M., Kayhoe, K., Yee H. K. C., 1998, ApJSS, 115,163
Lim, J., Ho, P. T. P., 1999, ApJ, 510, L7
Muchaey, J. S., Regan, M. W., 1997, ApJ, 482, L135
Véron-Cetty, M. P., Véron, P. A., 2000, A Catalogue of Quasars and Active Galactic Nuclei (ESO Sci. Rept.) (9th ed.; Garching:ESO)

STARBURST-AGN CONNECTION; REGULATED BY RADIATIVELY-SUPPORTED OBSCURING WALLS

Ken Ohsuga and Masayuki Umemura
Center for Computational Physics, University of Tsukuba, Japan.
ohsuga@rccp.tsukuba.ac.jp

Abstract To examine a novel model for obscuration of the AGN, whereby the AGN is hidden by the obscuring walls supported by radiation force due to a circumnuclear starburst, we attempt to confront the condition for the formation of the walls with observational data of existing AGNs. As a result, it is found AGNs are actually identified as type 2 in the case of satisfying the condition of the obscuring wall formation. Thus, the obscuration of the AGN by the radiatively-supported walls is a plausible mechanism which regulates the AGN type and links between the AGN type and starburst phenomena.

Keywords: AGNs, QSOs, Starburst, Obscuration, Radiative transfer, Evolution

1. Observational Data

A novel model for the obscuration of the AGN has been suggested by Ohsuga & Umemura (1999, 2001b), whereby the AGN is surrounded by the inner and outer obscuring walls and observed as type 2 AGN if the starburst is luminous and the AGN is relatively faint. To examine this picture, we apply it to existing AGNs in this paper.

The data of the AGN and the circumnuclear starburst are shown in Table 1 and 2. NGC 1068 and NGC 7469 are IRAS galaxies observed by Scoville et al. (1991), and the X-ray luminosities are measured by ROSAT (Pérez-Olea & Colina 1996). Here, we adopt the IR luminosities as the bolometric luminosities of the starburst regions, and the dynamical/gas masses as the masses of the starbursts. Also, we assume that the intrinsic AGN luminosities are about ten times as much as the X-ray luminosities. On the other hand, González Delgado et al. (1998), Maiolino et al. (1998), Schinnerer, Eckart & Tacconi (1998), Brotherton et al. (1999), and Ohsuga & Umemura (2001a) have estimated the bolometric luminosities and masses of the surrounding starburst regions based on the synthetic starburst model (see Table 2). The AGN luminosity is given by the X-ray observations or comparison with other Seyfert galaxy, NGC 1068. Since the nucleus of the Circinus is thought to be heavily obscured, we adopt a nuclear starburst as a central radiation source.

Table 1. The luminous infrared galaxies containing the Seyfert nucleus of the AGN

Name	Type	L_{IR} $(10^{11} L_\odot)$	L_X $(10^8 L_\odot)$	M_{gas} $(10^{10} M_\odot)$	M_{dyn} $(10^{10} M_\odot)$	L_{SB}/M_{SB} $(L_\odot/M_\odot)$
NGC 1068	Sy2	1.5	2.4	0.45	...	33
NGC 7469	Sy1	2.6	27	0.74	1.06	25

Table 2. The properties of the starbursts estimated by the synthetic starburst model and the bolometric luminosity of the AGN

Name	Type	M_{SB} $(10^9 M_\odot)$	L_{SB} $(10^{10} L_\odot)$	L_{AGN} $(10^{10} L_\odot)$	L_{SB}/M_{SB} $(L_\odot/M_\odot)$
Circinus	Sy2	0.25-1.0	1.1	0.2	11-40
I ZW 1	QSO/NLSy1	19	6.3	1.3	3.3
UN J1025-0040	QSO	70	40	40	5.7
Mrk 477	Sy2	4.9×10^{-2}	3.6	4.0 − 15	7.8×10^2
NGC 7130	Sy2	9×10^{-3}	1.1	0.4 − 3	1.2×10^3
NGC 5135	Sy2	9×10^{-3}	1.1	<2	1.2×10^3
IC 3639	Sy2	3×10^{-3}	0.3	0.3 − 1.3	10^3

2. Confrontation with observations

We plot the data of these galaxies on the diagram which represents the condition for the formation of obscuring walls in Fig. 1. Here, type 1 and type 2 AGNs are shown by open and filled circles, respectively. Also, the age of the starburst based on the simple stellar evolution model is shown in the upper abscissa (Ohsuga & Umemura 2001b). As shown in this figure, all Sy2s are located in the area of wall formation, and two QSOs are located in no wall area. Hence, it is confirmed that the radiation force can work to build up the inner and outer obscuring walls, which contribute to the obscuration of circumnuclear regions in Seyfert 2 galaxies. In the case of NGC 7469, it is expected that the outer wall forms whereas no inner wall is built up. NGC 7469 seems to have been shifted from type 2 to type 1 about $\sim 10^7$ yr ago.

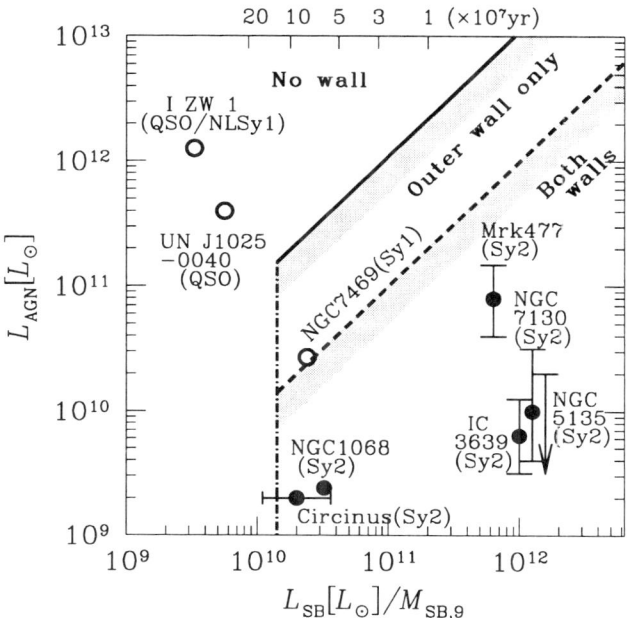

Figure 1. The observed properties of Seyfert galaxies in the $L_{SB}/M_{SB,9}$ - L_{AGN} diagram with the condition for the formation of the obscuring walls, where $M_{SB,9}$ is the mass of a starburst in units of $10^9 M_\odot$.

References

Brotherton, M. S., van Breugel, W., Stanford, S. A., Smith, R. J., Boyle, B. J., Miller, L., Shanks, T., Croom, S. M., Filippenko, A. V., 1999, ApJ, 520, L87
González Delgado, R. M., Heckman, T., Leitherer, C., Meurer, G., Krolik, J., Wilson, A. S., Kinney, A., Koratkar, A., 1998, ApJ, 505, 174
Maiolino, R., Krabbe, A., Thatte, N., Genzel, R., 1998, ApJ, 493, 650
Ohsuga, K., Umemura, M., 1999, ApJ, 521, L13
Ohsuga, K., Umemura, M., 2001a, submitted to A&A
Ohsuga, K., Umemura, M., 2001b, submitted to ApJ
Pérez-Olea, D. E., Colina, L., 1996, ApJ, 468, 191
Schinnerer, E., Eckart, A., Tacconi, L. J., 1998, ApJ, 500, 147
Scoville, N. Z., Sargent, A. I., Sanders, D. B., Soifer, B. T., 1991, ApJ, 366, L5

ROTATION CURVE AND MEAN STELLAR POPULATION OF THE I ZW 1 QSO HOST

J. Scharwächter[1], A. Eckart[1] and S. Pfalzner[1,2]
[1] *I.Physikalisches Institut, University of Cologne, Zülpicher Straße 77, 50937 Köln, Germany*
[2] *University of Jena, Schillergaesschen 3, D-07745 Jena, Germany*

Introduction

With a redshift of $z = 0.0611$ (Condon et al. 1985), corresponding to a distance of 244 Mpc ($H_0 = 75$ km s^{-1} Mpc^{-1}, $q_0 = 0$), I Zw 1 is one of the closest QSOs embedded in a gas-rich host galaxy disk. The optical disk has a diameter of about 32 kpc (26″.7) and shows spiral arms. Two bright knots in the east and the north of the disk are a nearby companion and a foreground star, respectively (Stockton 1982). The dynamics of the ISM in the I Zw 1 host galaxy have been investigated by Schinnerer et al. (1998) and Staguhn, Schinnerer & Eckart (2000) by means of millimetric CO line emission. We used their position-velocity diagrams to analyse the rotation curve with respect to dark matter (DM) content and mass-to-light ratio.

1. Analysis of the Rotation Curve

The observed rotation curve of I Zw 1, corrected for an inclination of 38°± 5° (Schinnerer et al. 1998), shows a steep rise up to a maximum circular velocity of about 360 km s^{-1} and flattens out for larger radii (Fig. 1). This typical shape suggests a high surface brightness galaxy (comp. de Blok et al. 1996). Various investigations (e.g. de Blok et al. 1997, Persic et al. 1996) on DM in spiral galaxies have revealed that the inner parts of the rotation curves of high surface brightness galaxies are generally well reproduced by the stellar mass contribution while in the case of lower surface brightness an increasing amount of DM is required. I Zw 1 should therefore exhibit no remarkable amount of DM within the optical disk. In order to confirm this result the observed rotation curve was compared with the one merely resulting from the contributions of the luminous matter in the host galaxy (Fig. 1). The latter was obtained

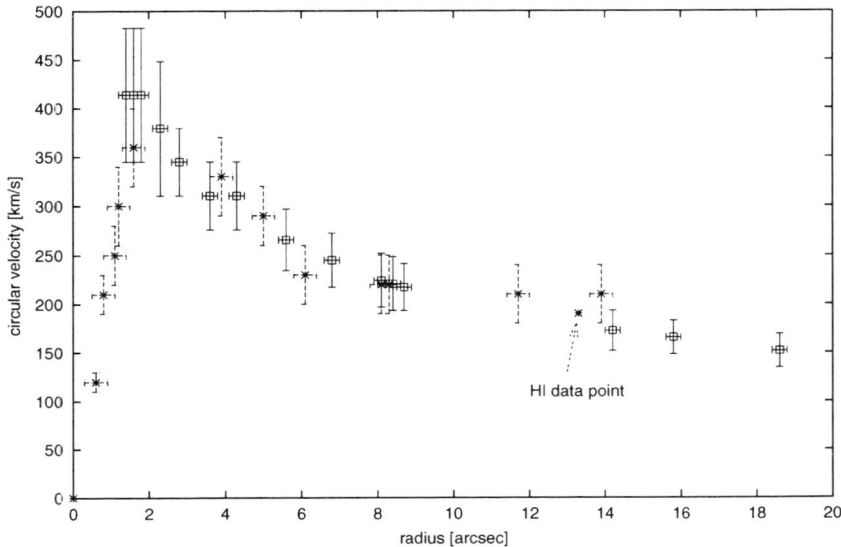

Figure 1. Comparison of the observed rotation curve (crosses) of the I Zw 1 host galaxy with the rotation curve (boxes) produced by luminous matter only.

from an R-band contour image of I Zw 1 (Hutchings & Crampton 1990), assuming a constant mass-to-light ratio, via

$$v = \sqrt{\frac{G\, M_{in}(r)}{r}} \propto \sqrt{\frac{G\, L_{in}(r)}{r}},$$

where G is the gravitational constant and $M_{in}(r)$ or $L_{in}(r)$ are the off-nuclear integrated mass or luminosity up to radius r. At least inside a radius of about 10″.0, no significant contribution of DM was required to reproduce the observed rotation curve. The proportionality factor results in a constant mass-to-light ratio of $\Upsilon = (2.77 \pm 0.28)\, M_\odot/L_\odot$ for the host galaxy of I Zw 1.

2. Mean Spectral Type of the I Zw 1 Host Galaxy

Besides the analysis of the rotation curve we also used another approach to determine the mass-to-light ratio or, equivalently, the mean spectral type of the I Zw 1 QSO host and the visual extinction of stellar light along the line of sight. For different values of the two fit parameters (spectral class or mass-to-light ratio and visual extinction) we computed the stellar $B - V$ color on the one hand and, assuming a uniform stellar population of the respective type, the expected I Zw 1 host galaxy's $B - V$ color on the other hand. The observed off-nuclear $B - V$ color of 0.6 − 0.7 mag (Hutchings & Crampton 1990) and its variation with visual extinction were used for further comparison. The stellar masses, absolute luminosities, L_{star}, and magnitudes, M_V^{star}, were taken from Lang (1992). Mass-to-light ratios of the galaxy were computed with the dynamical mass of $(3-7) \times 10^{11} M_\odot$ (Eckart et al. 1994) and an absolute luminosity of

$$L_{gal} = L_{star} \times 10^{-0.4 \times (M_V^{gal} - A_V - M_V^{star})}.$$

According to the flat shape of the broad band spectrum from the V-band to the R-band (Schinnerer et al. 1998), M_V^{gal} was approximated by the absolute off-nuclear R-band magnitude $M_R^{gal} = (-22.6 \pm 0.2)$ mag obtained by integration of the R-band contour image of Hutchings & Crampton (1990).

The best correspondence of all three $B - V$ colors is achieved at $A_V \approx 0$ with a mean spectral type of G9 V - K5 V and a mass-to-light ratio of $\Upsilon = (2.5 \pm 0.5) \, M_\odot/L_\odot$. We are aware that this method includes several inaccuracies, above all a strong dependence on the applied inclination of I Zw 1. However, within the error bars, the mass-to-light ratio agrees with the value obtained from the analysis of the rotation curve. The low mean visual extinction towards the disk is also supported by its blue colors and the fact that I Zw 1 is seen nearly face-on in the optical images (Bothun et al. 1984, Hutchings & Crampton 1990, McLeod & Rieke 1995).

References

Bothun, G. D., Heckman, T. M., Schommer, R. A., Balick, B., 1984, AJ, 89, 1293
Condon, J. J., Hutchings, J. B., Gower, A. C., 1985, AJ, 90, 1642
de Blok, W. J. G., McGaugh, S. S., van der Hulst, J. M., 1996, MNRAS, 283, 18
de Blok, W. J. G., McGaugh, S. S. 1997, MNRAS, 290, 533
Eckart, A., van der Werf, P. P., Hofmann, R., Harris, A. I., 1994, ApJ, 424, 627
Hutchings, J. B., Crampton, D., 1990, AJ, 99, 37
Lang, K. R., 1992, "Astrophysical Data: Planets and Stars", Springer-Verlag NY
McLeod, K. K., Rieke, G. H., 1995, ApJ, 441, 96
Persic, M., Salucci, P., Stel, F., 1996, MNRAS, 281, 27
Schinnerer, E., Eckart, A., Tacconi, L. J., 1998, ApJ, 500, 147
Staguhn, J., Schinnerer, E., Eckart, A., 2000, Proc. of IAU Symp. No.205, ASP Conf. Series, in press.
Stockton, A., 1982, ApJ, 257, 33

V

QSO AT HIGH z AS TRACERS OF STRUCTURES. GALAXY FORMATION.

Carlton Baugh (left) and George Miley (right)

SEMI-ANALYTIC GALAXY FORMATION: UNDERSTANDING THE HIGH REDSHIFT UNIVERSE

C. M. Baugh[1], A. J. Benson[2], S. Cole[1], C.S. Frenk[1] and C.G. Lacey[3]
[1]*Dept. of Physics, University of Durham, Durham, DH1 3LE, UK.*
[2]*California Institute of Technology, MC 105-24, Pasadena, CA 91125-2400, USA.*
[3]*SISSA, via Beirut 2-4, 34014 Trieste, Italy.*

Abstract There is now compelling evidence in favour of the hierarchical structure formation paradigm. Semi-analytic modelling is a powerful tool which allows the formation and evolution of galaxies to be followed in a hierarchical framework. We review some of the latest developments in this area before discussing how such models can help us to interpret observations of the high redshift Universe.

1. Hierarchical structure formation

The hierarchical structure formation paradigm is based upon the simple premise that large scale structure in the Universe results from the gravitational amplification of small, primordial density fluctuations. The origin of the fluctuations is uncertain, but one explanation is that they are quantum ripples boosted to macroscopic scales by inflation.

Many clear examples of interacting or merging galaxies, a key feature of hierarchical models, were presented during this meeting. Further convincing evidence for the paradigm can be derived by comparing the relative amplitudes of density fluctuations in Universe today with those present at some earlier epoch. The cosmic microwave background radiation is a snapshot of the distribution of photons and baryons just a few hundred thousand years after the Big Bang. Fluctuations in the temperature of the background radiation can be related to fluctuations in the distribution of baryons at the epoch of recombination, $z \sim 1000$. The inferred fluctuations are tiny, on the order of one part in a hundred thousand. However, if an additional component to the mass density of the Universe is included, weakly interacting cold dark matter, these fluctuations can subsequently develop into the large scale structure that we measure in the Universe today (Peacock et al. 2001).

An important challenge for theorists is to predict the formation and evolution of galaxies in a model universe in which the formation of structure in the dark matter proceeds in a hierarchical manner. Two powerful simulation techniques have been developed to address this issue: direct N-body or grid codes that follow the dynamical evolution of dark matter and gas, and semi-analytic codes that use a set of simple, physically motivated rules to model the complex physics of galaxy formation. These techniques have their advantages and disadvantages (e.g. limited resolution in case of N-body/grid based codes; the assumption of spherical symmetry for cooling gas in the semi-analytics), and so are complementary tools with which to attack the problem of galaxy formation. A preliminary study comparing the cooling of gas and merging of "galaxies" in a Smooth Particle Hydrodynamics simulation with the output of a semi-analytic code has shown that there is reassuringly good agreement between the results obtained using the two techniques (Benson et al. 2001a).

2. The Durham semi-analytic code

The past decade witnessed an explosion in observations of galaxies at high redshift, mainly as a result of new facilities such as the Hubble Space Telescope and the Keck telescopes in the optical, and the opening of other parts of the electromagnetic spectrum, e.g. the sub-millimetre, probed by the SCUBA instrument on UKIRT. In order to interpret these exciting new data, semi-analytic galaxy formation codes have been developed that model a wide range of physical processes. Below, I will outline the scheme developed by the Durham group and collaborators (Benson et al. 2000a, Cole et al. 2000, Granato et al. 2000). Similar codes have also been devised by other groups (e.g. Avila-Reece & Firmani 1998, Kauffmann et al. 1999, Somerville & Primack 1999).

The physical processes that play a fundamental role in hierarchical galaxy formation can be set out as follows (White & Rees 1978):

(i) The formation and merging of dark matter haloes, driven by gravitational instability. This process is completely determined by the initial power spectrum of density fluctuations and by the values of the cosmological parameters Ω, Λ and Hubble's constant.

(ii) The shock heating and virialisation of gas within the gravitational potential wells of dark matter haloes.

(iii) The cooling of gas in haloes.

(iv) The formation of stars from cooled gas. This process is regulated by the injection of energy into the cold gas by supernovae and stellar winds.

(v) The mergers of galaxies after their host dark matter haloes have merged.

There are a number of major improvements in the Cole et al. (2000) semi-analytic code over earlier versions: a more accurate technique is used in the Monte-Carlo generation of dark matter halo merger trees, the chemical enrichment of the ISM is followed, disk and bulge scale lengths are computed using a prescription based on conservation of angular momentum and the obscuration of starlight by dust is computed in a self-consistent fashion.

The semi-analytic model requires a number of physical parameters to be set. Some of these describe the background cosmology and are gradually being pinned down, for example, by measurements of supernovae brightnesses at high redshift or through the production of high resolution maps of the microwave background radiation. Other parameters refer to the prescriptions we adopt to model the physics of galaxy formation. Their values are set by reference to a subset of data on the local galaxy population, as explained by Cole et al.

The observational constraint to which we attach the most weight is the field galaxy luminosity function. Somewhat disappointingly, and in spite of much effort, this fundamental characterisation of the local galaxy distribution was not well known until this year. Fig. 5 of Cole et al. (2000) shows that any semblance of a consensus between the various determinations of this quantity prior to 2000 is lost even after moving just one magnitude faintwards of L_*. However, this situation is now changing beyond recognition. The 2dF Galaxy Redshift Survey (2dFGRS) and Sloan Digital Sky Survey are pinning down the field galaxy luminosity function to a high level of accuracy. The degree of improvement that is now possible with the 2dFGRS is readily apparent in Fig. 1. In this figure, we compare measurements obtained from the 2dFGRS with a representative determination of the luminosity function made from a redshift survey completed in the last millenium. For the first time, random errors in the luminosity function estimate are unimportant over a wide range of magnitudes.

The solid lines in Fig. 1 show the luminosity function of the Cole et al. model. The faint end is influenced by the strength of feedback in low mass haloes. The break at high luminosities is due to long cooling times in more massive dark matter haloes, which have higher virial temperatures and form more recently in hierarchical models. Assuming

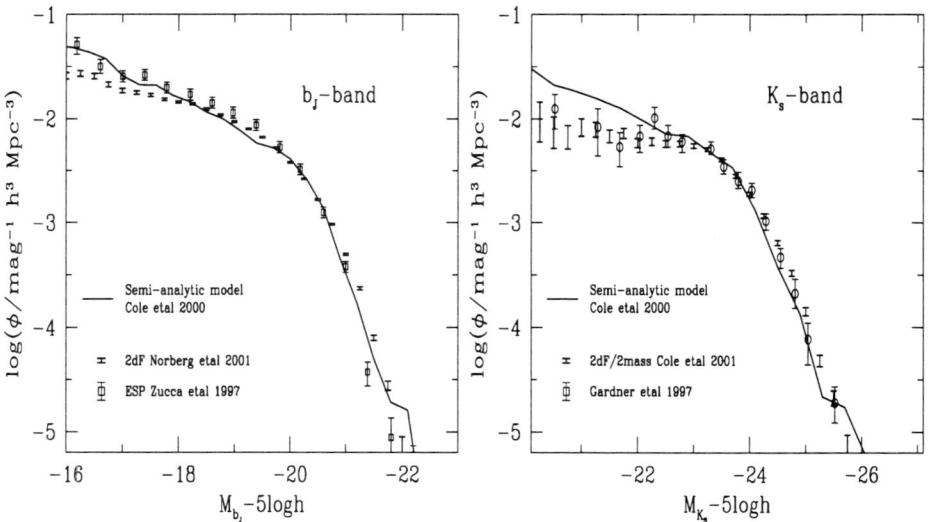

Figure 1. The field luminosity function in the b_J and K_s bands. The error bars without points show the luminosity functions estimated from the 2dF Galaxy redshift survey (the measurement in the right hand panel uses photometry from the near-infrared 2mass survey). The points with errorbars show a representative determination of the luminosity function from an earlier redshift survey. The lines show the luminosity function of the semi-analytic model of Cole et al. (2000).

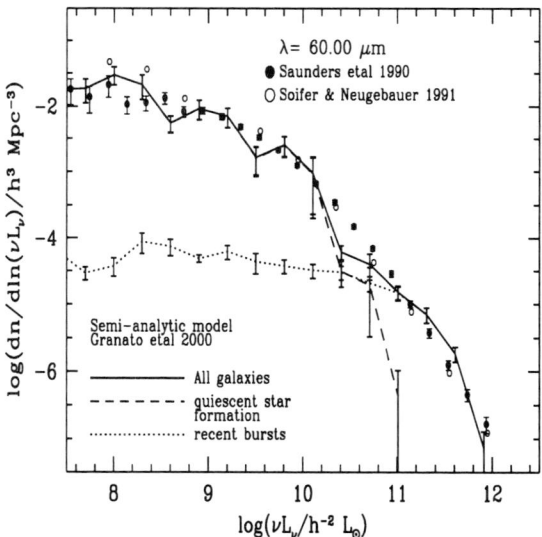

Figure 2. The $60\mu m$ luminosity function. The lines show the model predictions: the dashed line is the contribution of galaxies that are quiescently forming stars, the dotted lines correspond to galaxies that recently experienced or are undergoing a burst. The solid line is the total luminosity function.

a higher galaxy merger rate would depress the luminosity function at the faint end and weaken the break at the bright end. From a naive point of view, the model in Fig. 1 would be incorrectly dismissed as an abject failure due to an unacceptably large χ^2 value with reference to the 2dFGRS estimate of the luminosity function. However, it is important to appreciate that the parameters in the semi-analytic model are *physical* parameters. As such, they have a completely different meaning to the parameters that specify a Schechter function fit to these data, which is merely a convenient mathematical shorthand to describe the data points. We are not at liberty to chose any *ad hoc* combination of the parameters in the semi-analytic model. For example, changing the strength of feedback in order to reduce the slope of the faint end of the luminosity function also has an impact on the shape of the Tully-Fisher relation and upon the size of galactic disks.

In collaboration with Alessandro Bressan, Gian-Luigi Granato and Laura Silva, we have combined the semi-analytic model of Cole et al. with the spectro-photometric code of Silva et al. (1998), which treats the reprocessing of radiation by dust. The range of wavelengths spanned by the spectral energy distribution of model galaxies now extends from the extreme ultra-violet through the optical to the far-infrared, sub-millimetre and on to the radio (Granato et al. 2000). One highlight of this work is the reproduction of the observed smooth attenuation law for starbursts, starting from a dust mixture that reproduces the Milky Way extinction law which has a strong bump at 2000Å; this implies that the observed attenuation is strongly dependent on the geometry of stars and dust. In Fig. 2, we show the model predictions for the $60\mu m$ luminosity function. Above $\nu L_\nu \sim 10^{11} h^{-2} L_\odot$, the model luminosity function is dominated by galaxies undergoing bursts driven by mergers. This agrees with observations of ultra-luminous IRAS galaxies, which are all identified as being at some stage of the interaction/merger process (see Sanders' contribution).

2.1. Galaxy clustering at $z = 0$

One of the key science goals of the 2dF and SDSS redshift surveys is to produce definitive measurements of galaxy clustering over a wide range of scales for samples selected by various galaxy properties. In order to interpret the information encoded in the measured clustering, it is necessary to understand how galaxies illuminate the underlying distribution of dark matter. Progress has been made towards this end by marrying the semi-analytic galaxy formation technique with high

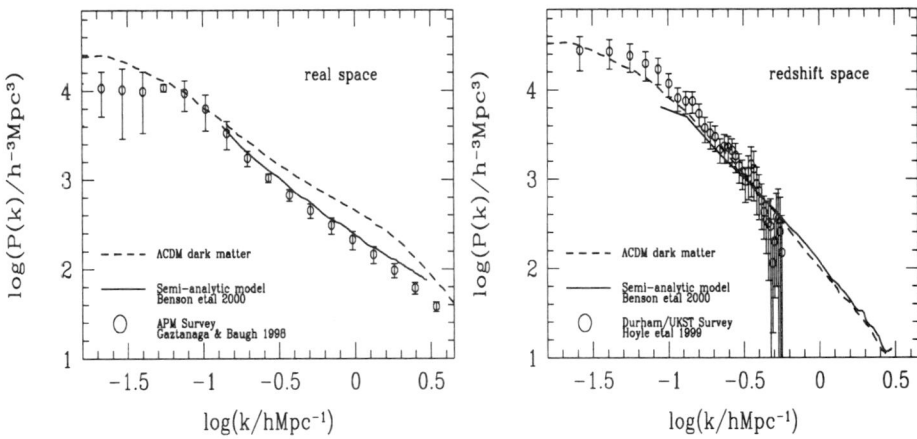

Figure 3. The power spectrum of galaxies at $z = 0$ in real space (left-hand panel) and redshift space (i.e. including the effects of peculiar motions - right-hand panel). The solid lines show the predictions for galaxies in the semi-analytic models, the dashed lines show the power spectrum of the underlying dark matter, which is a CDM universe with $\Omega_0 = 0.3$ and $\Lambda_0 = 0.7$. The points with errorbars show observational determinations of the power spectrum.

resolution N-body simulations of representative volumes of the universe (Kauffmann et al. 1999, Benson et al. 2000a,b, 2001b).

In the approach of Benson et al., the masses and positions of dark matter haloes are extracted from an N-body simulation using a standard group finding algorithm. The semi-analytic machinery is then employed to populate the dark haloes with galaxies. The central galaxy is placed at the centre of mass of the dark matter halo and satellite galaxies are placed on random dark matter particles within the halo, resulting in a map of the spatial distribution of galaxies within the simulation volume. Fig. 3 compares the power spectrum of bright, optically selected galaxies predicted by the semi-analytic model, with observational determinations and with the power spectrum of the dark matter. The left hand panel shows power spectra in real space. For $k \gtrsim 0.1 h\mathrm{Mpc}^{-1}$, the measured galaxy power spectrum has a lower amplitude than that of the dark matter in the popular ΛCDM model; the galaxies are said to be 'antibiased' with respect to the mass (Gaztañaga 1995). The semi-analytic model provides an excellent match to the data. This is particularly noteworthy as no additional tuning of parameters was carried out to make this prediction once certain properties of the local galaxy population, such as the field galaxy luminosity function, had been reproduced (see Cole et al. 2000 for a full explanation of how the model parameters are set). Furthermore, this level of agreement is not found for the galaxy clustering predicted in CDM models with $\Omega = 1$. The most important factor in shaping the predicted galaxy clustering amplitude is the way in which the efficiency of galaxy formation depends upon dark matter halo mass. This is illustrated by the variation in the mass to light ratio with halo mass shown by Fig. 8 of Benson et al. (2000a): for low mass haloes, galaxy formation is suppressed by feedback, whilst for the most massive haloes, gas cooling times are sufficiently long to suppress cooling.

The power of the approach of combining semi-analytic models with N-body simulations is demonstrated on comparing the left hand panel (real space) of Fig. 3 with the right hand panel, which shows power spectra in redshift space. Again, the same model gives a very good match to the observed power spectrum when the effects of peculiar motions are included to infer galaxy positions. However, the impression that one would gain about the bias between dark matter and galaxy fluctuations is qualitatively different; in redshift space galaxies appear to be unbiased tracers of the dark matter. The apparent contradiction between the implications for bias given by the panels of Fig. 3 can be resolved by turning back once more to the models. The pairwise velocity dispersion of model galaxies is lower than that of the dark matter, and as a result is in much better agreement with the observational determination of pairwise motions. Again, this difference is driven by a reduction in the efficiency of galaxy formation with increasing dark matter halo mass (Benson et al. 2000b).

2.2. The evolution of the galaxy distribution

Once the parameters of the semi-analytic model have been set by comparing the model output with a subset of data for the local galaxy population, firm predictions can be made regarding the evolution of the galaxy distribution (Benson et al. 2001b).

The properties of the distribution of galaxies and the way in which these properties evolve with redshift are intimately connected to the growth of structure in the dark matter, as illustrated by a sequence of high resolution pictures in Benson et al. (2001b) that show the evolution of galaxies and of the dark matter. An example of this is the morphology-density relation, namely the correlation of the fraction of early type galaxies with local galaxy density. The semi-analytic models reproduce the observed form of the morphology density relation at $z = 0$. Remarkably, essentially the same strength of effect is also predicted at

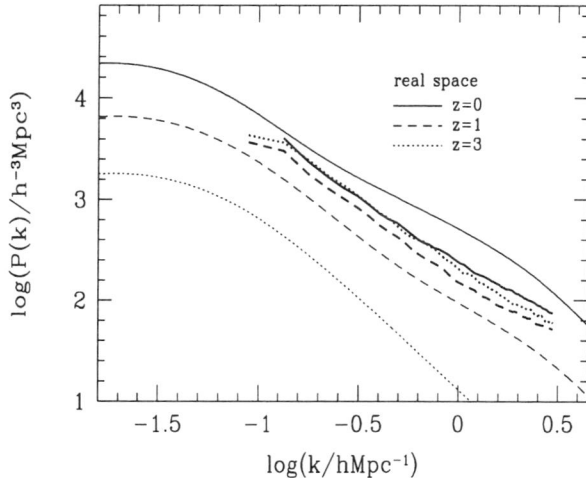

Figure 4. The evolution of galaxy clustering. The thick lines show the power spectrum predicted by the semi-analytic model for galaxies brighter than L_* (defined as $M_{b_J} - 5 \log h = -19.5$ in the rest frame b_J-band) at $z = 0, 1$ and 3. The light lines show the power spectrum of the dark matter at the same epochs, reading from top to bottom.

$z = 1$. The physical explanation for this result lies in the accelerated dynamical evolution experienced by galaxies that form in overdensities destined to become rich clusters by the present day.

A generic prediction of hierarchical clustering models is that bright galaxies should be strongly clustered at high redshift compared to the underlying dark matter (Davis et al. 1985). Fig. 4 shows the evolution of the power spectrum for galaxies and for dark matter in a ΛCDM universe. The amplitude of the dark matter power spectrum increases as fluctuations grow through gravitational instability. Between $z = 3$ and $z = 0$, the amplitude of the dark matter power spectrum increases by an order of magnitude on large scales. The shape of the dark matter power spectrum is significantly modified on small scales (high k) through nonlinear evolution of the density fluctuations – 'cross-talk' between fluctuations on different spatial scales. However, the amplitude and shape of the galaxy power spectrum show little change over the same redshift interval (Pearce et al. 2000, Benson et al. 2001b). The amplitude of the galaxy power spectrum drops by around 50% from $z = 3$ to $z = 1$, and by $z = 0$ it has been overtaken in amplitude by the mass power spectrum (Baugh et al. 1999). The clustering predictions can be readily explained. At $z = 3$, bright galaxies are only found in the most massive haloes in place at this time. Such haloes are much more strongly clustered than the underlying dark matter, hence the large difference in amplitude or bias between the galaxy and dark matter spectra at $z = 3$. The environment of bright galaxies becomes less exceptional as $z = 0$ is approached.

2.3. The formation and evolution of QSOs

The similarity in the general evolution of the global star formation rate per unit volume and of the space density of luminous quasars suggests a connection between the physical processes that drive the formation and evolution of galaxies and those that power quasars (see Dunlop's contribution). Spurred on by mounting dynamical evidence for the presence of massive black holes in galactic bulges (e.g. Magorrian et al. 1998), Guinevere Kauffmann and Martin Haehnelt have produced the first treatment to follow the properties of QSOs within a fully fledged semi-analytic model for galaxy formation (Kauffmann & Haehnelt 2000, Haehnelt & Kauffmann 2000).

The model of Kauffmann & Haehnelt assumes that black holes form during major mergers of galaxies, and that during the merger event, some fraction of the cold gas present is accreted onto the black hole to fuel a quasar. The qualitative properties of the observed quasar population are reproduced well by the model, including the rapid evolution in the space density of luminous quasars. There are three key features of the model responsible for the evolution in quasar space density between $z \sim 2$ and $z = 0$: (i) a decrease in the merger rate of objects in a fixed mass range over this interval, (ii) a reduction in the supply of cold gas from mergers, and (iii) an increase in the time-scale for gas accretion onto the black hole. The mass of cold gas available in mergers is reduced at low redshift because the star formation timescale in the model is effectively independent of redshift; at lower redshifts, gaseous disks have been in place for longer and a larger fraction of the gas has been consumed in quiescent star formation and so less gas is present in low redshift mergers (see Fig. 6 of Baugh, Cole & Frenk 1996). If the star formation timescale is allowed to depend upon the dynamical time, gas is consumed more rapidly in the disk and less gas is present in mergers at all redshifts.

The Kauffmann & Haehnelt model predicts strong evolution in the properties of QSO hosts with redshift, suggesting that quasars of a given luminosity should be found in fainter hosts at high redshift. This issue is just beginning to be addressed observationally (see, for example, the contributions of Ridgway and Kukula). At present, it is hard to reach any firm conclusions, though there is apparently little evidence for a strong trend in host luminosity with redshift.

The 2dF QSO redshift survey has recently reported measurements of the clustering in a sample of QSOs that is an order of magnitude larger than any previous sample (Hoyle et al. 2001). It should be relatively straight forward to obtain predictions for the clustering of quasars from the semi-analytic models to compare with these new data.

Acknowledgments

CMB would like to thank the organisers of this enjoyable meeting, for their hospitality and for providing financial support. We acknowledge the contribution of our GRASIL collaborators, Alessandro Bressan, Gian-Luigi Granato and Laura Silva to the work presented in this review. We thank Peder Norberg and the 2dF Galaxy Redshift Survey team for communicating preliminary luminosity function results.

References

Avila-Reese, V., Firmani, C., 1998, ApJ, 505, 37.
Baugh, C.M., Benson, A.J., Cole, S., Frenk, C.S., Lacey, C.G., 1999, MNRAS, 305, L21.
Baugh, C.M., Cole, S., Frenk, C.S., 1996, MNRAS, 283, 1361.
Benson, A.J., Cole, S., Frenk, C.S., Baugh, C.M., Lacey, C.G., 2000a, MNRAS, 311, 793.
Benson, A.J., Baugh, C.M., Cole, S., Frenk, C.S., Lacey, C.G., 2000b, MNRAS, 316, 107.
Benson, A.J., Pearce, F.R., Frenk, C.S., Baugh, C.M., Jenkins, A., 2001a, MNRAS, 320, 261.
Benson, A.J., Frenk, C.S., Baugh, C.M., Cole, S., Lacey, C.G., 2001b, MNRAS, submitted, astro-ph/0103092.
Cole, S., Lacey, C.G., Baugh, C.M., Frenk, C.S., 2000, MNRAS, 319, 168.
Cole, S. et al. (the 2dFGRS team), 2001, MNRAS, in press.
Davis, M., Efstathiou, G., Frenk, C.S., White, S.D.M., 1985, ApJ, 292, 371.
Granato, G.L., Lacey, C.G., Silva, L., Bressan, A., Baugh, C.M., Cole, S., Frenk, C.S., 2000, ApJ, 542, 710.
Gaztañaga, E., 1995, ApJ, 454, 561.
Gaztañaga, E., Baugh, C.M., 1998, MNRAS, 294, 229.
Haehnelt, M., Kauffmann, G., 2000, MNRAS, 318, L35.
Hoyle, F., Baugh, C.M., Shanks, T., Ratcliffe, A., 1999, MNRAS, 309, 659.
Hoyle, F., Outram, P.J., Shanks, T., Croom, S.M., Boyle, B.J., Loaring, N.S., Miller, L., Smith, R.J., 2001, MNRAS, submitted, astro-ph/0102163
Kauffmann, G., Colberg, J.M., Diaferio, A., White, S.D.M., 1999, MNRAS, 303, 188.
Kauffmann, G., Haehnelt, M., 2000, MNRAS, 311, 576.
Magorrian J. et al., 1998, AJ, 115, 2285
Peacock, J.A. et al., (the 2dFGRS team), 2001, Nature, 410, 169.

Pearce, F.R. et al., (the VIRGO consortium), 1999, ApJ, 521, L99.
Silva, L., Granato, G.L., Bressan, A., Danese, L., 1998, ApJ, 509, 103.
Somerville, R.S., Primack, J.R., 1999, MNRAS, 310, 1087.
White, S.D.M., Rees, M.J., 1978, MNRAS, 183, 341.

Masayuki Umemura and his daughter

Chris Impey

RADIATION-HYDRODYNAMICAL MODEL FOR THE QSO FORMATION

Masayuki Umemura
Center for Computational Physics, University of Tsukuba, Japan
umemura@rccp.tsukuba.ac.jp

Abstract Based on a novel radiation-hydrodynamical model for the formation of supermassive black holes, we propose a scenario for the QSO formation, whereby the evolution from ultraluminous infrared galaxies to QSOs is of direct consequence.

Keywords: QSO, Supermassive black holes, IRAS galaxies

Introduction

The paradigm that ultraluminous infrared galaxies (ULIRGs) could evolve into QSOs was proposed by a pioneering work by Sanders et al. (1988). By recent observations, the X-ray emission (Brandt et al. 1997) or Paα lines (Veilleux, Sanders & Kim 1999) intrinsic for active nuclei have been detected in roughly one fourth of ULIRGs. On the other hand, a lot of efforts have revealed that QSO host galaxies are mostly luminous and well evolved early-type galaxies (McLeod & Rieke 1995, Bahcall et al. 1997, Hooper, Impey & Foltz 1997, McLeod, Rieke & Storrie-Lombardi 1999, Brotherton et al. 1999, Kirhakos et al. 1999, McLure et al. 1999, McLure, Dunlop & Kukula 2000). These facts suggest that the QSO phenomena is closely related to the formation of the bulge.

In addition, the recent compilation of the kinematical data of galactic centers in both active galaxies and inactive ones has shown that a central 'massive dark object' (MDO), which is the nomenclature for a black hole (BH) candidate, correlates with the properties of galactic bulges. The demography of MDOs have revealed the following relations:
1) The BH mass exhibits a linear relation to the bulge mass with the ratio of $f_{\rm BH} \equiv M_{\rm BH}/M_{\rm bulge} = 0.001 - 0.006$ as a median value (Kormendy & Richstone 1995, Richstone et al. 1998, Magorrian et al. 1998,

Gebhardt et al. 2000a, Ferrarese & Merritt 2000, Merritt & Ferrarese 2001a).

2) The BH mass correlates with the velocity dispersion of bulge stars with a power-law relation as $M_{\rm BH} \propto \sigma^n$, $n = 3.75$ (Gebhardt et al. 2000b) or 4.72 (Ferrarese & Merritt 2000, Merritt & Ferrarese 2001a,b).

3) The $f_{\rm BH}$ tends to grow with the age of youngest stars in a bulge until 10^9 yr (Merrifield et al. 2000).

4) In disk galaxies, the mass ratio is significantly smaller than 0.01 if the disk stars are included (Salucci 2000, Sarzi et al. 2001).

5) For quasars the $f_{\rm BH}$ is of a similar level to that for elliptical galaxies (Laor 1998).

6) The $f_{\rm BH}$ in Seyfert 1 galaxies is not well converged, which may be considerably smaller than 0.01 (Wandel 1999, Gebhardt et al. 2000a) or similar to that for ellipticals (McLure 2001), while the BH mass-to-velocity dispersion relation in Seyfert 1 galaxies seems to hold good in a similar way to elliptical galaxies (Gebhardt et al. 2000a, Nelson 2000).

Comprehensively judging from all these findings on QSO hosts and supermassive BHs, it is likely that the formation of a QSO, a bulge, and a supermassive BH is mutually related. In this paper, based on a radiation-hydrodynamical mechanism for the formation of supermassive BHs, we propose a scenario for the QSO formation which predicts the evolution from ULIRGs to QSOs.

1. Formation of Supermassive Black Holes

A novel physical model which could account for the putative correlations between supermassive BHs and bulges is recently proposed by Umemura (2001), where the relativistic drag force by the radiation from bulge stars is considered. The radiation drag can extract angular momentum from gas and allow the gas to accrete onto the center (Umemura, Fukue & Mineshige 1997, 1998, Fukue, Umemura & Mineshige 1997). In an optically-thick regime, the radiation drag efficiency gets saturated due to the conservation of the photon number (Tsuribe & Umemura 1997). Given the total luminosity L_* in a uniform bulge, the mass accretion rate is estimated to be $\dot{M} = L_*/c^2$, which is numerically $\dot{M} = 0.1 M_\odot {\rm yr}^{-1} (L_*/10^{12} L_\odot)$ (see Umemura 2001 for the detail). This rate is comparable to the Eddington mass accretion rate for a black hole with $10^8 M_\odot$, $\dot{M}_{\rm Edd} = 0.2 M_\odot {\rm yr}^{-1} \eta^{-1} (M_{\rm BH}/10^8 M_\odot)$, where η is the energy conversion efficiency. Throughout this paper, the efficiency of 0.42 for an extreme Kerr black hole is assumed. The timescale of radiation drag-induced mass accretion is $t_{\rm drag} = 8.6 \times 10^7 {\rm yr} R_{\rm kpc}^2 (L_*/10^{12} L_\odot)^{-1} (Z/Z_\odot)^{-1}$, where $R_{\rm kpc} = R/{\rm kpc}$ with bulge radius R, and Z is the metallicity of gas. It is noted that the gas which is more abundant in metals accretes in a shorter timescale. The mass of an MDO is estimated by

$$M_{\rm MDO} = \int_0^t \dot{M} dt \simeq \int_0^t L_*/c^2 dt. \quad (1)$$

Next, we employ a simplest analytic model for bulge evolution. The star formation rate is assumed to be a Schmidt law, $S(t) = k f_g$. If we invoke the instantaneous recycling approximation, the star formation rate is given by $\dot{M}_*/M_b = k e^{-\alpha k t}$, where α is the net efficiency of the conversion into stars after subtracting the mass loss. The radiation energy emitted by a main sequence star is 0.14ε to the rest mass energy of the star, $m_* c^2$, where ε is the energy conversion efficiency of nuclear fusion from hydrogen to helium, which is 0.007. Thus, the luminosity of the bulge is estimated to be $L_* = 0.14\varepsilon k e^{-\alpha k t} M_b c^2$. By substituting this in (1),

$$M_{\rm MDO} = 0.14 \varepsilon \alpha^{-1} M_b (1 - e^{-\alpha k t}). \quad (2)$$

The term $M_b(1 - e^{-\alpha k t})$ represents just the stellar mass in the system which is $M_{\rm bulge}$ observationally. As a consequence, the MDO mass to bulge mass ratio is given by $M_{\rm MDO}/M_{\rm bulge} = 0.14\varepsilon \alpha^{-1} = 0.002 \alpha_{0.5}^{-1}$, where $\alpha_{0.5} = \alpha/0.5$. It should be noted that the final mass is basically determined by ε. If the BH accretion causes the nuclear activity, one should add the further radiative mass accretion induced by the nuclear luminosity $L_{\rm AGN}$. Finally, the BH mass to bulge mass ratio is predicted as

$$M_{\rm BH}/M_{\rm bulge} = 0.14 \varepsilon \alpha^{-1} (1-\eta)^{-1} = 0.003 \alpha_{0.5}^{-1}. \quad (3)$$

This is just comparable to the observed ratio.

If the BH mass is determined by the present mechanism, the BH mass to velocity dispersion relation is naturally understood in the context of a cold dark matter (CDM) cosmology. Supposing the bulge is a virialized system, then $GM_{\rm tot}/R = \sigma^2$ and $R \approx 0.5 R_{\rm max} \propto M_{\rm tot}^{1/3}(1+z_{\rm max})^{-1}$, where $R_{\rm max}$ is the radius at the maximum expansion epoch $z_{\rm max}$. If a CDM cosmology is assumed, then $(1+z_{\rm max}) \propto M_b^{-\beta}$, where $\beta \simeq 1/6$ around $M_b = 10^{12} M_\odot$, almost regardless of the cosmological parameters (Bunn & White 1997). Combining all these relations, we find $M_{\rm BH} \propto \sigma^n$ with $n = 6/(2-3\beta)$, which is 4 for $\beta = 1/6$. This result is just corresponding to the inferred relation between the BH mass and the stellar velocity dispersion.

The radiation drag efficiency would be strongly subject to the effect of geometrical dilution (Umemura, Fukue & Mineshige 1998). If the system is spherical, the emitted photons are effectively consumed within the system, whereas a large fraction of photons can escape from a disk-like system and thus the drag efficiency is considerably reduced. This may be the reason why $f_{\rm BH}$ is observed to be significantly smaller than 0.01.

2. QSO Formation

In this section, the QSO formation is addressed based on the radiation-hydrodynamical formation of supermassive BHs. Here, the MDO is distinguished from a supermassive BH, although they are often used in the same meaning. In the present picture, the MDO includes not only a supermassive BH but also a massive disk composed of the mass accreted by radiation drag. If the mass accretion onto a BH is limited by an order of the Eddington rate, the BH mass grows according to

$$M_{\rm BH} = M_0 e^{\nu t/t_{\rm Edd}}, \qquad (4)$$

where ν is the ratio of the BH accretion rate to the Eddington rate, $\nu = \dot{M}_{\rm BH}/\dot{M}_{\rm Edd}$, and $t_{\rm Edd}$ is the Eddington time-scale, $t_{\rm Edd} = 1.9 \times 10^8$ yr. The M_0 is the mass of a seed BH, which could be a remnant of a massive population III star with $10^{2-5} M_\odot$ (Carr, Bond & Arnett 1984, Nakamura & Umemura 2001). Since the MDO mass is given by (2), it exceeds $M_{\rm BH}$ in an early stage and stays greater than $M_{\rm BH}$ until a time $t_{\rm cross}$ when (4) equals (2). After $t_{\rm cross}$, the BH mass must be limited by (2). The timescale $t_{\rm cross}$ is estimated to be $t_{\rm cross} \approx 4\nu^{-1} t_{\rm Edd} \approx 10^9 \nu^{-1}$ yr. Recently, McLeod, Rieke & Storrie-Lombardi (1999) estimate the Eddington ratio to be $\nu \approx 0.1$ for QSOs.

However, before $t_{\rm cross}$ the interstellar gas may be blown out by a galactic wind. Kodama & Arimoto (1997) argue that the color-magnitude relation of bulges can be reproduced if a galactic wind sweeps away the gas at the epoch of a few 10^8 yr. After the galactic wind epoch $t_{\rm w}$, the bulge would evolve passively without the star formation episodes.

With considering $t_{\rm cross}$ and $t_{\rm w}$, we propose a possible scenario for the QSO formation, which is schematically shown in Fig. 1. Here, the system is assumed to be optically thick before $t_{\rm w}$ and become transparent at $t_{\rm w}$. At $t < t_{\rm w}$, the mass of stellar component in the bulge, $M_{\rm bulge}$, increases with continuous star formation and the MDO mass ($M_{\rm MDO}$) grows in proportion to $M_{\rm bulge}$ by radiation drag-induced mass accretion. The bulge luminosity (L_*) gradually decreases with decaying star formation rate according to the consumption of gaseous materials. This optically-thick bright phase may correspond to a ULIRG. In this

Radiation-Hydrodynamical Model for the QSO Formation 311

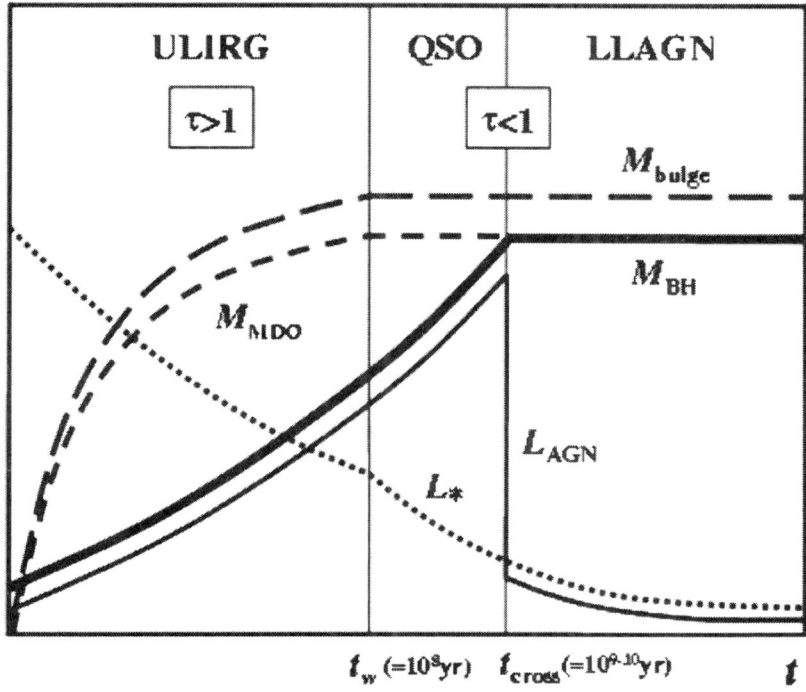

Figure 1. A schematic illustration of the present scenario for the QSO formation. The abscissa is time and the ordinate is arbitrary. $M_{\rm bulge}$ is the mass of stellar component in the bulge. $M_{\rm MDO}$ is the mass of the massive dark object (MDO). $M_{\rm BH}$ is the mass of the supermassive BH. L_* and $L_{\rm AGN}$ are the bulge luminosity and the black hole accretion luminosity, respectively.

picture, a ULIRG harbors a more or less active nucleus. At $t_{\rm w}$, the radiation drag-induced mass accretion practically stops owing to the reduced efficiency of radiation drag due to the small optical depth. But, $M_{\rm MDO}$ is still greater than $M_{\rm BH}$ and thus $M_{\rm BH}$ continues to grow until $t_{\rm cross}$. The AGN should brighten according to the growth of $M_{\rm BH}$ if the Eddington rate is constant. At $t_{\rm cross}$, all gas of the MDO falls onto the BH and then $M_{\rm BH}/M_{\rm bulge} = 0.003\alpha_{0.5}^{-1}$ is achieved. Simultaneously, the AGN luminosity ($L_{\rm AGN}$) drops abruptly, because the radiation drag-induced accretion becomes weak in the optically-thin passive evolution phase and also the energy conversion efficiency of an ADAF is proportional to $\dot{M}/\dot{M}_{\rm Edd}$. Such a nucleus would be a low luminosity AGN (LLAGN) as suggested by Kawaguchi & Aoki (2001). As a result, there is time delay between L_* and $L_{\rm AGN}$, and the AGN luminosity exhibits a peak around $t_{\rm cross}$. This peak phase may correspond to QSO phenomena. Finally,

M_{BH}/M_{bulge} is predicted to increase with L_{AGN} or age until t_{cross} and with the metallicity of the gas.

References

Bahcall, J. N. et al., 1997, ApJ, 479, 642
Brandt, W. N. et al., 1997, MNRAS, 290, 617
Brotherton, M. S., et al., 1999, ApJ, 520, L87
Bunn, E. F., White, M., 1997, ApJ, 480, 6
Carr, B. J., Bond, J. R., Arnett, W. D., 1984, ApJ, 277, 445
Davies, R. L. et al., 1983, ApJ, 266, 41
Ferrarese, L., Merritt, D., 2000, ApJ, 539, L9
Fukue, J., Umemura, M., Mineshige, S., 1997, PASJ, 49, 673
Gebhardt, K. et al., 2000a, ApJ, 539, L13
Gebhardt, K. et al., 2000b, ApJ, 543, L5
Hooper, E. J., Impey, C. D., Foltz, C. B., 1997, ApJ, 480, L95
Kawaguchi, T., Aoki, K. 2001, preprint
Kirhakos, S. et al., 1999, ApJ, 520, 67
Kodama, T., Arimoto, N., 1997, A&A, 320, 41
Kormendy, J., Richstone, D., 1995, ARA&A, 33, 581
Krolik, J. H., 2001, ApJ, in press (astro-ph/0012134)
Laor, A., 1998, ApJ, 505, L83
Laor, A., 2001, ApJ, in press (astro-ph/0101405)
McLeod, K. K., Rieke, G. H., 1995, ApJ, 454, L77
McLeod, K. K., Rieke, G. H., Storrie-Lombardi, L. J., 1999, ApJ, 511, L67
McLure, R. J. et al., 1999, MNRAS, 308, 377
McLure, R. J., Dunlop, J. S., Kukula, M. J., 2000, MNRAS, 318, 693
McLure, R. J., 2001, in this volume
Magorrian, J. et al., 1998, AJ, 115, 2285
Merrifield, M. R., Forbes, Duncan A., Terlevich, A. I., 2000, MNRAS, 313, L29
Merritt, D., Ferrarese, L., 2001a, MNRAS, 320, L30
Merritt, D., Ferrarese, L., 2001b, ApJ, 547, 140
Nakamura, F., Umemura, M., 2001, ApJ, 548, 19
Nelson, C. H., 2000, ApJ, 544, L91
Richstone, D. et al., 1998, Nature, 395A, 14
Salucci, P. et al., 2000, MNRAS, 317, 488
Sanders, D. B. et al., 1988, ApJ, 325, 74
Sarzi, M. et al., 2001, ApJ, in press (astro-ph/0010240)
Tsuribe, T., Umemura, M., 1997, ApJ, 486, 48
Umemura, M., 2001, preprint
Umemura, M., Fukue, J., Mineshige, S., 1997, ApJ, 479, L97
Umemura, M., Fukue, J., Mineshige, S., 1998, MNRAS, 299, 1123
Veilleux, S., Sanders, D. B., Kim, D.-C., 1999, ApJ, 522, 139
Wandel, A., 1999, ApJ, 519, L39

GRAVITATIONALLY LENSED QUASAR HOST GALAXIES

Chris Impey
Steward Observatory, University of Arizona, Tucson, AZ 85721, USA
cimpey@as.arizona.edu

Hans-Walter Rix
Max-Planck-Institut für Astronomie, Königstuhl 17, D-69117 Heidelberg, Germany
rix@mpia-hd.mpg.de

Brian McLeod
Harvard-Smithsonian Center for Astrophysics, 60 Garden Street, Cambridge, MA 02138, USA
bmcleod@cfa.harvard.edu

Chien Peng, Charles Keeton
Steward Observatory, University of Arizona, Tucson, AZ 85721, USA
cyp,ckeeton@as.arizona.edu

Chris Kochanek, Emilio Falco, Joseph Lehár*
Harvard-Smithsonian Center for Astrophysics, 60 Garden Street, Cambridge, MA 02138, USA
ckochanek,efalco,jlehar@cfa.harvard.edu

José A. Muñoz
Instituto Astrofísica de Canarias, E-38200 La Laguna, Tenerife, Spain

---

*Currently at the Center for Genome Research, One Kendall Square, Cambridge, MA 02139

Abstract   Gravitational lensing improves the contrast of a quasar host galaxy because the extended emission is stretched away from the nucleus at constant surface brightness. As a result, the lensing technique is sensitive to low luminosity hosts at high redshift, using samples that are unlikely to be biased by host galaxy properties. In data from the CASTLES project, lensed host light is detected in roughly 2/3 of the cases. Hosts of radio-quiet quasars have modest luminosity ($L < L_*$) at $z \sim 2$, and they are 2-5 times fainter than hosts of radio-loud quasars at the same epoch. The comparison with low redshift, radio-quiet hosts suggests rapid early growth of quasar black holes.

Keywords: Quasars, Gravitational lensing, Host galaxies

1. Using Nature's Telescope

Approximately 1 in 300 quasars is closely enough aligned with an intervening massive galaxy to be multiply imaged. "Nature's Telescope" helps us to see host galaxies, because gravitational lensing magnifies both the nuclear quasar emission and the surrounding host galaxy. For the unresolved nucleus this magnification simply results in a flux increase without observable changes in the spatial source extent. In contrast, the extended host emission is "stretched away" from the nucleus by lensing at a constant surface brightness. If the PSF falls off with radius $\propto r^{-3}$, as it does at large radii for the Hubble Space Telescope (HST), the net effect is that gravitational lensing strongly changes the local AGN-host contrast in favor of the extended light.

Everything comes with a price, and the increased host galaxy contrast that results from gravitational lensing comes at the expense of more complex modelling. First, some terminology. As light rays travel through the universe, they are subject to focusing by intervening concentrations of mass (for details and diagrams, see Blandford & Narayan 1992). The rays converge at a conjugate point, where the size is zero and the magnification is infinite. The locus of conjugate points is a caustic sheet. The source plane intersects caustic sheets at caustic lines, and the images of caustic lines are critical lines. In the source plane, the tangential caustic separates 5-image regions from 3-image regions (the odd image highly demagnified and is never observed). For an extended source, or a point source quasar plus surrounding host galaxy, the light is sheared into arcs. For a source that entends over all four folds of the tangential caustic, the arcs join up to form a complete Einstein ring.

Fig. 1 gives a visual sense of how host galaxy light is imaged into an Einstein ring. The lens in this case is a singular isothermal ellipsoid (SIE) model. The trace of intensity along the Einstein ring forms a "pattern" that can only be understood in terms of a lens model. The ring

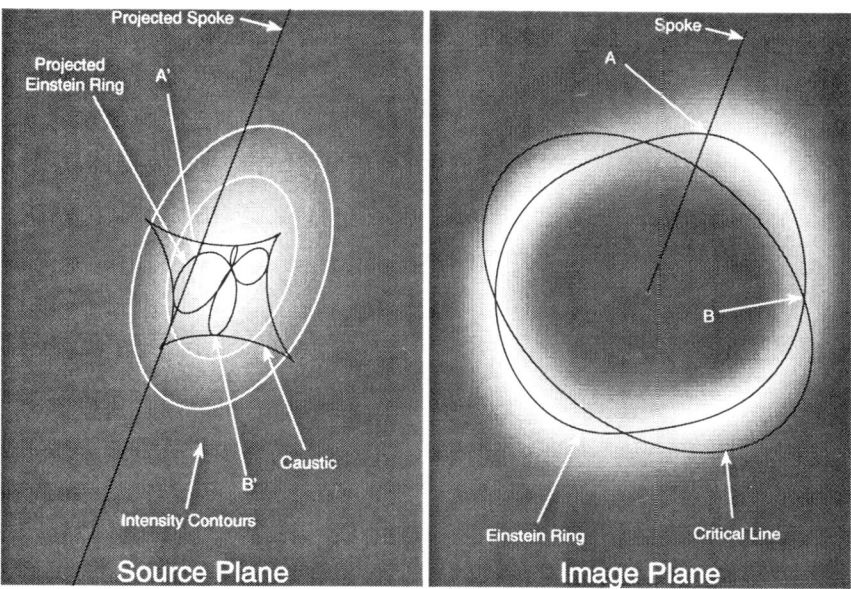

Figure 1. Ring formation by an SIE lens, from Kochanek, Keeton & McLeod (2001). An ellipsoidal source on the left (i.e. a host galaxy) is magnified by a factor of 2.5 into an Einstein ring on the right. The tangential caustic in the source plane and the critical line in the image plane are superimposed. Points A and B in the image plane map back to points A' and B' in the source plane.

in the image plane corresponds to the four-lobed pattern in the source plane. Intensity maxima in the Einstein ring mark the four images of the quasar point source. Intensity minima occur where the Einstein ring crosses the critical line, representing the four positions in the host galaxy where the caustic is tangential to the intensity contours. The intensity at an arbitrary position along the ring corresponds to the spoke in the image plane of Fig. 1, projecting to a host galaxy intensity at the point in the source plane where the spoke tangents the intensity contours.

Thus, the host galaxy intensity is measured at different positions relative to the quasar point source, and the intensity maps in a complex way into the image plane. With the spatial resolution of HST, the typical Einstein ring radius of 0.5-1 arcsecond corresponds to 50-100 resolution elements and that same number of independent measures of the host galaxy light. The modelling assumes only that the host is ellipsoidal, is roughly centered on the point source, and has intensity falls monotonically with distance from the point source. Optical Einstein rings are easier to interpret than radio rings, where the surface brightness distri-

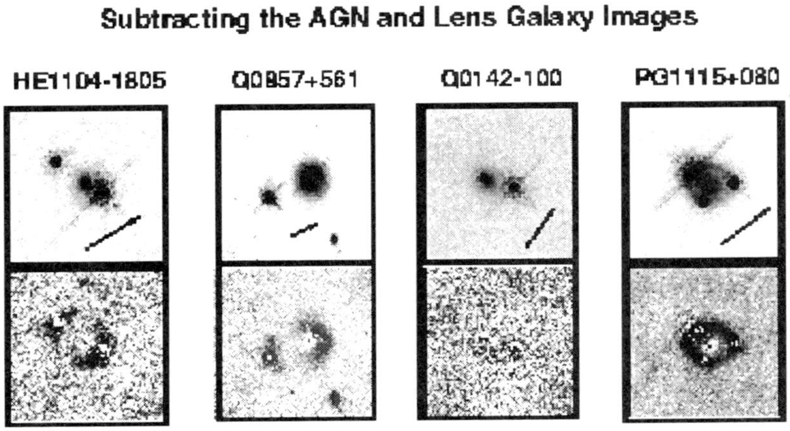

Subtracting the AGN and Lens Galaxy Images

Figure 2. Four examples of lensed quasar host galaxies from the CASTLES survey. The top panels are reduced H-band images; the lower panels show the images after subtracting and modelling the lens and quasar images. The source redshifts are $z = 2.32$ (1104−1805), 1.41 (0957+561), 2.72 (0142−100), and 1.72 (1115+080).

bution of the emission in the jets and lobes is poorly known a priori. In this initial work we will only model the total intensity.

2. The CASTLES Survey

Using the data from the CfA-Arizona Space Telescope Lens Survey (CASTLES, e.g. Kochanek et al. 2000, Lehár et al. 2000), we have begun a study of the host galaxies around lensed quasars. As the lens magnification is physically unrelated to the source structure, our sample of lensed quasars should not be biased towards more or less luminous hosts compared to other quasar samples of similar (de-magnified) luminosity. The identification of lenses depends in practice only on the nuclear or radio properties; therefore we should not expect the selection procedure to impact the stellar host galaxy luminosities.

We have initially restricted our attention to the "doubles" with optically bright AGN, where the de-magnified host fluxes can be reliably modelled, plus PG 1115+080 (Impey et al. 1998), and the physical pair Q2345+007(A/B). The sample is 12 objects from $z = 0.96$ to 4.5. The analysis is based on F160W imaging with NIC2 (except for NIC1 imaging of B 1030+074). The colors, and hence the constraints on stellar populations of the host galaxies, will be presented in a later paper. Details of the observations are given in Lehár et al. (2000); effective exposure times were typically one orbit or 30-40 minutes. Although the

NIC2 PSF morphology is quite complex, experiments with well-chosen PSF stars (McLeod et al. 2001) have shown that systematic PSF errors are usually at the 1% level or lower. All reduction steps were carried out with the software package NICRED, and images of sample systems before and after subtraction of the lens and the point sources are shown in Fig. 2. After decomposition, the SIE models provide a good description of the image and lens geometry for most systems.

3. Initial Results

We represented the unlensed source system by a point lensed source at the center of a simple host galaxy model (round, with a half-light radius of $R_{1/2} = 3h_{0.8}^{-1}$ kpc and an exponential profile) and match the corresponding image to the data. The de-magnified magnitudes are shown as a function of redshift in Fig. 3. The demagnified quasar point source luminosities ($M_B = -24$ to -27) are representative of the luminous QSO population at that epoch. We did not force the host flux to be positive definite, yet only for Q0142−100 (UM673) did the best fit result in a negative flux, implying a non-detection. It is very unlikely that we have underestimated the host luminosity by 1–3 magnitudes; image simulations show that such bright hosts would appear dramatically in the original un-subtracted images. Fig. 2 shows two SFHs that lead to L_* galaxies at the present epoch: in one, all stars formed at high redshift and simply faded subsequently (E-type), in the other, star formation is continuous, with SFR $\sim \exp(-t/t_{\text{Hubble}})$, similar to the Milky Way. These findings are consistent with a picture where luminous, mostly radio-quiet QSOs at all epochs live in typical galaxies (headed to become L_*) with unexceptional, Milky Way-like star formation histories.

4. Implications

As the right panel of Fig. 3 illustrates, for a given nuclear luminosity, most distant hosts are intrinsically fainter than even the least luminous low-redshift hosts. The difference becomes even larger, if one interprets the host luminosity in terms of a stellar host mass. The stellar populations at $z \sim 2$ are considerably younger than the host populations at $z < 0.5$ and therefore have lower M/L even in the H-band; for a Sb-type star formation history the M/L difference between $z = 2$ and $z < 0.5$ is indicated by the arrow. Clearly, for a given nuclear luminosity the distant host galaxies have less stellar mass than the $z < 0.5$ hosts. One conceivable explanation is that high-z QSOs produced luminosity more efficiently, while $M_{BH}/M_* \sim 0.003$ held at all epochs. In this case a smaller host galaxy, with a smaller BH, could sustain a higher central

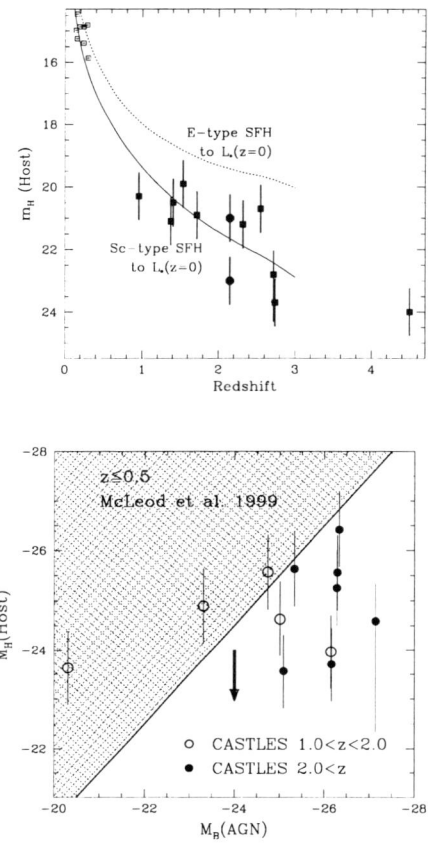

Figure 3. The left panel shows the H-band host magnitudes from the CASTLES survey, with low redshift data from McLeod et al. (1999). The lines give models that lead to L_* galaxies at the present epoch. The right panel compares nuclear luminosity to stellar host luminosity, with a limiting line at $M_{BH}/M_* \sim 0.003$ and $L \sim 0.5 L_{Edd}$.

luminosity. However, after accounting for the H-band M/L evolution, we would be led to conclude that most $z > 1$ QSOs radiate at super-Eddington luminosities, if the same $M_{BH}/M_* \sim 0.003$ relation were to hold. Instead, it appears more plausible to infer that the central BHs grew faster at early epochs than the surrounding hosts, in which case, a less massive and less luminous host galaxy contains a very massive BH and so can sustain a relatively more luminous nucleus.

Acknowledgments

Support for CASTLES was provided by NASA through grant numbers GO-7495, GO-7887 and GO-8804 from STScI. We would also like to thank Steve Beckwith for supporting the CASTLES project.

References

Blandford, R.D., Narayan, R., 1992, ARA&A, 30, 311
Impey, C.D. et al., 1998, ApJ, 509, 551
Kochenek, C.S. et al., 2000, ApJ, 535, 692
Kochanek, C.S., Keeton, C.R., McLeod, B.A., 2001, ApJ, 547, 50
Lehár, J. et al., 2000, ApJ, 536, 584
McLeod, B.A., Rieke, G.H., Storrie-Lombardi, L.J., 1999, ApJ, 511, L67
McLeod, K.K., McLeod, B.A., 2001, ApJ, 546, 782

Sebastian Rabien

Marek Kukula and Susan Ridgway

HOST GALAXIES OF LOW LUMINOSITY QUASARS AT INTERMEDIATE REDSHIFT

Sebastian Rabien and Matthew Lehnert
Max-Plank-Institut für extraterrestrische Physik, Postfach 1312, D-85741 Garching, Germany

Abstract

We report on observations of a sample of 18 faint radio-quiet quasars with redshifts from 0.8 to 2.7. This faint sample probes the luminosity range roughly at to 2 magnitudes higher than the QSO/Seyfert 1 divide at redshifts ranges little studied to date with such a sample. These QSOs have luminosities matching samples observed at low redshift, so we can compare the host measurements without effects that arise from quasar luminosity dependence. And to further facilitate the comparison with low redshift QSOs, the observations using the ESO NTT were carried out using filter bandpasses specifically chosen to observe at approximately rest-frame B or V in the optical and to avoid strong emission lines. We measured the properties of the quasars and the underlying host galaxies with PSF-subtraction and global modeling of the galaxies. We find that most of the QSO hosts can be fit with a "spiral-like" light profile, and that the low luminosity QSOs ($<M_B$ (nucleus)$>$ = -22) have hosts which have average magnitudes of $\approx L^*$ galaxies, while the more luminous nuclei ($<M_B$ (nucleus)$>$ = -24.5) are typically about 2 L^* galaxies, both results being independent of the redshift of the QSO. Broadly speaking, these results are consistent with more massive black holes being hosted by more massive galaxies, no evolution in the fueling rate with redshift or nuclear magnitude, and when combined with other known facts about the QSO population, favors a picture where most galaxies harbor black holes with few fueling episodes as opposed to a small subset of galaxies harboring black holes with many, longer-lived episodes of activity.

Keywords: Galaxies: quasar hosts

Introduction

Until recently, studies of quasars at high and intermediate redshift have almost exclusively focused on the most extreme radio-loud members of the AGN family. Such objects are the rarest within the active nuclei galaxies: for every radio-loud quasar with a radio flux sufficient to place it in the 4C radio catalog, there are of order 10^3 less radio bright quasars with similar absolute magnitudes at the same redshift. Since the samples studied so far have been flux limited, there is a strong correlation between redshift and quasar luminosity. Thus if z-dependent properties are found, it is not possible to determine whether these trends are more related to quasar luminosity or to cosmic epoch.

On the other hand we know that in the "quasar epoch" at z≈2–3, the amount of emission provided by quasars had most likely a profound impact on the early evolution of the baryonic component of the Universe. The strong and rapid evolution of the quasar population is certainly one of the most remarkable phenomena in astronomy, and it is hard to resist speculating that this evolution is connected in some way with the formation and evolution of "normal" galaxies (e.g., Kauffmann & Haehnelt 2000).

To attack the possible connection between QSOs, the development of black holes, and the population of normal galaxies, to make a comparison of the host galaxies of powerful AGN over a wide range of redshifts, and to attempt to understand the dominate population of QSOs, we observed exclusively a small sample of faint ($B_J \approx 20 - 22$) radio-quiet quasars covering the redshift range 0.8–2.7.

1. Observations

The observed quasars are all chosen to have low to medium luminosities, with apparent magnitudes in the range $B_J \approx 20 - 22$, from the survey of faint QSOs from Zitelli et al. (1992); Boyle et al. (1991); Marano, Zamorani, Zitelli (1988). The largest number of observed QSOs lies around $z \approx 0.8$ and ≈ 1.2 with smaller numbers in the other redshift bins of ≈ 1.8 and ≈ 2.7 (only 2 QSOs). All observations were carried out at the ESO NTT telescope on La Silla. To match the filter bandpasses to the rest-frame optical wavelengths at the different redshifts, two instruments were used: for the optical we used SUSI2 and for the near infrared imagery we used SOFI. To keep the data free of strong emission lines, like [OIII] or Hα we used I-band for $z \approx 0.8$ QSOs, J for $z \approx 1.2$ QSOs, H for $z \approx 1.8$ QSOs, and K_s for the for $z \approx 2.7$ QSOs. A list of the observed objects and their redshifts as well as other observational parameters are given in Table 1. The total integration times ranged from one to two hours with the highest redshift sources getting preferentially the longest integration times. During the scheduled observing time, only the best seeing condition were used due to the difficulty in using ground-based images to detect hosts, and thus in the present data set, the average seeing is measured to be 0.6″, with a range from 0.5″to 0.8″.

Objects and observational parameters

Object name	RA (B1950.0)	Dec (B1950.0)	z	Int	Filter	Date
MZZ 9854	03 11 59.8	−55 31 56	2.706	7110	K_s	09/20/99
MZZ 5250	03 13 49.8	−55 20 31	1.192	1680	J	09/20-21/99
SGP 2:16	00 50 28.2	−29 07 42	0.852	1620	H	09/21-22/99
MZZ 1246	03 14 36.7	−55 05 24	1.132	3480	J	09/21/99
MZZ 921	03 15 36.7	−55 01 58	1.972	4320	H	09/21/99
QS M3:36	22 02 01.5	−19 01 46	0.873	3510	H	09/22/99
MZZ 2994	03 14 40.7	−55 13 49	2.735	6840	K_s	09/22/99
MZZ 9554	03 12 51.1	−55 37 02	1.821	2250	H	09/22/99
QS F5:40	03 37 01.4	−44 28 47	0.90	3600	I	02/10/00
QS F1X:66	03 41 29.1	−45 16 12	0.90	3600	I	02/10/00
QM B2:23	10 42 19.1	+00 44 01	0.857	3600	I	02/10/00
QN Y2:09	12 35 04.5	+00 31 57	0.820	3600	I	02/10/00
QN Y2:16	12 35 20.2	+00 33 07	0.866	3600	I	02/10/00
QN Z4:24	15 18 00.0	+03 01 21	0.826	3600	I	02/10/00
F861:134	12 39 27.8	−00 25 49	1.12	3600	J	02/12/00
QS F5:25	03 38 06.1	−44 18 32	1.762	3600	H	02/13/00
QN B2:28	10 41 28.4	+00 56 46	1.194	3600	J	02/13/00
F855:155	10 44 28.7	−00 04 14	1.789	3600	H	02/13/00
QN Y2:19	12 35 22.8	+00 17 28	1.122	3600	J	02/13/00

Table 1. Col. (1) — Source designation. Col. (2) — Right Ascension at Epoch B1950.0. Col. (3) — Declination of the source at Epoch B1950.0. Col. (4) — redshift of the source. Col. (5) — total integration time in seconds for filter listed in col. (6). Col. (6) — Filter used in observation. The optical data were taken using SUSI2 on the NTT while the near-IR images were taken with SOFI on the NTT. Col. (7) — Date of observations (UT) in the format MM/DD/YY.

2. Data reduction and PSF subtraction

The data were reduced in a standard way. The reduction included flat field correction, bad pixel removal on both the single and dithered frames, and sky subtraction. Both the optical and near-IR data were offset between each exposure.

We applied a search algorithm to extract all point-like sources which could be candidates for a point-spread-function (PSF) reference. Depending on the individual filter and how deep the final combined image was, about 5–20 point sources could be found in each field (≈ 4 arcmin^2). These point sources were then selected by their relative brightness and distance from the QSO. The goal was to find as many possible point sources in the same field as the QSO to act as a suitable reference PSF in order to give a robust comparison of the QSO and PSF and to guage accurately the uncertainty in the contribution of the host to the image of the QSO.

For the PSF subtraction we developed an algorithm that is capable of taking into account all the selected point sources and selecting the best of them to construct the final PSF. As a first step in finding the most appropriate PSF, we fit each possible PSF star over a grid of points that is 20 times finer than the original sampling, and where each fitted PSF adds information about the total PSF sub-structure. We then subtract this combined PSF from all point sources after making the obviously necessary sub-pixel shift. Some of the sources then showed a non-smooth or non-zero residual, which means that they are either affected by PSF variations across the field (e.g., at the edge of the fields) or they are previously unrecognized extended sources. The procedure is then repeated ignoring the PSF stars that had significant residuals or sub-structure in the first step. The PSF so constructed becomes the reference for the analysis of the QSO and for all objects a minimum of 5 suitable PSF references were found.

Applying the PSF for a subtraction to the QSO needs extreme accuracy in minimizing the differences in the PSF and QSO centers of their light profiles (sub-pixel shifting). Our algorithm uses a smoothness test in a grid of sub-pixel positions around the measured centroids of the QSO and the PSF. In addition, we tested a simpler version that just takes the standard deviation of the residual, and which gives the same result as the χ^2 test. After having obtained the best position we try to get the best scaling factor for the PSF in a several step process. First, we subtract the PSF with increasing scaling. At every choice of relative scaling, we measure the integrated flux in many annuli around the center, which gives a very sensitive estimate of small deviations from a point source. The next step is estimate an error on the parameters in every ring, which is done by applying the same procedure to point sources in the field.

The final scaling for the subtraction can only be achieved by putting in a model: one has to assume a certain light profile of the underlying galaxy. We constructed artificial galaxies assuming that they follow either a spheroid dominated or disk-like profile. The artificial object is then convolved with the observed PSF and combined with a point source. The radius as well the scaling of this galaxy profile can then be varied and the same measurement procedure as for the QSO can be applied to every variation, until the observed data is matched with the smallest deviation relative to the uncertainties. With this procedure we

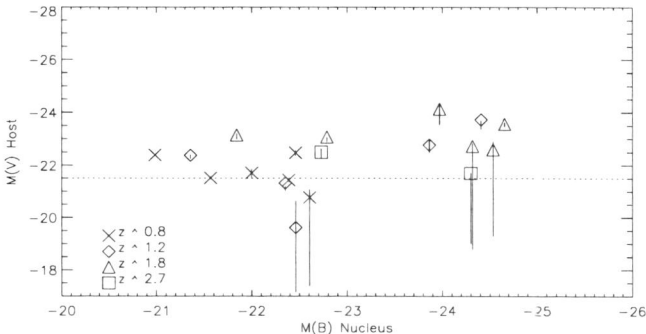

Figure 1. Host and nuclear absolute magnitudes of all observed QSOs. The magnitudes are in rest-frame V and B for the host and nucleus respectively for easy comparison with the results of the simulations of Kauffmann & Haehnelt 2000, and the data that has been obtained by Ridgway et al. (2000 and in this volume). The B–V color for conversion has been assumed to 0.4, the cosmology used is $q_0=0.5$ and $H_0=50$. The horizontal dashed line shows the position of a typical L_* galaxy.

were able to determine the scaling parameter for the PSF, the possible uncertainty due to PSF variations and noise in the images and, if the profile was sufficiently extended and high S/N, to discriminate between an elliptical or disk like profile.

3. Host properties

Using the procedure outlined above, we estimated the contribution (and the corresponding uncertainty) of the underlying host galaxy in each final image. The measured magnitudes were converted to absolute magnitudes assuming a cosmology with $q_0=0.5$ and $H_0=50$. In Fig. 1 we summarize our results by plotting M(V) host against M(B) core, assuming a B–V color of 0.4. There are some cases where the uncertainties from the estimation procedure are too large to provide a reliable value for the host magnitude. For these cases, we take the best fit result as an upper limit and these are so denoted in Fig. 1. We find that the low luminosity QSOs ($<M_B$ (nucleus)$>$ = -22) have hosts which have average magnitudes of $\approx L^*$ galaxies, while the more luminous nuclei ($<M_B$ (nucleus)$>$ = -24.5) are typically about 2 L^* galaxies, both results being independent of the redshift of the QSO. Such a trend is consistent with the data of Ridgway et al. (2000) (see also Ridgway this volume) and for the more luminous QSOs, with Kukula et al. (2000) (see also Kukula this volume).

Interestingly, nearly all of the host galaxies we have measured in the range $z \approx 0.8$–1.2 are fit clearly better with a "disk-like" profile. For the higher z objects the profile shape of the extended emission becomes more ambiguous, since depth of the images is insufficient to detect much emission at large projected distances from the QSO nucleus and the physical spatial resolution is considerably more coarse. In addition, while we find some possible companions to some of the QSOs, their environments are not particularly rich. The numbers are consistent with the hosts being field galaxies or lying in small groups.

4. Conclusions

The addition of our results to those in the literature allows us to discuss the various relationships within the QSO host population. We have found that more luminous QSO nuclei appear typically to be hosted by more luminous galaxies and this does not change significantly with redshift. The nature of this result suggests that the fueling rate must not be a strong function of redshift and is thus consistent with more massive black holes being hosted by more massive galaxies (they follow a "Magorrian-type relation"; Magorrian et al. 2000). If we combine the fact that low luminosity QSOs are the dominant population of powerful AGN which at high redshifts have relatively large co-moving space densities and that such QSOs appear to be hosted by quite ordinary galaxies without many luminous galaxies nearby, our results favors a picture where most galaxies harbor black holes with few fueling episodes as opposed to a small subset of galaxies harboring black holes with many, longer-lived episodes of activity. This is consistent with recent evidence of the ubiquity of supermassive black holes in nearby galaxies (Kormendy & Richstone 1995; Magorrian et al. 2000) – including the Milky Way (Genzel & Eckart 1996).

References

Boyle, B. J., Fong, R., Shanks, T., Peterson, B. A., 1991, MNRAS, 243, 1
Genzel, R., Eckart, A., 1996, Nature, 383, 415
Kauffmann G., Haehnelt M., 2000, MNRAS, 311, 579
Kormendy, J., Richstone, M., 1995, ARA&A, 33, 581
Kukula, M. J. et al., 2000, astro-ph/0010007
Magorrian, J. et al., 1998, AJ, 115, 2285
Marano, B., Zamorani, G., Zitelli, V., 1988, MNRAS, 232, 111
Ridgway S., Heckman T., Calzetti D., Lehnert M., 2000, astro-ph/0011330
Zitelli, V., Mignoli, M. Zamorani, G., Marano, B., Boyle, B. J., 1992, 256, 349

A NICMOS IMAGING STUDY OF QUASAR HOST GALAXY EVOLUTION

Marek J. Kukula, J. S. Dunlop, R. J. McLure, L. Miller, W. J. Percival, S. A. Baum and C. P. O'Dea

Abstract We summarise the results of a major HST/NICMOS programme designed to investigate the cosmological evolution of quasar host galaxies from z=2 to the present day. At $z = 1$ we have been able to establish host-galaxy luminosities and scalelengths and, apart from simple passive stellar evolution, we find little difference between these galaxies and the elliptical hosts of low-redshift quasars of comparable nuclear output - implying that the hosts are virtually fully assembled by $z = 1$. At $z = 2$ the host galaxies are harder to detect, but we have been able to derive host luminosities. We find evidence that the difference between RQQ and RLQ hosts appears to grow with increasing redshift, with the RLQ hosts remaining consistent with passive evolution while the RQQ hosts appear to show a (modest) drop in mass. Possible physical causes of this effect are briefly discussed.

Introduction

Thanks largely to the efforts of the Hubble Space Telescope (*HST*), our understanding of the host galaxies of quasars in the local universe has improved imeasurably over the last decade. Quasar hosts at low redshifts ($z \leq 0.4$) are now known to be large, luminous and typically bulge-dominated systems (e.g. Disney et al. 1995, Bahcall et al. 1997, McLeod et al. 1999). Indeed, in our own *HST* R-band study of quasars at $z \simeq 0.2$) we find that the hosts of all quasars more luminous than $M_V = -24$ are massive ellipticals with $L \geq 2L^\star$, irrespective of the quasar's radio power (McLure et al. 1999, Dunlop et al. 2000). Complementary ground-based spectroscopy shows that these host galaxies have old, well-established stellar populations, similar to those of nearby inactive elliptical galaxies (Hughes et al. 2000, Nolan et al. 2000). In retrospect such results can perhaps be viewed as an inevitable consequence of the correlation between black-hole mass and spheroid mass suggested for nearby galaxies by Maggorian et al. (1998). A quasar with $M_V < -24$

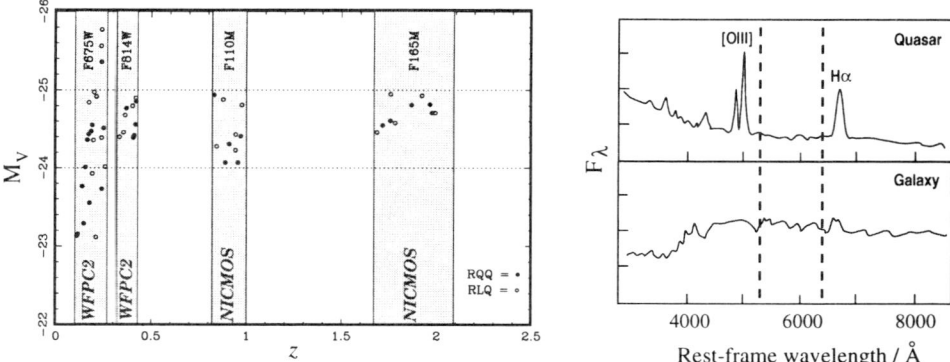

Figure 1. Absolute V magnitude versus redshift for the quasars in our *HST* studies, including our earlier work on quasars at $z \simeq 0.2$. In each case we match our choice of WFPC2 or NICMOS filter to the redshift of each object so that our images always sample the rest-frame V-band continuum of the quasar, whilst avoiding strong emission lines in the quasar spectrum.

requires a black hole of $> 10^9 M_\odot$ which, according to the m_{bh}:m_{sph} relation, can only be housed in a spheroid with $m_{sph} > 5 \times 10^{11} M_\odot$ - in other words, a massive elliptical galaxy.

However, at high redshifts ($z \geq 1$) the situation is less clear. Recent models based on hierarchical galaxy formation suggest that quasars of a given luminosity should be found in hosts of progressively smaller mass at higher redshifts (Kauffmann & Haehnelt 2000). But ground-based observations of high-z quasar hosts have not necessarily supported such a picture, finding evidence for very luminous hosts around quasars at redshifts of $z \sim 2.5$ (e.g. Aretxaga et al. 1998). However, these observations have tended to concentrate on the most luminous quasars for which there are no counterparts at low redshift, making unbiased comparisons very difficult.

1. A NICMOS study of high-redshift quasar hosts

With the installation of the NICMOS camera in 1997, *HST* gained the ability to follow the rest-frame optical continuum of more distant quasars as it is shifted into the near-infrared, allowing the telescope for the first time to make detailed studies of quasar hosts out to high redshifts. Our own 60-orbit *HST* program uses NICMOS to do just this, extending our existing low-z study to $z \simeq 2$ (Fig. 1).

Our program is designed to determine the host-galaxy properties of *both* RLQs and RQQs from $z \simeq 2$ to the present day in a genuinely unbiased manner. Its key features are: (i) sufficient HST-based angular resolution to allow a meaningful attempt at determining galaxy scalelengths at all redshifts; (ii) the use of quasar samples with the same characteristic absolute magnitude ($-24 > M_V > -25$) at all redshifts; (iii) our insistence that the radio-quiet quasars selected for study are known to lie below a definite radio-luminosity threshold; (iv) filter selection, which when coupled with careful sample redshift constraints, guarantees line-free imaging at the same rest-wavelength ($\sim$ V-band) (removing concerns about emission-line contamination, and obviating the need for k-corrections); (v) the extraction of host-galaxy parameters using an identical modelling approach at all redshifts, minimising potential surface-brightness bias and concerns over aperture corrections; (vi) the use of properly-sampled, high-dynamic-range PSFs derived from observations of stars through the same filters, and on the same regions of the relevant detectors, as the quasar images.

Full details of this study can be found in Kukula et al. (2001). Here, we summarise the results.

Quasar hosts at $z \simeq 1$: Analysis of our NICMOS images shows that the hosts of both RQQs and RLQs at $z \simeq 1$ are large ($r_e \simeq 10$kpc), luminous elliptical galaxies which lie on the same Kormendy relation as that deduced for 3CR radio galaxies at comparable redshift. The size scales of both the RLQ and RQQ hosts seem to have changed little, if at all, between $z \simeq 0.2$ and $z \simeq 1$. Moreover, as Fig. 2 demonstrates, the typical luminosity of the RLQ hosts has increased between $z \simeq 0.2$ and $z \simeq 1$ by an amount that is perfectly consistent with pure passive evolution of a mature stellar population - in other words with a galaxy which forms at high redshift ($z \sim 5$) and undergoes little active star formation thereafter. The implication is that the RLQ host masses are constant between redshifts of 0.2 and 1. By comparison, the typical luminosity of the RQQ hosts at $z \simeq 1$ appears basically unchanged. Because the effect of passive evolution is to cause a given stellar population to increase in luminosity between redshifts of 0.2 and 1, there is therefore a suspicion that the RQQ hosts at $z \simeq 1$ are *less massive* than low-z RQQs of comparable nuclear output. A similar trend has been noted in other samples (e.g. Örndahl et al., this volume). However, in our $z \simeq 1$ sample the significance of this result is marginal, and within a Λ-dominated cosmology (Fig. 2, RH panel) our results are certainly consistent with passive stellar evolution, in which the host masses of both class of quasar remain unchanged between $z \simeq 0.2$ and $z \simeq 1$.

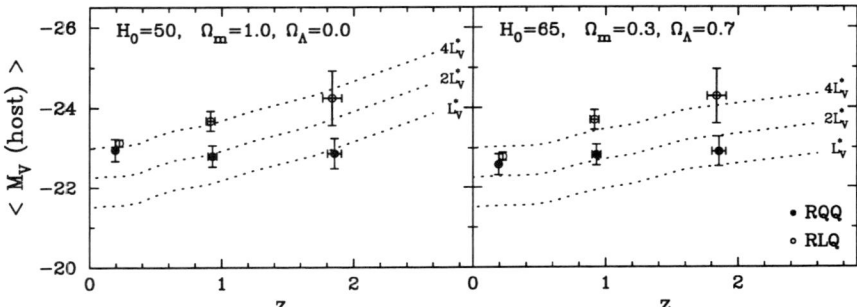

Figure 2. Mean absolute V-band magnitude versus mean redshift for the host galaxies of the RLQs (open circles) and RQQs (filled circles) in the current study. Also shown is the subset of 5 RLQs and 7 RQQs from our previous study of quasars at $z \sim 0.2$ which have total (host + nuclear) luminosities in the same range as the high-redshift samples ($-24 \geq M_V \geq -25$). Error bars show the standard error on the mean. The dotted lines show the luminosity evolution of present day L^*, $2L^*$ and $4L^*$ elliptical galaxies, assuming a formation epoch of $z = 5$ with a single rapid burst of star formation followed by passive evolution thereafter. LH panel: assuming a cosmology with $H_0 = 50$ km s^{-1} Mpc^{-1}, $\Omega_m = 1.0$ and $\Omega_\Lambda = 0.0$. RH panel: $H_0 = 65$ km s^{-1} Mpc^{-1}, $\Omega_m = 0.3$ and $\Omega_\Lambda = 0.7$.

Quasar hosts at $z \simeq 2$: The quasar hosts at $z \simeq 2$ are harder to detect and consequently to model, but we have been able to extract host-galaxy luminosities with sufficient confidence to demonstrate that the hosts of RLQs continue to brighten exactly as expected under the hypothesis of pure passive evolution (Fig. 2) implying that their masses remain roughly constant. In contrast, the hosts of RQQs again seem little changed in luminosity, and are also systematically smaller in size compared to those at $z \simeq 1$. This apparently growing radio-loud/radio-quiet host-mass gap is mirrored in other studies of radio galaxies (Lacy et al. 2000) and radio-quiet quasars (Ridgway et al. 1999, this volume, Rix et al. 1999) at comparable redshift, lending further support to its reality. However, after allowance for brightening due to passive evolution, the inferred drop in RQQ host mass is still relatively modest and is in fact consistent with the black-hole/spheroid population being essentially unchanged out to $z \simeq 2$, but with the $z \simeq 2$ black holes being fueled at close to the Eddington limit, compared to the more modest fueling rates inferred for typical nearby radio-quiet quasars.

2. Discussion

One explanation for a growing discrepancy between RLQ and RQQ host masses with increasing redshift might be that the additional criterion of radio-loudness ensures a black hole mass - and hence host spheroid mass - above a fixed threshold. This was apparently found to be the case at $z \simeq 0.2$ by Dunlop et al. (2001). If this is true then concentrating on radio-loud objects will blind us to underlying trends in galaxy evolution because at each redshift we are only selecting those galaxies which have already attained a certain critical mass. This means that it is the masses of RQQ hosts which may offer a more representative test of the predictions of hierarchical growth of massive galaxies.

However, to date the interpretation of our data has relied heavily on the assumption of passive stellar evolution. If we find that more active star formation is underway in these galaxies, the evidence for a substantial decrease in mass for the RQQ hosts (and perhaps even the RLQ hosts) would then become compelling. We thus need to obtain some basic information on the stellar composition of these galaxies. This issue will be addressed by a forthcoming HST study to obtain rest-frame $U - V$ colours for the galaxies in our sample.

References

Aretxaga I., Terlevich R. J., Boyle B. J., 1998, MNRAS, 296, 643
Bahcall J. N., Kirhakos S., Saxe D. H., Schneider D. P., 1997, ApJ 479, 642
Disney M. J. et al., 1995, Nature, 376, 150
Dunlop J. S., McLure R. J., Kukula M. J., Baum S. A., O'Dea C. P., Hughes D. H., 2001, MNRAS, submitted
Hughes D. H., Kukula M. J., Dunlop J. S., Boroson T., 2000, MNRAS, 316, 204
Kauffmann G., Haehnelt M., 2000, MNRAS, 311, 576
Kukula M. J., Dunlop J. S., McLure R. J., Miller L., Percival, O'Dea C. P., 2001, MNRAS, in press
Lacy M., Bunker A. J., Ridgway S. E., 2000, AJ, 120, 68
Magorrian J. et al., 1998, AJ, 115, 2285
McLeod K. K., Rieke G. H., Storrie-Lombardi L. J., 1999, ApJ, 511, L67
McLure R. J., Kukula M. J., Dunlop J. S., Baum S. A., O'Dea C. P., Hughes D. H., 1999, MNRAS, 308, 377
Nolan L. A., Dunlop J. S., Kukula M. J. et al., 2000, MNRAS, in press
Ridgway S., Heckman T., Calzetti D., Lehnert M., 1999 (astro-ph/9911049)
Rix H. -W. et al., 1999 (astro-ph/9910190)

Matt Jarvis

Andrea Cattaneo

THE RADIO GALAXY $K-z$ RELATION TO $z \sim 4.5$

Matt J. Jarvis[1,2], Steve Rawlings[2], Steve Eales[3]
Katherine M. Blundell[2] and Chris J. Willott[2]
[1] *Sterrewacht Leiden, The Netherlands*
[2] *University of Oxford, UK*
[3] *University of Wales College of Cardiff, UK*

Abstract Using a new radio sample, 6C* designed to find radio galaxies at $z > 4$ along with the complete 3CRR and 6CE sample we extend the radio galaxy $K-z$ relation to $z \sim 4.5$. The 6C* $K-z$ data significantly improve delineation of the $K-z$ relation for radio galaxies at high redshift ($z > 2$). Accounting for non-stellar contamination, and for correlations between radio luminosity and estimates of stellar mass, we find little support for previous claims that the underlying scatter in the stellar luminosity of radio galaxies increases significantly at $z > 2$. This indicates that we are not probing into the formation epoch until at least $z \gtrsim 3$.

1. Why radio galaxies?

Radio galaxies provide the most direct method of investigating the host galaxies of quasars if orientation based unified schemes are correct. The nuclear light which dominates the optical/near-infrared emission in quasars is obscured by the dusty torus in radio galaxies, therefore difficulties surrounding the PSF modelling and subtraction are not required to determine the properties of the underlying host galaxy. Unfortunately compiling samples of radio loud AGN is a long process, because of the radio selection there is no intrinsic optical magnitude limitation, making follow-up observations extremely time consuming, especially when dealing with the faintest of these objects. However, low-frequency selected radio samples do now exist with the completion of 3CRR (Laing, Riley & Longair 1983) along with 6CE (Eales et al. 1997, Rawlings et al. 2001) and the filtered 6C* sample (Blundell et al. 1998, Jarvis et al. 2001a,b).

We can now use these radio samples to investigate the underlying stellar populations through the radio galaxy $K-z$ Hubble diagram.

2. Previous radio samples and the $K-z$ Hubble diagram

There has been much interest in the $K-z$ relation for radio galaxies in the past decade. Dunlop & Peacock (1993) using radio galaxies from the 3CRR sample along with fainter radio sources from the Parkes selected regions demonstrated that there exists a correlation between radio luminosity and the K-band emission. Whether this is due to a radio luminosity dependent contribution from a non-stellar source or because the galaxies hosting the most powerful radio sources are indeed more massive galaxies has yet to be resolved. Eales et al. (1997) confirmed this result and also found that the dispersion in the K-band magnitude from the fitted straight line increases with redshift. This result, along with the departure to brighter magnitudes of the sources at high redshift led Eales et al. to conclude that we are beginning to probe the epoch of formation of these massive galaxies. Using the highest redshift radio galaxies from ultra-steep samples of radio sources van Breugel et al. (1998) found that the near infrared colours of radio galaxies at $z > 3$ are very blue, consistent with young stellar populations. They also suggest that the size of the radio structure is comparable with the size of the near infrared region, and the alignment of this region with the radio structure is also more pronounced at $z > 3$. Lacy et al. (2000) using the 7C-III sample found evidence that the hosts of radio galaxies become more luminous with redshift and are consistent with a passively evolving population which formed at high redshift ($z > 3$). Thus, all of this work points to a radio galaxy population which formed at high redshift and has undergone simple passive evolution since. However, all of these studies were made with only a few high-redshift ($z > 2$) sources. With the 6C* sample we are now able to probe this high redshift regime with increased numbers from samples with well-defined selection criteria.

3. The 6C* filtered sample

The 6C* sample is a low-frequency radio sample ($0.96\text{Jy} \leq S_{151} \leq 2.00\,\text{Jy}$) which was originally designed to find radio sources at $z > 4$ using filtering criteria based on the radio properties of steep spectral index and small angular size. The discovery of 6C*0140+326 at $z = 4.41$ (Rawlings et al. 1996) and 6C*0032+412 at $z = 3.66$ (Jarvis et al. 2001a) from a sample of just 30 objects showed that this filtering was indeed effective in finding high-redshift objects. Indeed, the median redshift of the 6C* sample is $z \sim 1.9$ whereas for complete samples at similar flux-density levels the median redshift is $z \sim 1.1$ (Willott et al. in prep.). We can now use this sample to push the radio galaxy $K - z$ diagram to high redshift ($z > 2$) where it has not yet been probed with any significant number of sources (e.g. Eales et al. 1997, van Breugel et al. 1998, Lacy et al. 2000). Fig. 1 shows the radio luminosity-redshift plane for the 3 samples used in this analysis.

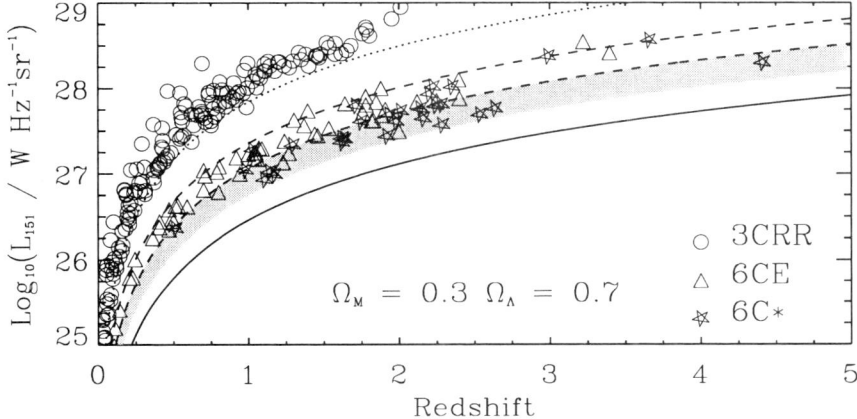

Figure 1. Rest-frame 151 MHz luminosity (L_{151}) versus redshift z plane for the 3CRR (circles), 6CE (triangles) and 6C* (stars) samples. The rest-frame 151 MHz luminosity L_{151} has been calculated according to a polynomial fit to the radio spectrum (relevant radio data from Blundell et al. 1998). The curved lines show the lower flux-density limit for the 3CRR sample (dotted line; Laing et al. 1983) and the 7CRS (solid line; Blundell et al., in prep, Willott et al., in prep). The dashed lines correspond to the limits for the 6CE sample (Rawlings et al. 2001) and the shaded region shows the 6C* flux-density limits (all assuming a low-frequency radio spectral index of 0.5). Note that the area between the 3CRR sources and 6CE sources contains no sources, this is the area which corresponds to the absence of a flux-density limited sample between the 6CE ($S_{151} \leq 3.93$ Jy) and 3CRR ($S_{178} \geq 10.9$ Jy) samples. The reason why some of the sources lie very close to or below the flux-density limit of the samples represented by the curved lines is because the spectral indices lie very close to or below the assumed spectral index of the curves of $\alpha = 0.5$.

4. Emission-line contamination

The most-luminous sources at high redshift may be contaminated by the bright optical emission lines redshifted into the infrared. This is particularly true for sources in radio flux-density limited samples. The high redshift sources in these samples are inevitably some of the most luminous, and we also know there is a strong correlation between low-frequency radio luminosity and emission-line strength (e.g. Rawlings

& Saunders 1991, Willott et al. 1999, Jarvis et al. 2001a) which will increase the contribution to the measured K−band magnitudes from the emission-lines in the most radio luminous sources.

To subtract this contribution we use the correlation between [OII] emission-line luminosity $L_{[OII]}$ and the low-frequency radio luminosity L_{151} from Willott (2000), where $L_{[OII]} \propto L_{151}^{1.00\pm0.04}$. Then by using the emission-line flux ratios for radio galaxies (e.g. McCarthy 1993) we are able to determine the contribution to the K−band magnitude from all of the other emission-lines. This is illustrated in Fig. 2 where the emission-line contamination to the K−band flux is shown for various radio flux-density limits and a range of redshifts.

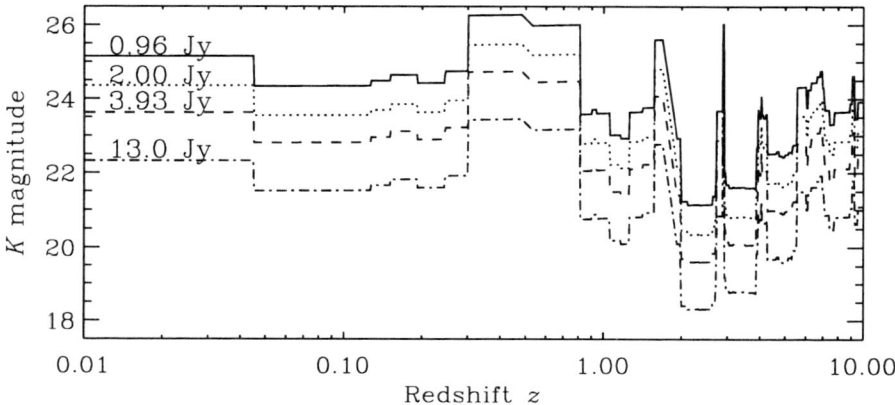

Figure 2. Emission line contribution to the K−band magnitudes for various radio flux-densities assuming the power-law relation of $L_{[OII]} \propto L_{151}^{1.00}$.

5. The $K - z$ relation

Following the prescription of Eales et al. (1997) we split the data into three redshift bins: $z < 0.6$, the redshift above which the alignment effect is readily observed (e.g. McCarthy 1993); $0.6 \leq z \leq 1.8$, the medium-redshift bin to compare 3CRR and 6CE/6C* sources at the same redshift; and $z > 1.8$, the redshift above which there are no 3CRR sources. A least-squares fit line to the K−band magnitudes is plotted on the $K - z$ diagram for the three samples used in our analysis (Fig 3). The main results from this plot are:

(i) We confirm the previous results of Dunlop & Peacock (1993) and Eales et al. (1997) and find a correlation between radio luminosity and near infrared magnitude. Sources from the new 6C* sample occupy a similar range in K−band magnitudes as the 6CE sources, which are

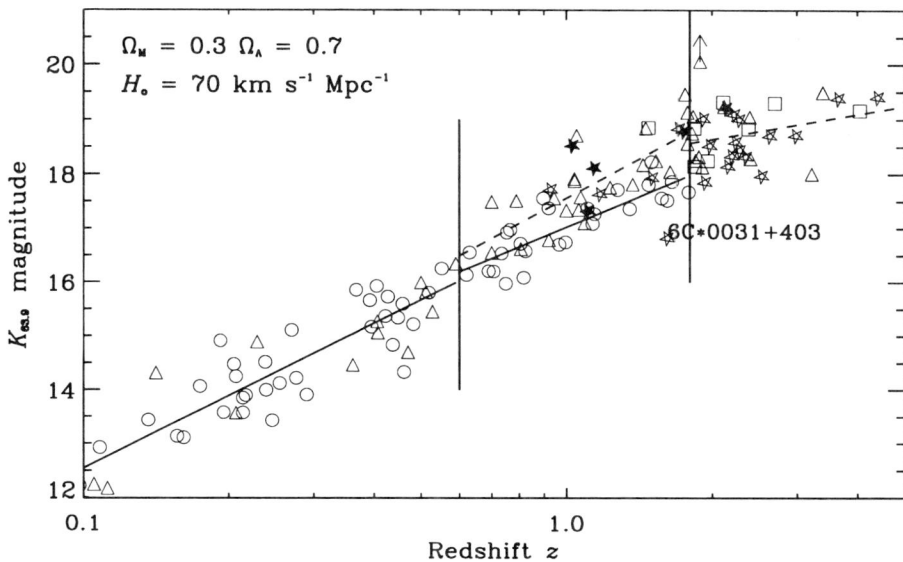

Figure 3. The $K - z$ Hubble diagram for radio galaxies from the 3CRR (circles), 6CE (triangles), 6C* (stars) and 7C-III (squares) samples. $K_{63.9}$ denotes the K-band magnitude within a comoving metric aperture of 63.9 kpc (c.f. Eales et al. 1997, Jarvis et al. 2001b). The two vertical lines show the redshift above which the alignment effect begins to be seen ($z = 0.6$) and the higher redshift at which we chose to split the data beyond which there are no 3CRR sources ($z = 1.8$). The solid lines are the fits to the 3CRR data points at $z < 0.6$ and $0.6 < z < 1.8$. The dashed line is the fit to the 6CE and 6C* sources at $0.6 < z < 1.8$ and $z > 1.8$. 6C*0031+403 probably has an AGN component contributing to the K-band magnitude and is labelled. The filled stars represent the objects in 6C* which do not yet have completely secure redshifts.

a factor of two brighter in radio luminosity than the 6C* sources. The 3CRR sources, which are a factor of six brighter than the 6CE sources occupy K-band magnitudes approximately 0.7 magnitudes brighter than the 6CE/6C* sources at $0.6 < z < 1.8$ (A Mann-Whitney U-test shows that the distribution in magnitudes between the 6CE/6C* and 3CRR sources are different at $> 99.9\%$ level).

(ii) We find that the dispersion in the K-band magnitudes for the 3CR sources decrease from the $z < 0.6$ bin ($\sigma = 0.52$) to the $0.6 \leq z \leq 1.8$ bin ($\sigma = 0.36$). This may be because of a non-stellar component linked to the radio luminosity becoming more important at the higher redshifts as the sources are necessarily more radio luminous than their low-redshift counterparts and such a luminosity dependent effect would reduce the observed dispersion. However, an alternative explanation in which these extreme objects can only exist in specific physical conditions also needs to be explored. It is not inconceivable that the most luminous radio galaxies need some of the narrowest set of conditions to form and exist, whereas the lower luminosity radio galaxies may be able to form and exist in a broader range of physical environments.

(iii) The dispersion in the $K - z$ diagram does not increase from $z < 0.6$ ($\sigma = 0.57$) out to $z \sim 3$ ($\sigma = 0.51$) for the sources in the 6CE and 6C* samples. This is in opposition to results in which the dispersion was found to increase, which led previous authors to conclude that we are probing the epoch of formation at $z \sim 2$. The lack of an apparent increase in dispersion toward high redshift may be due to the possibility outlined above in which non-stellar emission may be contributing to the $K-$band flux at these high redshifts, and thus high luminosities. Alternatively, the lack of scatter may be informing on the lack of ongoing star formation at these redshifts, and would mean that the period of star formation has come to an end. If this scenario is correct then it does fit in with other results concerning the epoch at which these massive galaxies first formed. First, Archibald et al. (2001) have found that the dust mass in radio galaxies appears to increase with redshift, at least out to $z \sim 3$, thus implying that the majority of star formation activity in these galaxies is occurring at high redshift. Second, the discovery of six extremely red objects at $1 < z < 2$ in the 7C redshift survey Willott et al. (2001, these proceedings) with inferred ages of a few Gyrs implies that the bulk of their stellar population formed at $z \simeq 5$.

References

Archibald E.N. et al., 2001, MNRAS, in press (astro-ph/0002083)
Blundell K.M. et al., 1998, MNRAS, 295, 265
Dunlop J.S., Peacock J.A., 1993, MNRAS, 263, 936
Eales S.A., Rawlings S., Law-Green D., Cotter G., Lacy M., 1997, MNRAS, 291, 593
Jarvis M.J. et al., 2001a, MNRAS, submitted
Jarvis M.J. et al., 2001b, MNRAS, submitted
Lacy M., Bunker A.J., Ridgway S.E., 2000, AJ, 120, 68
Laing R.A., Riley J.M., Longair M.S., 1983, MNRAS, 204, 151
McCarthy P.J., 1993, ARA&A, 31, 639
Rawlings S., Eales S.A., Lacy M., 2001, MNRAS, 322, 523
Rawlings S. et al., 1996, Nature, 383, 502
Rawlings S., Saunders R., 1991, Nature, 349, 138
van Breugel W.J.M. et al., 1998, ApJ, 502, 614
Willott C.J., 2000, to appear in Proc. "AGN in their Cosmic Environment", Eds. B. Rocca-Volmerange & H. Sol, EDPS Conf. Series (astro-ph/0007467)
Willott C.J., Rawlings S., Blundell K.M., 2001, MNRAS, in press (astro-ph/0011082)
Willott C.J., Rawlings S., Blundell K.M., Lacy M., 1999, MNRAS, 309, 1017

WHAT FUELS AGNS?

Andrea Cattaneo
University of Cambridge, Institute of Astronomy
Madingley Road, Cambridge CB3 0HA, UK
cattaneo@ast.cam.ac.uk

Abstract

This contribution presents the results of semi-analytic simulations to test the hypothesis that quasars are fuelled by merger-driven inflows. It also discusses the limitations of this type of modelling and shows how different implementations can affect the simulated luminosity function.

1. Scenarios

High resolution images of quasar hosts, the detection of supermassive black holes in nearby galaxies and the presence of AGN features in ULIRGs spectra have corroborated the idea that quasars and galaxy formation are related phenomena. On the theoretical side, Nulsen & Fabian (2000), Kauffmann & Haehnelt (2000) and Cattaneo (2001) have begun to include cooling, galactic mergers and star formation in models of the quasar population (see Baugh's review in this volume). A main goal of this program is to test different assumptions for the fuelling mechanism. The last two papers considered a scenario where quasars are fuelled by inflows in major mergers (see the reviews by Stockton 1999 and Barnes 1999, and the references therein). This scenario relates quasars to ULIRGs and to the formation of elliptical galaxies. Nulsen & Fabian proposed an entirely different picture, in which quasars are fuelled with hot gas from cooling flows.

2. The fraction of accreted gas

The simplest way of modelling the merger scenario is to assume that at each major merger a fraction of the gas in the merging galaxies flows to the centre and forms or refuels a supermassive black hole. In fact, the fraction of accreted gas will have a complicated dependence on halo and galactic properties (dynamical and kinematical structure, abundance of cold gas, past merging history, star formation history, angular momentum) and will also depend on the physics of the accretion and emission mechanism. If the accreted mass is proportional to the gas mass, then the accreted mass becomes a smaller fraction of the galactic mass at low redshift, where mergers become increasingly gas-poor (Fig. 1). This result was predicted by Cattaneo, Haehnelt & Rees (1999) and confirmed observationally by Merrifield, Forbes & Terlevich (2000), who found that the ratio of black hole mass to bulge mass is higher in bulges with older stellar populations.

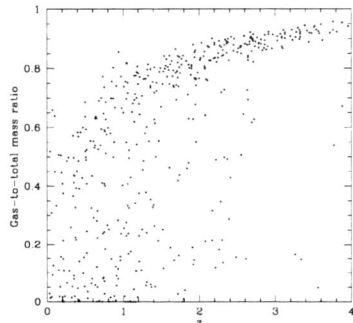

Figure 1. Gas to total mass ratio in simulated mergers as a function of redshift. High redshift mergers typically involve discs that are 80-90% gas. Low redshift mergers are mostly stellar mergers. Reproduced from Cattaneo (2001).

3. The light curve

Converting the accretion of a given gas mass into a luminosity curve $L(t)$ entails a number of non-trivial steps:

$$\Delta M_{\rm accr} \rightarrow \dot{M}_{\rm accr}(t) \rightarrow L_{\rm bol}(t) = \epsilon \dot{M}_{\rm accr} c^2 \rightarrow \nu f(\nu).$$

Converting the accreted mass $\Delta M_{\rm accr}$ into an accretion rate $\dot{M}_{\rm accr}$ requires specifying not only an accretion time $t_{\rm accr}$, but also a law that determines the dependence of $\dot{M}_{\rm accr}(t)$ with time (e.g. a decreasing exponential). The regime of accretion is specified by $\dot{m} \equiv \dot{M}_{\rm accr}/\dot{M}_{\rm Edd}$, where $\dot{M}_{\rm Edd}$ is the accretion rate required to sustain emission at the Eddington luminosity. Different choices for the dependence of $\dot{M}_{\rm accr}$ with t can substantially change the luminosity function (Fig. 2) and the conditions of accretion (Fig. 3). If most of the energy dissipated in the accretion disc is radiated, then $0.057 < \epsilon < 0.432$. But if $\dot{m} \leq 0.01$, a transition to ADAFs can occur, and if the accretion rate becomes super-Eddington, most of the photons may be trapped in the inflowing plasma and pushed into the event horizon. In both cases ϵ can drop much below 0.1. Finally some of the photons may be absorbed and reemitted in different parts of the spectrum by gas or dust in the host galaxy. The effects of this reprocessing are essentially indistinguishable from those of lowering ϵ. The shortcut solution is to treat ϵ as a free parameter including all sources of obscuration. However this gives a systematic error if the fraction of absorbed light correlates with the luminosity.

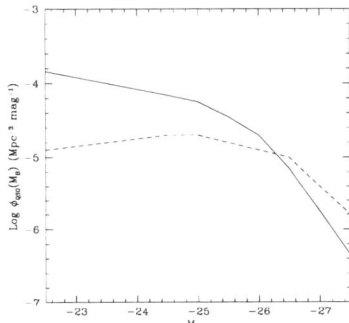

Figure 2. Simulated luminosity functions of quasars at $z = 2.5$ for two different models. In both cases the growth of black holes in major mergers starts with an initial accretion rate of $\dot{M}_{\rm accr} = 3 \times 10^{-10} M_{\rm gas} {\rm yr}^{-1}$ and last for a characteristic time of $t_{\rm accr} = 3 \times 10^7$ yr. However, in one case (solid line) the accretion rate decreases as $\dot{M}_{\rm accr} \propto \exp(-t/t_{\rm accr})$ while in the other (dashed line) $\dot{M}_{\rm accr} \propto (1+t/t_{\rm accr})^{-2}$.

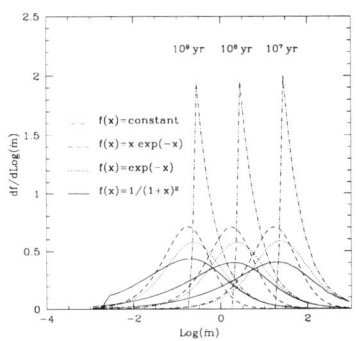

Figure 3. Accreted mass per interval of $\dot{m}$ for $t_{\rm accr} = 10^7, 10^8, 10^9$ yr. Four different growth curves are considered: 1) $\dot{M}_{\rm accr}$ =constant (dash-dotted lines); 2) $\dot{M}_{\rm accr} = (t/t_{\rm accr}) \exp(-t/t_{\rm accr})$ (dashed lines); 3) $\dot{M}_{\rm accr} = \exp(-t/t_{\rm accr})$ (dotted lines); 4) $\dot{M}_{\rm accr} = (1+t/t_{\rm accr})^{-2}$ (solid lines). A growth curve with a longer tail distributes the accretion over a larger range of $\dot{m}$. From Cattaneo (2001).

4. The quasar luminosity function

In spite of the uncertainties discussed in the last Section, Kauffmann & Haehnelt and I agree that the reduced abundance of cold gas is the main reason for the drop in counts of bright quasars at low redshift. Moreover, despite the modelling freedom, it was very difficult to obtain enough bright objects for both of us. The steep decline of the simulated function at high luminosities is attributable to the presence of an upper limit to the mass of a galaxy which one can form through cooling. If this interpretation is correct and the result is not the consequence of a systematic error (see Section 3), then a plausible explanation is that mergers are not the only mechanism for fuelling quasars.

References

Barnes, J.E., 1999, in *The evolution of galaxies on cosmological timescales*, J.E. Beckman and T.J. Mahoney, eds., ASP Conference Series vol. 187, p. 293

Cattaneo, A., 2001, MNRAS, accepted

Cattaneo A., Haehnelt, M., Rees, M.J., 1999, MNRAS, 308, 77

Kauffmann, G., Haehnelt, M., 2000, MNRAS, 311, 576

Nulsen, P.E.J., Fabian, A.C., 2000, MNRAS, 311, 346

Stockton, A., 1999, in *Galaxy interactions at low and high redshift*, J.E. Barnes and D.B. Sanders, eds., IAU Symp. 186, p. 311

VLT–ISAAC IMAGING OF THREE RADIO LOUD QUASARS AT $Z \sim 1.5$

R. Falomo
Osservatorio Astronomico di Padova, Padova, Italy; falomo@pd.astro.it

J.K. Kotilainen
Tuorla Observatory, University of Turku, Piikkiö, Finland; jarkot@astro.utu.fi

A. Treves
Università dell'Insubria, Como, Italy

Abstract

We present high spatial resolution near–infrared images, obtained with Very Large Telescope, of three radio-loud quasars (RLQ) at z∼ 1.5 as a part of a study aimed at investigating the QSO-galaxy connection for a sample of RLQ and radio-quiet quasars (RQQ) in the redshift range $1 < z < 2$. We are able to clearly detect the host galaxy in two quasars and marginally in the third one. The host galaxies appear compact (average scale-length $\sim$ 4 kpc) and very luminous (average M(H) = −27.6±0.1). They are ∼2.5 mag. more luminous than the typical galaxy luminosity (M*(H) = −25.0±0.2). Compared with recent results on RQQ at similar z we find that the difference of host galaxies luminosity is more apparent than at low redshift. All three quasars have at least one close companion galaxy at a projected distance < 50 kpc, assuming they are at the same redshift.

Keywords: Galaxies: active – Galaxies: nuclei – Infrared: galaxies – Quasars: general

1. Observations and Results

We present high spatial resolution near–infrared H-band (1.65 μm) observations, of three radio-loud quasars at $z \sim 1.5$. These are part of a larger matched sample of quasars (both radio loud and radio quiet) in the redshift range 1 to 2 of a program to investigate QSO host galaxy evolution. The observations were taken in Service mode in October 1999 with ISAAC on UT1 of ESO VLT during excellent seeing conditions (< 0.5 arcsec). For two quasars (PKS 0000-177 and PKS 0348-120) the host galaxy is clearly resolved while for the third one (PKS 0402-362) there is only a marginal detection of the extended emission.

Fig. 1 shows the central portion of the images of the three quasars together with their PSF-subtracted images. After the subtraction of the scaled PSF, two out of the three quasars images exhibit clearly extended emission and suggest the presence of knotty structure.

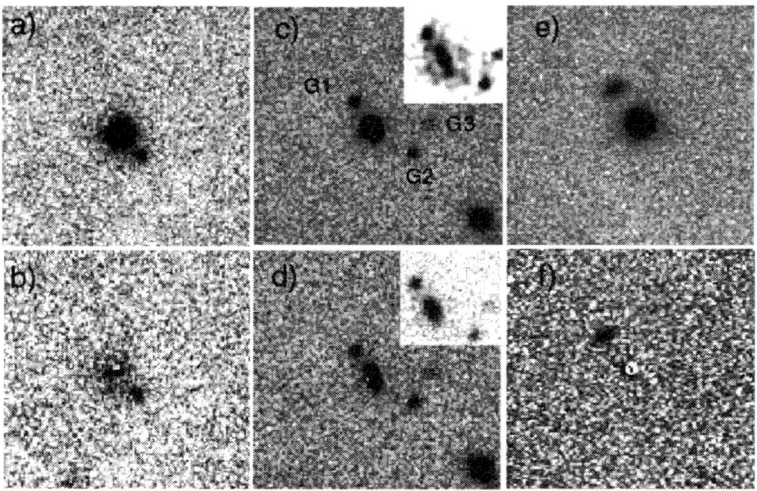

Figure 1. H-band images of the three quasars before (top) and after (below) subtraction of a scaled PSF. From left to right: a), b) PKS 0000-177, c), d) PKS 0348-120 and e), f) PKS 0402-362. The size of the image shown is $\sim 30''$ across. North is to the top and east to the left. The inset in panel c) shows the result of a deconvolution of the image. The inset in the panel d) yields a different grey-scale of the central portion of the PSF subtracted image in order to enhance the knot structure.

From modeling of the luminosity profiles of these host galaxies (see example in Fig. 2) we find they are consistent with massive ellipticals the luminosity of which exceeds by ~ 2.5 mag. the typical galaxy luminosity ($M^*(H) = -25$). The hosts of these radio loud AGNs are therefore similar to those of low redshift objects when pure passive luminosity evolution is considered. On the other hand the host galaxies appear more compact and possibly in denser environments than their counterpart at low z. All three quasars have at least one close companion galaxy at a projected distance < 50 kpc from the quasar.

These observations lead support to the ongoing idea (see e.g. Kukula et al. 2000) that galaxies hosting radio loud quasar (RLQ) are well

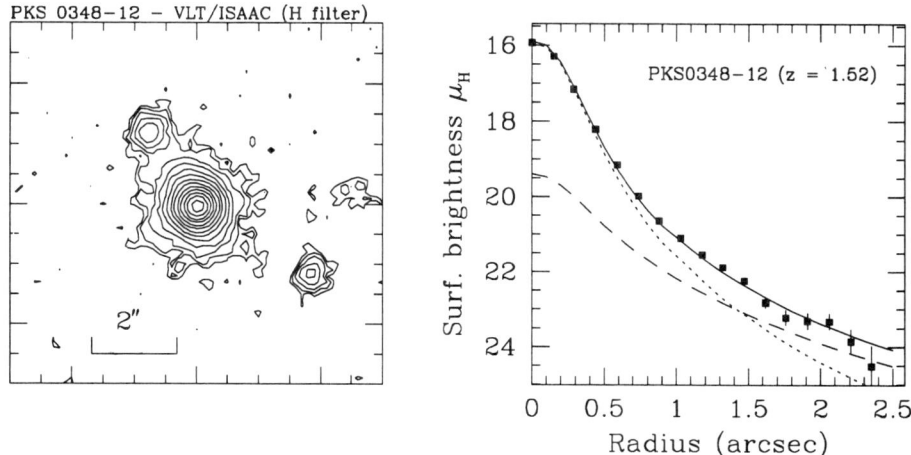

Figure 2. Observations of the quasar PKS 0348-120 at $z = 1.520$. **Left panel:** Contour plot of the H-band image (1 hour integration time; seeing $0.45''$ FWHM). The elongated host galaxy is clearly resolved and has a complex environment. **Right panel:** The radial luminosity profile (filled squares) of the object (after removing the companion) is superimposed on the fitted model (solid line) consisting of the PSF (dotted line) and a de Vaucouleurs bulge (dashed line).

formed at high redshift consistently with observations of high z radio galaxies (e.g. Eales et al. 1997). Comparison with hosts of radio quiet quasars (RQQ) suggests the latter are sistematically less luminous (Ridgway et al. 2000) and that the difference between RLQ and RQQ hosts increases with z. Our results taken together with the other observations of quasar hosts at high z appear difficult to reconcile with quasar models based on pure luminosity evolution and also with hierarchical models of galaxy/AGN formation (e.g. Kauffmann & Hähnelt 2000) that predict a significant drop of host luminosity at high z. A full discussion of these observations is reported in Falomo et al. 2001.

References

Eales, S., Rawlings,S., Law-Green et al., 1997, MNRAS, 291, 593
Falomo, R. Kotilainen, Treves, A., 2001, ApJ, 547 124.
Kauffmann G., Hähnelt M., 2000, MNRAS, 311, 576
Kukula M.J. et al., 2000, MNRAS, in press. (see also these proc.)
Ridgway,S., Heckman,T., Calzetti,D. et al., 1999, Lifecycles of Radio Galaxies (eds. J.Biretta et al.), New Astronomy Reviews (also these proc.)

Renato Falomo

Björn Kuhlbrodt and Margrethe Wold

TWO-DIMENSIONAL MODELING OF AGN HOST GALAXIES

Björn Kuhlbrodt[1], Lutz Wisotzki[2] and Knud Jahnke[3]

[1,3] *Hamburger Sternwarte;* [2] *Universität Potsdam*

[1] BKuhlbrodt@uni-hamburg.de, [2] lutz@astro.physik.uni-potsdam.de, [3] KJahnke@uni-hamburg.de

Abstract We present a method to model the two-dimensional light distribution of Quasar and Seyfert 1 host galaxies by multi-component fitting. The method has been tested on a large set of simulations, showing its ability to recover total fluxes and morphological properties within the observational errors. We describe the current applications for several well-defined AGN samples

To determine the properties of AGN host galaxies it is necessary to separate stellar light from the light of the AGN. One way to do this is modeling the light distributions of the single components.

1. Analytic PSF representation

Before modeling the light distributions the point-spread function (PSF) with which the images will be convolved has to be determined. Choosing a star as a representation of the PSF is only possible if the PSF is invariable with position on the CCD. Instead we fit an analytical model to all stars in the field. We can thus determine an essentially noise-free PSF by averaging the fit parameters, or modeling the spatial variations over the field. Using an analytical model we also circumvent the problem of shifting the PSF centroid to the same subpixel position as the AGN.

This PSF can now be used for the numerical convolution. Here sampling plays an important role. In an area where the gradient is large, the function value taken at the pixel center may not be a good approximation for the average function value for the pixel. One way to avoid this, is to divide the pixel into smaller subpixels. Optimally this is done only for the pixels with large gradients. We use an adaptive subpixeling grid, ensuring that the function gradient in each subpixel is smaller than a specified value.

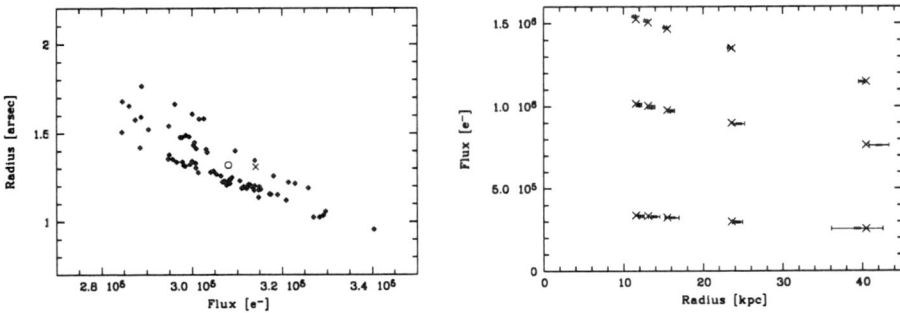

Figure 1. Left: The first set of simulations. Each dot represents the result of one fit, the cross is the average of the fitted values and the circle is the input value. 1σ of the average is 4% for the flux and 16% for the radius. Right: The second set of simulations. The crosses represent the input values, dots the fitted values with 1σ errorbars. For SOFI on ESO NTT the radii correspond to $z = 0.1, 0.2, 0.4, 0.6, 1.2$. Larger redshifts were not simulated as the apparent radius does not increase beyond that.

2. Fitting

The input models we used have two components (nucleus and standard exponential disk or de Vaucouleurs spheroid) or three components (nucleus plus a superposition of disk and bulge). Fitting those models to the data is done by a downhill-simplex χ^2 minimization. For this method no derivatives are needed which are not available in our case because of the numerical convolution. Foreground stars, background galaxies and image artifacts can be masked and do not contribute to the fit.

The minimization is done in multiple runs, using the result of each run as initialization for the next while lowering the convergence tolerance in each pass.

3. Simulations

We consider two sets of simulated AGN hosts:

- An elliptical galaxy with half-light radius $r_{1/2} = 10$ kpc, $M_J = -25$ at $z = 0.6$ as seen with a 4m-class telescope with a nucleus five times brighter than the host galaxy. The galaxy was shifted to different positions on the image pixel grid and artificial Poisson noise was added at a level typical for an exposure time of 50 seconds. The PSF is $0''\!.8$ FWHM compared to $r_{1/2} = 1''\!.33$ for the galaxy.

- The second set extends the first by adopting three different scale lengths and three different flux relations from 5:1 to 1:1 (AGN to

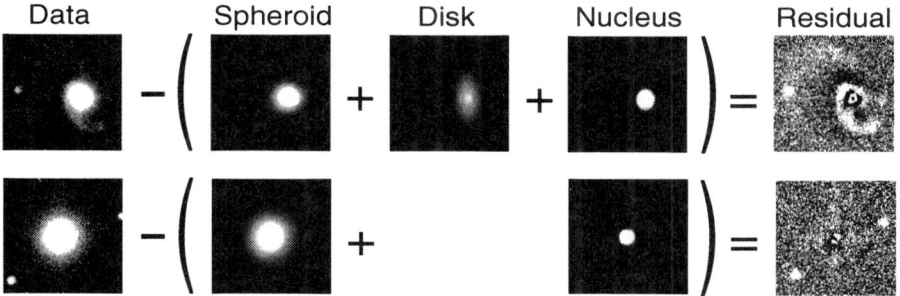

Figure 2. Example of a decomposition.

host galaxy). Those models were noisyfied at the same level as in the first set, but with fixed position.

The fitted parameters usually fit the input values very well. For the two most important parameters (flux und radius) this can be seen in detail in Fig. 1.

4. Applications

A first application was the investigation of a sample of $z < 0.05$ Seyfert galaxies taken from a subsample of 1700 deg^2 of the Hamburg/ESO survey. As the objects were relatively nearby and spacially well resolved, all objects have been fitted with a three-component model. Some objects with apparently faint disk component were also fit with only a spheroid.

With the three-component fits we looked for correlations between the nuclear luminosities and the bulge of the host galaxy, found e.g. by McLure et al. (1999). We too found such a relation but steeper than McLure and we also found a relation between nucleus and disk. A detailed analysis of this sample will be presented in a future Paper (Kuhlbrodt et al., in prep.).

Further applications include the investigation of a multicolor (BVRI-JHK) sample by K. Jahnke (Jahnke et al., these proceedings) and a high redshift sample taken with VLT by L. Wisotzki.

References

McLure R. J., Dunlop J. S., Kukula M. J. et al., 1999, MNRAS, 308, 377

Ichi Tanaka

David Floyd

SUPERCLUSTERING OF GALAXIES TRACED BY A GROUP OF QSOS AT $Z = 1.1$

Ichi Tanaka, Toru Yamada, Edwin L. Turner and Yasushi Suto
National Astronomical Observatory of Japan (IT)
itanaka@optik.mtk.nao.ac.jp

Abstract
We present the result of a wide-area ($48' \times 9'$) imaging survey of faint galaxies toward the 1338+27 field where an unusual concentration of five QSOs at $z \sim 1.1$ is known to exist. We have detected a significant clustering signature of faint red galaxies with $I > 21$ and $R - I > 1.2$ over a scale extending to $\sim 20 h_{50}^{-1}$ Mpc at $z \sim 1.1$. The $R - I$ colors of these galaxies and the angular correlation function analysis strongly suggests that these galaxies are indeed at the redshift of the group of five QSOs. Our new K'-band images for three selected density peaks also supported the idea. Since this group of five QSOs are indeed a part of much larger ($\sim 70 h_{50}^{-1}$ Mpc) clustering of 23 QSOs discovered by Crampton et al. (1988), the whole size of the detected structure may be as large as the local "Great Wall" structure.

Introduction

High-redshift QSOs are often considered as a possible tracer of the large-scale structure in the Universe. However, there are few observational studies of how they are related with each other. The large group of 23 QSOs at $z = 1.1$ in the 1338+27 field, which was discovered by Crampton et al. (1988), is the suitable system to see whether QSOs and a large-scale structure are actually related with each other or not.

1. The Detection of the Possible Superstructure

Motivated above, we started the survey for a postulated superstructure using the ARC 3.5-m telescope at the Apache Point Observatory (APO). Using the "scan mode" of the SPIcam CCD imager, we con-

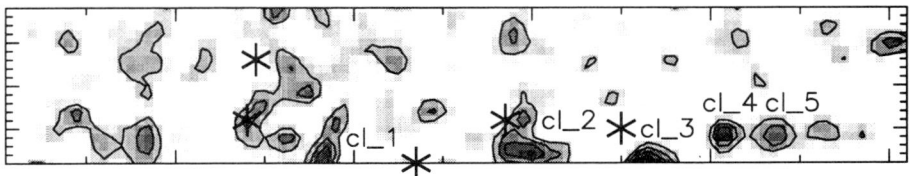

Figure 1. The density map of selected $z \sim 1$ galaxy candidates in the $48' \times 9'$ region around the QSO group. The lowest contour is 1σ excess with the interval of 1σ. Asterisks are the position of member QSOs at $z \sim 1.1$.

structed quite homogeneous image of the $48' \times 9'$ region around the subgroup of five QSOs in the field in R and I bands. The nominal completeness limit of the catalog is about $I \sim 23.5$ magnitude.

To detect the signal of the galaxy excess at $z \sim 1.1$, we have applied the color and magnitude selection "filters" to our catalog to enhance the contrast of the signal. Fig. 1 shows the result of the density excess analysis. We can see several density excess ($> 2\sigma$) regions lying mostly in the lower half of the figure where the four of five QSOs are. We do not think that such concentration of possible clusters and unusual concentration of five QSOs commonly occuring in the lower-half of the survey area is a mere coincidence. Rather, they could be physically associated system. The idea is checked and strongly supported as true by the color-magnitude diagram, the luminosity function, and the angular correlation function analysis (Tanaka et al. 2001).

2. New Near-Infrared Data for Selected Regions

To constrain the redshift of the cluster signal more strongly, we already took the K'-band data of the $1'.5 \times 1'.5$ area around three density peaks using the GRIM II Near-IR camera on APO ARC 3.5-m telescope. Unfortunately the weather conditions were bad so we can only reach $K' \sim 19$ and the errors in color measurement were rather large. Still, we can detect a number of faint ($K > 17$) and red ($I - K = 3 - 4$) galaxies there.

Fig. 2 shows the composite two-color diagram of the three fields. In the figure, circles, triangles and boxes each refer to different density peaks. Although the errorbars are rather large, we can see that there are many galaxies whose colors are $R - I = 1 \sim 1.5$ and $I - K = 3 \sim 4$. Such very red color is a typical of passively-evolving galaxies at $z \sim 1.1$ (see Tanaka et al. 2000). This strongly suggests that the fields contain

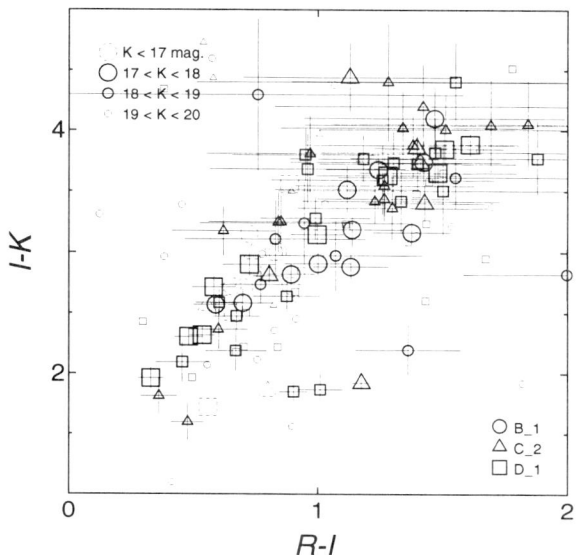

Figure 2. The $R - I$ versus $I - K$ color-color diagram for the three highest-density regions in Fig. 1. Circles, triangles, boxes each represents the data for B_1(cl_1), C_2(cl_2), D_1(cl_3 in Fig. 1), respectively. The size of each symbol represents the K'-band magnitudes. We did not add the errorbars to the data points of $K > 19$ to avoid confusion.

cluster systems of old galaxies at $z = 1.1$. The most important thing is that the color distribution of these red galaxies for different symbols cannot distinguish with each other, which suggests that they are all at the similar redshifts of $z = 1.1$.

These data suggest that the group of QSOs actually acts as a marker of the large-scale structure. Recently we got the new $i'\&z'$ data for much wider area around the QSO concentration using Subaru 8.2-m telescope. The data will enable us to reveal the existence of the largest-scale ($\sim 70 h_{50}^{-1}$Mpc) structure in the young Universe.

References

Crampton, D., Cowley, A.P., Schmidtke, P.C., Janson, T., Durrell, P., 1988, AJ, 96, 816

Tanaka, I., Yamada, T., Aragón-Salamanca, A., Kodama, T., Miyaji, T., Kouji Ohta, Arimoto, N., 2000, ApJ, 528, 123

Tanaka, I., Yamada, T., Turner, E. L., Suto, Y., 2001, ApJ, 547, 521

VI

FUTURE PROSPECTS WITH NEW INSTRUMENTATION

Xavier Barcons (left) and José Miguel Rodríguez Espinosa (right)

From left to right, Fausto Vagnetti, Dario Trèvese, Valentina Zitelli and Birgit Kelm

UNCOVERING AGN WITH X-RAY OBSERVATIONS

Xavier Barcons
Instituto de Física de Cantabria (CSIC-UC)
E-39005 Santander, Spain
barcons@ifca.unican.es

Abstract In this paper I review how X-ray observations can help to uncover Active Galactic Nuclei (AGN), with particular emphasis on hidden AGNs. After introducing the X-ray view of AGN, I'll show how unified models for the X-ray background require the existence of AGNs of all luminosities with various levels of (gas) absorption. Then I discuss the search for type 2 QSOs in X-ray selected samples and the relation of X-ray absorption to optical obscuration. Finally, I report on the current status of medium and deep X-ray surveys carried out with the *Chandra* and *XMM-Newton* observatories. The implications for the hosts of AGN emerging from these surveys are outlined.

Keywords: Active Galactic Nuclei, X-ray emission, Absorption, Surveys

Introduction

According to the standard AGN model, the accretion disk around a massive black hole is heated at $\sim 10^5$ K emitting mostly UV radiation. This radiation is, however, Compton upscattered to the X-ray band by relativistic electrons that surround the disk. X-ray radiation is therefore at the very heart of the AGN phenomenon and it provides direct information about the central engine. Although reprocessing of this primary X-ray emission certainly takes place in AGN, X-ray observations provide the best view to deepen our knowledge of the physics and geometry of the innermost part of AGN.

Besides that, in many (probably most) cases only hard X-rays can escape from the obscured nucleus. Unless observed at high photon energies these obscured AGN might display a non-active appearance. This is why X-ray observations, particularly at hard photon energies, are crucial to uncover hidden AGN.

In this paper I review some aspects of our current knowledge of AGN in the X-ray band, with particular emphasis on hidden AGN.

1. X-ray emission from AGN

It is currently believed that X-ray emission in AGN originates close to the accretion disk. The UV radiation emitted by the disk is Compton up-scattered by hot electrons surrounding the disk. Depending on whether this electron gas is close to thermal equilibrium or not, the resulting X-ray spectrum will have a high energy exponential cutoff (reflecting the maximum energy that electrons can transfer via Compton scattering to the UV photons) or not. If this energy cutoff is present it happens at photon energies $E > 100$ keV (as suggested, e.g., by Madejski et al. 1995 for IC4329A), and therefore a power-law is a good approximation for the primary X-ray spectrum up to fairly high energies.

Radio Loud AGN are known to have different spectral characteristics, in particular the underlying X-ray spectrum is significantly flatter than for Radio Quiet AGN (Wilkes & Elvis 1987). It has been suggested that the X-rays we receive are scattered in the radio lobes and that beaming might play an imporatant role in detremining the X-ray spectrum. In this paper I'll concentrate mostly on Radio-Quiet AGN.

The X-ray spectrum of AGN

The X-ray photons generated by the disk's atmosphere suffer reprocessing before they can escape the surroundings of the AGN. In particular a fraction of these photons (depending on the geometry) will be directed towards the disk again and experience reflection. This has various effects on the X-ray spectrum (George & Fabian 1991), most importantly the Fe K-band reflection lines (in particular the Kα line at 6.4 keV for neutral Fe) and a Compton reflection hump above ~ 20 keV which results from a combination of absorption at low energies and Compton downscattering at high energies. A soft excess resulting from direct UV radiation from the disk itself may also be seen.

Absorption and the unified AGN model

Absorption features are also often seen as this X-ray radiation goes through atomic gas. The most prominent absorption edges in the X-ray regime are those of OVII, OVIII and Fe K. Photoelectric absorption (cross section $\propto E^{-3}$, where E is the photon energy) dominates the effect of absorption when the absorbing gas is neutral, up to equivalent column densities of $N_{HI} \sim 10^{24.5}$ cm^{-2}. At higher column densities, Compton scattering strongly suppresses X-rays up to ~ 30 keV. It is clear that ROSAT (0.1–2 keV) is rather insensitive to AGN with absorption in excess to 10^{22} cm^{-2} and that even *Chandra* and *XMM – Newton* (0.2–10 keV) will miss most of the sources with column densities above 10^{23} cm^{-2}.

In the unified AGN model, the spectroscopic AGN type depends mostly on orientation (Antonucci 1993). That means that type 2 AGN are viewed through different amounts of molecular gas and dust (the "torus") while type 1 AGN have a relatively unobstructed line of sight. If atomic gas, which is ultimately responsible for X-ray absorption, is well mixed with the obscuring material in the torus, then the X-ray spectra of Seyfert 2s should display larger amounts of absorption than the X-ray spectra of Seyfert 1s. This general trend was noted by Awaki et al. (1991). Moreover Risaliti et al. (1999) show how "intermediate" Seyfert types tend to display "intermediate" absorbing columns. The general picture proposed by the unified AGN model is therefore consistent with X-ray observations.

2. The extragalactic X-ray background

The cosmic X-ray background (XRB) was one of the first discoveries of X-ray astronomy (Giacconi et al. 1962). Its spectrum is well fitted by a thermal bremsstrahlung model with temperature $kT \approx 30$ keV (Marshall et al. 1980) which means that most of its energy content resides at that energy. Below 10 keV the XRB intensity can be approximated by a power law $I_E \propto E^{-\alpha}$ with $\alpha = 0.4$. The XRB is isotropic within a few per cent at high galactic latitudes on scales of degrees.

X-ray surveys of different depths have been performed in trying to identify the sources that produce the XRB. AGN were known to be an important constituent of the XRB, but their spectrum appeared too steep (typically $\alpha \sim 1$ in front of the $\alpha = 0.4$ of the XRB below 10 keV). Within the unified AGN model Setti & Woltjer (1989) proposed that type 2 AGN should have a harder X-ray spectrum due to absorption than type 1 AGN, and therefore it could be possible to produce a combined flat ($\alpha = 0.4$) XRB spectrum as a sum of steep type 1 spectra and a flatter type 2 spectra.

The details of this XRB paradigm have been worked out by Madau et al. (1994), Comastri et al. (1995) and Gilli et al. (1999) among others. All these models assume a mixture of unabsorbed and absorbed (with a distribution for the column density) AGN, together with some redshift evolution. The simplest models assume a constant distribution of absorption with redshift, although this might be too simplistic (see, e.g., Comastri 2001 for a review).

An unavoidable prediction of the unified AGN model for the XRB is the existence of heavily absorbed AGN (including QSOs) at all redshifts and luminosities. The identification of absorbed with type 2 AGN, as suggested by the unified model itself, has lead to a search for type 2 QSOs within samples of absorbed X-ray sources (see below).

Another prediction is that most of the energy produced by accretion in AGN goes unnoticed in the X-ray band due to absorption. Fabian & Iwasawa (1999) assumed that the 30 keV peak in the spectral energy distribution of the XRB measures the unabsorbed emission and conclude that $\sim 85\%$ of the total energy produced by AGN is absorbed and re-radiated at some other wavelength. Indeed that very high fraction of absorbed energy is a challenge for a simple toroidal geometry of the obscuring material (the central source should only see $\sim 15\%$ of the sky unobscured in a typical AGN), and therefore other geometries with larger covering solid angles might be necessary (Fabian et al. 1998).

3. Searching for absorbed/obscured AGN

Where are the type 2 QSOs?

Absorbed AGN should manifest themselves in terms of hard X-ray spectra, as it is the case of their low-luminosity version (Seyfert 2 galaxies). Attempts to find such objects at high redshift with *ROSAT* data yielded a handful of candidates with optical/UV narrow emission lines (Almaini et al. 1995 at $z = 2.35$, Ohta et al. 1996 at $z = 0.9$, Boyle et al. 1998 at $z = 0.6$, Barcons et al. 1998 at $z = 1.23$). Only the 2 of these objects at the highest redshifts have QSO luminosities ($L_X > 10^{44}\,\mathrm{erg\,s^{-1}}$) and in fact the one at $z = 2.35$ has a broad Hα emission line (Georgantopoulos et al. 1999). A survey of hard X-ray selected *ROSAT* sources (Page, Mittaz & Carrera 2000) yields a majority of broad-line AGN (Page et al. 2001).

In the deepest survey carried out by *Rosat* in the Lockman Hole, a few faint ($R > 24$) sources have been found with extremely red colours ($R - K > 6$), which suggest heavy obscuration (Lehmann et al. 2000, Hasinger 2001). Photometric redshift techniques can then be used (as the nucleus is likely to be very obscured) to identify these objects which turn out to be consistent with moderate redshift normal galaxies. Hasinger (2001) has pointed out that the heavily obscured AGN NGC 6240 has a remarkably similar SED to that of some of these very red sources found in the Lockman Hole. NGC 6240 is a starburst ultraluminous IRAS galaxy where *BeppoSAX* has discovered a very luminous ($> 10^{44}\,\mathrm{erg\,s^{-1}}$) nucleus obscured by a column density $N_{HI} \sim 2 \times 10^{24}\,\mathrm{cm^{-2}}$. Perhaps these sources are the prototype of type 2 QSOs.

Absorbed vs obscured AGN

A naive assumption that has been made throughout is that AGN which are absorbed in the X-ray band are also obscured in the optical and vice-versa. A one-to-one correspondence between gas absorption and dust obscuration assumes a perfect mixture of both components which is likely to be wrong. The sample of hard X-ray sources studied by Page, Mittaz & Carrera (2000), which is selected to favour X-ray absorption, has very little evidence for obscuration in the optical.

For a galactic dust to gas ratio and standard extinction curve, the visual extinction and HI column density should be related by $A_V = N_{HI}/2 \times 10^{21}\,\mathrm{cm}^{-2}$, but the values found in Seyfert 2s are significantly lower than this (Maccacaro et al. 1982). Maiolino (2001) shows that there is an important dust deficit in type 2 objects which is more pronounced in high-luminosity AGN. Among the possible interpretations for this are dust sublimation near the central engine (Granato, Danese & Franceschini 1997) and a non-standard extinction curve. Other prototypical examples of dust-poor AGN include BAL QSOs which constitute $\sim 10\%$ of the optical population but are practically absent in X-ray surveys (see below), possibly due to the fact that the gas ejected by the QSO does not contain much dust.

In conclusion, the optical classification of AGN in terms of their optical spectrum (the type 1/type 2 classes) does not map on a one-to-one basis into the X-ray band to an unabsobed/absorbed classification.

Where does the absorbed energy go?

As $\sim 85\%$ of the energy generated by accretion is absorbed, its re-emission at other wavelengths has to lead to a quite prominent energy contribution. Unless dust is completely absent, its nature is likely to regulate the final destination of this vast amount of energy. One place to look for that is in the MIR, as there appears to be a good correlation between the hard X-ray luminosity and MIR luminosity in local AGN (Barcons et al. 1995). This correlation has been extended to much fainter sources by Alexander et al. (2001).

As the QSO SED peaks in the FIR, it is tempting to assume that most of that energy would be re-radiated at FIR/submm wavelengths. Almaini, Lawrence & Boyle (1999) computed that 10-20% of the FIR background and of the SCUBA source counts could be contributed by AGN (i.e. X-ray selected sources). However, with the advent of *Chandra* there have been searches for direct correlations between faint X-ray sources and submm sources, which invariably yielded negative results (Fabian et al. 2000, Severgnini et al. 2000). It appears therefore that the estimate by Almaini, Lawrence & Boyle (1999) is rather optimistic and that the X-ray and submm source populations are different. That strongly favours the previous alternative of most of the energy being re-radiated at $\sim 10\mu m$ rather than at $\sim 100 - 1000\mu m$.

4. X-ray surveys with hard X-ray telescopes

The launch of *Chandra* and *XMM-Newton* in 1999 has opened a new era in the search of hard X-ray emitting sources, after pioneering work with *ASCA* and *BeppoSAX*. Deep surveys now reach sensitivities of $\sim 10^{-15}$ erg cm^{-2} s^{-1} or fainter in the 2–10 keV band and source densities of several thousands of sources per square degree.

The Chandra and XMM-Newton deep surveys

There have been reports of various *Chandra* deep surveys, of which I'll discuss two. The first one (Mushotzky et al. 2000, Barger et al. 2001) has found very puzzling results when resolving 75% of the 2–10 keV XRB into three classes of sources: type 1 AGN and QSOs, fairly bright early-type galaxies (which might contain an AGN but show no sign of it in the optical) and very faint sources ($R > 26$). The type 1 QSOs have steep X-ray spectrum, and both the bright galaxies and the faint counterparts have a much harder spectra suggesting heavy absorption. Crawford et al. (2001) have found NIR counterparts for these very faint sources implying very red $R - K$ colours (as in the faint sources of the Lockman Hole discussed in Section 3.1) which probably means that most (if not all) of the 2–10 keV X-ray sources host an active nucleus. It is remarkable, concerning the topic of these proceedings, the variety of appearances that X-ray selected AGN can have in the optical.

Giacconi et al. (2001) have spectroscopically identified a dozen sources in the *Chandra* deep field south, with an even distribution of type 1 and type 2 AGN. In that small sample they also have identified the first X-ray selected BAL QSO, which means that these type of sources that escaped X-ray surveys with previous softer or less sensitive X-ray instruments are going to appear at some flux level in sensitive hard X-ray surveys.

XMM-Newton has also carried out its first ~ 100 ksec exposure on the Lockman Hole (Hasinger et al. 2001), reaching a flux limit similar to that of the *Chandra* exposures in the 2–10 keV band. A total of ~ 150 sources (of which 100 appear in the 2–10 keV band and 60 in the 5–10 keV band) have been found within the ~ 30 arcmin diameter field of view of the EPIC cameras. The most obvious result from the preliminary analysis results from a hardness ratio study where it is seen that $\sim 40\%$ of the sources show some sign of absorption (e.g., hard X-ray colours) in agreement with the unified AGN models for the XRB. An additional 200 ksec exposure on the same target field has been scheduled, which will allow detailed X-ray spectroscopic studies of very faint X-ray sources.

The AXIS XMM-Newton medium sensitivity survey

The large field of view of the EPIC cameras on board XMM-Newton ensures that in a typical exposure of a few tens of ks, 30-150 serendipitous X-ray sources will be discovered. Given the enormous potential of the XMM-Newton mission, the European Space Agency appointed a Survey Science Centre (SSC) to catalogue and identify a large number of these serendipitous sources (Watson et al. 2001), among other tasks. The identification programme of the XMM-Newton serendipitous sky survey sources (the XID programme) has been designed in two steps: a core programme where large representative samples of serendipitous sources will be fully identified by means of archival searches and optical/infrared imaging and spectroscopy and an imaging programme by which a significant fraction of XMM-Newton target fields will be imaged in the optical and near infrared. The hope is that the core programme will be used to train a statistical identification process based on multi-colour optical/infrared imaging and X-ray information, which will be subsequently applied to all XMM-Newton fields for which there is multi-colour imaging.

AXIS (An XMM-Newton International Survey) is a scientific observational programme being carried out at the Observatorio del Roque de los Muchachos in La Palma (Canary Islands, Spain) which constitutes the backbone of the XID programme. Besides providing a major boost in the multi-band wide-field imaging component of the XID programme, AXIS aims at spectroscopically identifying several hundreds of sources in each one of the following 3 samples: a high-galactic latitude "medium" sample with a 0.5-4.5 keV flux limit of $\sim 10^{-14}\,\mathrm{erg\,cm^{-2}\,s^{-1}}$, a high-galactic latitude "bright" sample with flux limit $\sim 10^{-13}\,\mathrm{erg\,cm^{-2}\,s^{-1}}$ and a Galactic plane sample. Within the first few months of observations, significant progress has been achieved in the "medium" sample which is on what I report here.

So far AXIS has identified ~ 80 sources, of which ~ 60 are at galactic latitude $\mid b \mid > 20°$. In the galactic plane, stars with active coronae dominate, but at high galactic latitude type 1 AGN dominate. As this is by no means a complete sample in its current status, I'd like to emphasize the variety of sources found and especially the variety of optical appearances that the hosts of these X-ray sources (which in most cases contain AGN) exhibit.

First, among the 40 type 1 AGN found, 2 of them are seen to be BAL QSOs at $z = 0.8$ and $z = 1.8$. The suspicion raised by the *Chandra* deep surveys (Giacconi et al. 2001) that BAL QSOs should start to populate X-ray selected samples is confirmed here at medium fluxes. It is too soon to establish what fraction of AGN will be BALs, but from this preliminary work it appears to be of the order of several per cent.

In agreement with previous *Rosat* surveys, we also find 8 narrow-line X-ray emitting galaxies. Most (if not all) of these galaxies are powered by AGN as demonstrated by their high X-ray luminosity ($L_X > 10^{42}\,\mathrm{erg\,s^{-1}}$) and will fall within the Seyfert 1.8 to Seyfert 2 category. In one case ($z = 1.2$) we find a luminous AGN which we tentatively classify as a type 2 QSO. The ratio of type 2 to type 1 AGN is indeed uncertain, but at the flux limit of the medium sample it appears to be of the order of 20%, in agreement with our expectations within the unified AGN model for the XRB.

Finally, we also find a few cases (6) where the optical counterpart to an X-ray source is an absorption line galaxy at moderate redshift ($z < 1$). The origin of the X-ray emission in such cases is difficult to assess (as in the cases reported by Mushotzky et al. (2000) in the *Chandra* deep surveys), but we are probably witnessing a deeply hidden active nucleus in these galaxies.

5. Conclusions

The high-galactic latitude X-ray sky is populated mostly with AGN, some (or most) of which are absorbed and/or obscured. X-ray observations at high photon energies are probably the best way to find these hidden AGN. The current hard X-ray observatories *Chandra* and *XMM-Newton* reach sensitivities where more than ~ 1000 AGN per square degree can be found out to large redshifts. It should be noted, however, that even *Chandra* and *XMM-Newton* are only sensitive up to photon energies of the order of 10 keV, whilst 30–40 keV would be needed to see the unaborbed energy produced by AGN.

Optical identification of these X-ray selected AGN exhibits a variety of hosts, from normal type 1 AGN, to type 2 AGN and also galaxies with no (optical) trace of nuclear or star formation activity. The X-ray view of AGN is then showing a very rich variety of AGN hosts.

Acknowledgments

I'm grateful to the XMM-Newton Survey Science Centre XID working group (http://xmmssc-www.star.le.ac.uk) and to the AXIS collaboration (http://www.ifca.unican.es/~xray/AXIS) for allowing me to review our results prior to publication. I acknowledge financial support to my host (the Spanish Scientific Research Council CSIC) and environment (the University of Cantabria UC) organisations when national agencies were not supporting my research. I'm also glad to be finally able to acknowledge partial financial support for this work to the spanish Ministry of Science and Technology under project AYA2000-1690.

References

Alexander, D.M. et al., 2001, ApJ, in press (astro-ph/0101546)
Almaini, O., Boyle, B.J., Griffiths, R.E., Shanks, T., Stewart, G.C., Georgantopoulos, I., 1995, MNRAS, 277, L31
Almaini, O., Lawrence, A., Boyle, B.J., 1999, MNRAS, 305, L59
Antonucci, R., 1993, ARA&A, 31, 473
Awaki, H., Koyama, K., Inoue, H., Halpern, J.P., 1991, PASJ, 43, 195
Barcons, X., Franceschini, A., De Zotti, G., Danese, L., Miyaji, T., 1995, ApJ, 455, 480
Barcons, X., Carballo, R., Ceballos, M.T., Warwick, R.S., González-Serrano, J.I., 1998, MNRAS, 301, L25
Barger, A.J., Cowie, L.L., Mushotzky, R.F., Richards, E.A., 2001, ApJ, in press (astro-ph/0007175)
Boyle, B.J., Almaini, O., Georgantopoulos. I., Blair, A.J., Stewart, G.C., Griffiths, R.E., Shanks, T., Gunn, K.F., 1998, MNRAS, 297, L53
Comastri, A., Setti, G., Zamorani, G., Hasinger, G., 1995, A&A, 296, 1
Comastri, A., 2001, In Stellar Endpoints, AGN and the diffuse X-ray background, in press (astro-ph/0003437)
Crawford, C.S., Fabian, A.C., Gandhi, P., Wilman, R.J., Johnstone, R.M., 2001, MNRAS, in press (astro-ph/0007456)
Fabian, A.C., Barcons, X., Iwasawa, K., Almaini, O., 1998, MNRAS, 297, L11
Fabian, A.C., Iwasawa, K., 1999, MNRAS, 303, L34
Fabian, A.C., Smail, I., Iwasawa, K., Allen, S.W., Blain, A.W., Crawford, C.S., Ettori, S., Ivison, R.J., Johnstone, R.M., Kneib, J.-P., Wilman, R.J., 2000, MNRAS, 315, L8
George, I.M., Fabian, A.C., 1991, MNRAS, 249, 352
Georgantopoulos, I., Almaini, O., Shanks, T., Stewart, G.C., Griffiths, R.E., Boyle, B.J., Gunn, K.F., 1999, MNRAS, 305, 125
Giacconi, R., Gursky, H., Paolini, F., Rossi, B., 1962, Phys Rev Lett, 9, 439
Giacconi, R., Rosati, P., Tozzi, P., Nonino, M., hasinger, G., Norman, C., bergeron, J., Borgani, S., Gilli, R., Gilmozzi, R., Zheng, W., 2001, ApJ, submitted (astro-ph/0007240)
Gilli, R., Risaliti, G., Salvati, M., 1999, A&A, 347, 424
Granato, G.L., Danese, L., Franceschini, A., 1997, ApJ, 486, 147
Hasinger, G., 2001, In ISO Surveys of a dusty Universe. D. Lemke, M. Stickel, K. Wilke (eds), in press (astro-ph/0001360)
Hasinger, G. et al. 2001, A&A, 365, L45
Lehmann, I., Hasinger, G., Schmidt, M., Gunn, J.E., Schneider, D.P., Giacconi, R., McCaughrean, M., Trümper, J., Zamorani, G., 2000, A&A, 354, 35
Maccacaro, T., Perola, G.C., Elvis, M., 1982, ApJ, 257, 47
Madau, P., Ghisellini, G., Fabian, A.C., 1994, MNRAS, 270, L17

Madejski, G.M., Zdziarski, A.A., Turner, T.J., Done, C., Mushotzky, R.F., Hartman, R.C., Gehrels, N., Connors, A., Fabian, A.C., Nandra, K., Celotti, A., Rees, M.J., Johnson, W.N., Grove, J.E., Starr, C.H., 1995, ApJ, 438, 672

Maiolino, R., 2001, In Stellar endpoints, AGN and the diffuse X-ray background, in press (astro-ph/0007473)

Marshall, F.E., Boldt, E.A., Holt, S.S., Miller, R.B., Mushotzky, R.F., Rose, L.A., Rothschild, R.E., Serlemitsos, P.J., 1980, ApJ, 235, 4

Mushotzky, R.F., Cowie, L.L., Barger, A.J., Arnaud, K.A., 2000, Nature, 404, 459

Ohta, K., Yamada, T., Nakanishi, K., Kohno, K., Akiyama, M., Kawabe, R., 1996, Nature, 382, 426

Page, M.J., Mittaz, J.P.D., Carrera, F.J., 2000, MNRAS, 318, 1073

Page, M.J., Mittaz, J.P.D., Carrera, F.J., 2001, MNRAS, submitted

Risaliti, G., Gilli, R., Maiolino, R., Salvati, M., 2000, A&A, 357, 13

Setti, G., Woltjer, L., 1989, A&A, 224, L21

Severgnini, P., Maiolino, R., Salvati, M., Axon, D., Cimatti, A., Fiore, F., Gilli, R., La Franca, F., Marconi, A., Matt, G., Risaliti, G., Vignali, C., 2000, A&A, 360, 457

Watson, M.G. et al. 2001, A&A, 365, L51

Wilkes, B.J., Elvis, M., 1987, ApJ, 323, 243

NEW GROUND BASED FACILITIES IN QSO RESEARCH

The GTC

José M. Rodríguez Espinosa
Instituto de Astrofísica de Canarias, Spain

Abstract
New ground based observing opportunities are becoming, or about to become, available to astronomers for QSO research. These, combined with state of the art focal plane instruments, provide unprecedented sensitivity for detecting faint surface brightness features. During the talk I will take the liberty of talking about one of these new large telescope facilities currently being built in Spain, and will discuss some of the advantages for QSO research offered by these new facilities.

Keywords: Large Telescopes

Introduction

Astronomy is entering an era of great observational opportunities. The advent of new large telescopes combined with state of the art focal plane instruments represent an unprecedented leap forward in the history of Astronomy.

These new telescopes are meeting the most demanding requirements of present day observational astronomy. A brief summary of those requirements shows that for morphology studies of low surface brightness features, and for subtracting the point sources in searching for extended emission, a good characterization of the telescope and Point Spread Function (PSF) is required. This is tantamount to meeting strict criteria for telescope image quality.

Spectroscopy of stellar populations in host galaxies, understanding the chemical evolution of faint compact dwarf galaxies or the determination of the history of star formation rate in the Universe, all demand large collecting apertures combined with excellent image quality. Last but not least, one would wish to measure scattered light off the dusty AGN environment.

These and many other similar observations will benefit enormously from the large increase in collecting aperture that the new generation of large ground based telescopes offers to observational astronomers.

In particular, the Spanish GTC telescope with its 10 metre collecting aperture is endeavoring to achieve high image quality and low emissivity for infrared observations.

In what follows, I will take the opportunity of this conference being held in Granada to show the work that is been done in Spain for the construction of the GTC.

1. The GTC status

The GTC Project is a Spanish initiative lead by the IAC (Instituto de Astrofísica de Canarias), for the construction of a segmented 10 meter mirror telescope at the ORM (Observatorio del Roque de los Muchachos), in the Island of La Palma. The GTC Project was approved early in 1996 and is fully funded by the Central Government of Spain and the Regional Government of the Canary Islands.

The Conceptual Design of the project was finished during the second half of 1997 (Alvarez et al. 1997) The GTC is currently planned to reach first Light at the end of 2002, with scientific operations (Day One) starting by the end of 2003.

All major contracts have been awarded along 1999 and 2000. By now, nearly 2/3 of the originally estimated budget has been committed, the project remaining well within the 1997 estimates.

The GTC Project Board, advised by the GTC SAC (Scientific Advisory Committee), has approved funds for initiating the Continuous Development Program. This allows for the development of new instruments and telescope upgrades after Day One. This program is a key element along the GTC lifetime to maintain it at the forefront of science. Additionally, an Adaptive Optics Program has been funded to provide AO capabilities to the GTC no later than 2005.

The GTC is taking steps to achieve its main science objectives, namely excellent image quality, high operational efficiency and reliability (Rodríguez Espinosa et al. 1999), both on time and budget. Moreover, the GTC is also set to achieve its full capabilities in the shortest possible time after Day One.

2. Construction work

The site work was initiated in October 1999 by ACS, one of the major Spanish Civil Work companies, who was awarded the GTC Civil Work and Auxiliary Equipment contract in September'99. The site (see Fig. 1) work has progressed steadily, having included earth moving for site preparation and the access road to this site from the current observatory road. The foundations for the telescope pier were carefully laid out and the pier is now rising and will be finished within the current year. The progress of this work can be monitored through a live web camera available in the project web pages (http://www.gtc.iac.es). Fig. 2 shows a recent view of this site work. The GTC Civil Work will continue until the end of 2001 although the cylinder building should be ready to receive the Dome by March 2001.

Figure 1. ORM aerial view. The hatched mark shows the GTC site.

The GTC enclosure and auxiliary buildings will be situated at an altitude of 2267 meters in the ORM. They will occupy a surface of about 2400 square meters over a ground platform of 5000 square meters. Three different areas will be built, namely, the telescope enclosure, an annex building, where the telescope control room, direct services for the telescope and office space will be housed, and an auxiliary building, well separated from the telescope area, to facilitate the exhaustion of the heat produced by certain equipment.

The GTC Dome is being built by a group of Spanish companies, encompassing engineering, steel construction and assembly. The contract was awarded in July'99. The Dome will be tested at the factory at the beginning of 2001 to reduce the risk during the final assembly at site where it will be erected afterwards. Two rows of windows situated around the GTC Dome plus an additional row of windows situated around the concrete base will provide a total of 228 square meters of open surface to facilitate the natural ventilation of the telescope enclosure (Serrano et al. 2000). A forced ventilation system will be available for ventilating the telescope chamber when the natural ventilation is not sufficient. Fig. 3 shows the current GTC Dome design.

Figure 2.   Current status of the GTC site work.

Figure 3.   Current status of the GTC Dome construction.

The GTC Telescope structure fabrication has recently been initiated. Factory test will be done in August'01 and the erection at the ORM is planed for the first half of the year 2002. Fig. 4 shows a picture of the Telescope construction at the factory in Tarragona, Spain.

Figure 4. Main yoke of the telescope being built at the factory.

The GTC Optics is also progressing with the fabrication of the primary segments in Germany and subsequent polishing and testing of these blanks to be performed by REOSC (Paris, France). The strict figure requirements of the GTC in order to meet the image quality criteria demand for the polishing a precision of the order of 11 nm rms. The first polished segments will be delivered in February'02 and they will continue to arrive so that by the end of the year 2003 all the segments, including the six spare ones will be delivered.

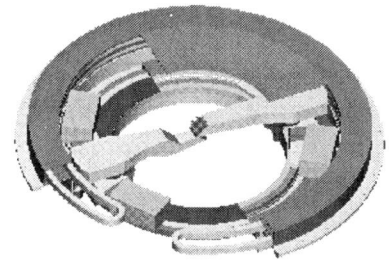

Figure 5. 3D image of the Acquisition, Guiding and Calibration Module.

The Acquisition, Guiding and Calibration Modules for the two Nasmyth foci are being constructed in Belgium. These modules (Fig. 5) will perform not only the more standard functions of telescope acquisition and guiding but also the function of primary mirror calibration on tip, tilt and phasing (Devaney et al. 2000). These modules will be installed at the telescope at the end of the year 2002, just before First Light.

3. Science Instruments

Two science instruments have been selected to be the first GTC instruments covering a wide range of wavelength from the visible to the mid-IR with imaging and low resolution capabilities. These are OSIRIS and CANARI-CAM. OSIRIS is multi-object low resolution spectrometer being built by the IAC, Spain, and the Institute of Astronomy in Mexico, under the leadership of Dr. Jordi Cepa (IAC). With a field of view of 8.0 x 8.0 arcminutes squared, it will be an ideal instrument for determining the dynamics of galaxy clusters. Provided with large format arrays, capable of charge shuffling, and tuneable filters, it can be used advantageously for low surface brightness applications, where an optimal reduction of sky noise is required. The OSIRIS expected limiting magnitudes for S/N 5 and 0.5 arcsecond seeing, can be seen in the following table:

Table 1. Osiris limiting magnitudes (S/N = 5 in 0.5 arcsecond seeing)

Exp Time (s)	U	B	V	R	I
100	24.3	25.9	25.5	25.4	24.6
3600	26.2	27.8	27.4	27.3	26.5

CANARICAM is a mid-IR imager and low-resolution spectrometer, including polarisation and coronagraphy capabilities. Dr. C. Telesco is building the instrument at the University of Florida. It will be an upgraded version of T-Recs, a similar instrument being built by this team for Gemini. CANARICAM should be able to provide diffraction limited images and slit spectroscopy in the 8 to 25 micron range.

CANARICAM is an ideal instrument for detecting evolved (4-5 Gyr) Giant Planets in the N band. Fig. 7 shows that such a measurement would be nearly impossible in the J and K bands with the current technology.

A third instrument, EMIR, is being designed as a second-generation instrument capable of multi-objects spectroscopy up to the K band. EMIR will use a cold mask exchange mechanism, which combined with

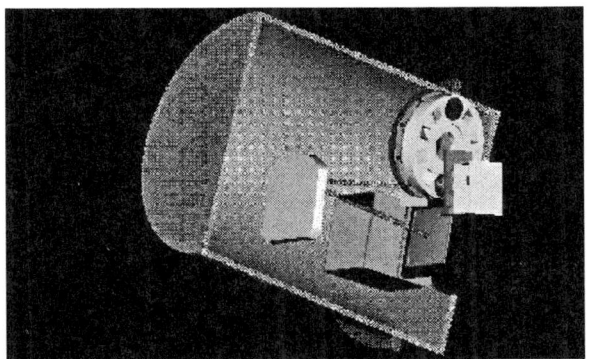

Figure 6. 3D rendition of OSIRIS.

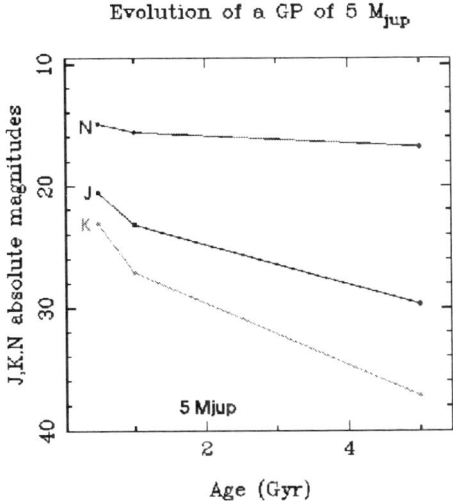

Figure 7. Expected absolute magnitudes of Jupiter type Giant Planets in the near and mid-IR bands. (Diagram courtesy of M. Rosa Zapatero-Osorio).

its large field of view (6 x 6 arcminutes squared) will be ideal for studying the history of the star formation rate in the Universe. Dr. F. Garzón from the IAC is the EMIR PI. EMIR is a collaboration of the IAC, the UCM (Madrid), and the Observatoire de Midi-Pyrenees (France).

4. Final considerations

The GTC is within a year of entering the assembly phase. At that time all the subsystems will be put together and the first functional tests will be performed. This is an exciting period indeed as many hardware pieces are being produced contrary to the time through here in which most work was on paper.

It is worth mentioning that as of today and with 75% of the budget either spent or committed in firm, fix price contracts, the project is mostly on schedule. None of the science requirements have been compromised, much to the credit of the team of engineers that have worked in the design and manufacture of the GTC.

Finally, let me mention that the GTC has also an international dimension. In relation to that, the University of Florida and Mexico have each signed a Memorandum of Understanding for participation in the GTC at the 5% level. The final agreement should be signed in a short time. The participation of these future partners goes beyond a mere participation in the observing time in return for some cash. The participation has been set up in both cases as an wide collaboration involving student and post-doctoral exchanges, the set up of instrumental programmes, etc.

References

Alvarez, P. et al., 1997, "Gran Telescopio CANARIAS. Conceptual design". GTC Project Document GEN/STMA/0012-L.

Rodríguez Espinosa J.M. et al., 1999, "The GTC: An advanced 10M telescope for the ORM", Astrophysics and Space Science 263, 355-360

Serrano J. et al., 2000, "The GTC Dome entering into manufacturing", SPIE's International Symposium on Astronomical Telescopes and Instrumentation, p. 4004-26

Devaney N. et al., 2000, "Guacamole: the GTC guiding, acquisition and calibration module", SPIE's International Symposium on Astronomical Telescopes and Instrumentation, p. 4003-17

Author Index

Baker, J.C., 45, 119
Barcons, X., 357
Barr, J.M., 119
Baugh, C.M., 295
Baum, S.A., 327
Benson, A.J., 295
Beuzit, J.L., 101
Blundell, K.M., 113, 333
Boisson, C., 241
Boschetti, C., 209
Boyle, B., 215
Braatz, J.A., 223
Bremer, M.N., 101, 119
Bressan, A., 171, 261
Cattaneo, A., 339
Chini, R.S., 199
Chuo, H., 191
Cole, S., 295
Combes, F., 185
Cotton, W.D., 65
Croom, S., 215
Downes, D., 265
Dultzin-Hacyan, D., 273, 277
Dunlop, J., 3, 21, 27, 327
Eales, S., 333
Eckart, A., 265, 289
Evans, A.S., 177
Falcke, H., 223
Falco, E., 313
Falomo, R., 13, 95, 343
Feretti, L., 65
Finn, R.A., 133
Focardi, P., 209, 229
Franceschini, A., 171
Frayer, D.T., 177
Frenk, C.S., 295
Fried, J., 39
Giovannini, G., 65
González Delgado, R.M., 247
Granato, G.L., 261
Haas, M., 199
Hardcastle, M., 127
Heidt, J., 39, 51
Henkel, C., 223

Hooper, E.J., 133
Hopp, U., 39
Ho, P.T.P., 191, 281
Hutchings, J.B., 71
Impey, C.D., 133, 313
Jahnke, K., 83, 89, 347
Jarvis, M.J., 333
Jiménez Bailón, E., 269
Joly, M., 241
Jäger, K., 39
Keeton, C., 313
Kelm, B., 209, 229
Kim, D.-C., 165
Klaas, U., 199
Knapen, J.H., 235
Kochanek, C., 313
Kotilainen, J.K., 95, 343
Kreysa, E., 199
Krongold, Y., 273, 277
Kuhlbrodt, B., 83, 89, 347
Kukula, M., 327
Lacey, C.G., 295
Lacy, M., 33
Laine, S., 235
Lara, L., 65
Lehnert, M., 321
Lehár, J., 313
Lemke, D., 199
Leon, S., 185
Letawsky, M., 215
Liao, W., 191, 281
Lilje, P.B., 33
Lim, J., 185, 191, 281
Loaring, N., 215
Marcaide, J.M., 65
Marquez, I., 65, 101
Marziani, P., 273, 277
Mas Hesse, J.M., 269
McLeod, B., 313
McLure, R.J., 21, 27, 327
Meisenheimer, K., 199
Miller, L., 21, 215, 327
Monnet, G., 101
Muñoz, J.A., 313

Müller
 S.A.H., 199
Nilsson, K., 39, 51
O'Dea, C.P., 327
O'Dowd, M., 13
Ohsuga, K., 285
Pelat, D., 241
Peletier, R.F., 235
Peng, C., 313
Percival, W.J., 21, 327
Pesce, J.E., 13
Petitjean, P., 101
Pfalzner, S., 289
Poggianti, B., 171
Pursimo, A., 51
Pérez Garcia, A.M., 255
Rabien, S., 321
Rawlings, S., 113, 333
Rix, H.-W., 313
Rodriguez Espinosa, J.M., 255, 367
Rowan-Robinson, M., 151
Ryś, S., 137
Rönnback, J., 61
Sanders, D.B., 165, 177, 205
Santos Lleó, M., 269
Scarpa, R., 13, 55
Schade, D., 215
Scharwächter, J., 289
Serjeant, S., 33

Shanks, T., 215
Sheinis, A.I., 141
Shlosman, I., 235
Sillanpää, A., 51, 59
Silva, L., 261
Smith, R., 215
Surace, J.A., 177, 205
Sutorius, E., 39
Suto, Y., 351
Takalo, L., 51
Tanaka, I., 351
Theodore, B., 101
Treves, A., 13, 343
Trèvese, D., 145
Turner, E.L., 351
Umemura, M., 285, 307
Urry, C.M., 13, 55
Vagnetti, F., 145
Van-Trung, D., 185
Veilleux, S., 165
Venturi, T., 65
Wen, S., 191
Willot, C.J., 113, 333
Wilson, A.S., 223
Wisotzki, L., 83, 89, 347
Wold, M., 33
Yamada, T., 107, 351
Zitelli, V., 209
Örndahl, E., 61, 89